Responsible Implementations of Generative AI for Multidisciplinary Use

Loveleen Gaur
University of South Pacific, Fiji & Taylor's University, Malaysia

Published in the United States of America by
    IGI Global
    701 E. Chocolate Avenue
    Hershey PA, USA 17033
    Tel: 717-533-8845
    Fax: 717-533-8661
    E-mail: cust@igi-global.com
    Web site: https://www.igi-global.com

Library of Congress Cataloging-in-Publication Data

CIP Pending
ISBN: 979-8-3693-9173-0
EISBN: 979-8-3693-9175-4

British Cataloguing in Publication Data
A Cataloguing in Publication record for this book is available from the British Library.

All work contributed to this book is new, previously-unpublished material.
The views expressed in this book are those of the authors, but not necessarily of the publisher.

Table of Contents

Detailed Table of Contents

Chapter 1
    D. Elavarasi, Mount Zion College of Engineering and Technology, India
    *M. S. Ramadevi, Mount Zion College of Engineering and Technology,
        India*
    *Jayson K. Jayabarathan, Mount Zion College of Engineering and
        Technology, India*
    S. Robinson, Mount Zion College of Engineering and Technology, India

Artificial intelligence (AI) has the capability for machines to learn from experience, change their inputs and perform actions as if they were human. To begin with, AI researchers focused on primitive algorithms that have predefined rules. There are also a number of shortcomings when it comes to these algorithms like interpretability problems, inadequate data sources, computation resource, data description and quality, ethical consideration, overfitting and underfitting computational burden, data collection and bias, high probability of error, lack of enough trained instances no causality as well as reproducibility issues among others. This chapter will guide you through understanding of Generative AI by discussing fundamental algorithms and models used in powering this game-changing technology. In this, it investigates into the basics by using some generative algorithms like probability-based models, VAEs, GANs and autoregressive models.

Chapter 2

Syed Ibad Ali, Parul Institute of Engineering and Technology, Parul
University, Vadodara, India
Mohammad Shahnawaz Shaikh, Parul Institute of Engineering and
Technology, Parul University, Vadodara, India
Smita Shahane, Ajeenkya D.Y. Patil University, Pune, India
Kamini Sharma, Indus University, India
Kiran Macwan, Parul Institute of Engineering and Technology, Parul
University, Vadodara, India

A type of artificial intelligence known as "generative AI" is capable of creating new text, pictures, audio, and video material on its own. Filling in the gaps in the metaverse's evolution, generative AI offers creative methods for content creation in the metaverse. Products like ChatGPT have the ability to improve search results, change how information is generated and presented, and open up new markets. points for the traffic on the internet. This is anticipated to have a substantial influence on conventional search engine offerings, speeding up industry innovation and modernization. In addition to providing an overview of the technologies and potential uses of generative AI in the development of metaverse technology, this article provides advice on how to make generative AI more useful for producing creative material.

Chapter 3

Bodhibrata Nag, Indian Institute of Management Calcutta, Kolkata,
India

AI ethics focuses on the systematic study of what is right and wrong, with the aim of providing principles to resolve ethical dilemmas. AI products, such as deepfakes, have raised concerns about their potential to disrupt traditional industries, devalue human work, and threaten intellectual property rights. AI ethics are intertwined with the need for an understanding of potential risks and benefits. We can categorize AI ethics into principles-based, processes-based, and ethical consciousness approaches. Key themes emerging from AI ethics include transparency, justice, fairness, non-maleficence, responsibility and accountability, privacy, beneficence, freedom and autonomy, trust in AI, dignity, sustainability, and solidarity. The development of AI ethics requires defining universally applicable guidelines and promoting global collaboration. Collaboration between industry, academia, and the public is critical for detecting and evaluating AI-generated content, addressing the social and economic impacts of AI displacements, and building public trust in AI technologies.

Artificial intelligence (AI) technologies are advancing at a rapid pace, which has created many potential and benefits for a variety of industries. But there are also moral questions about the creation and application of AI systems due to their growing autonomy and complexity. This study examines the fundamentals of AI ethics with an emphasis on privacy, responsibility, and openness as the cornerstones that uphold moral AI practices in order to allay these worries. Making AI systems comprehensible and explicable to users and stakeholders is emphasised by transparency. It entails giving concise descriptions of the decision-making procedures, data sources, and potential biases of AI systems. Contrarily, accountability describes the obligations of people and institutions engaged in the creation and application of AI systems. It entails making certain that AI systems are applied in a just, fair, and responsible way and that individuals in charge of their creation and application be held accountable for any unfavourable effects.

This chapter explores the critical need for robust accountability management in the rapidly evolving domain of generative artificial intelligence (GenAI). It examines the complexities and ethical implications of AI-driven decisions and content creation, highlighting the challenges posed by increased autonomy and sophisticated AI outputs. A holistic GenAI accountability framework is introduced. By analyzing ethical dilemmas, legal implications, and societal impacts, the chapter underscores the importance of transparency, fairness, and adaptability in AI governance. It provides actionable recommendations for AI developers, businesses, policymakers, educators, and the public to ensure responsible GenAI development and deployment. Emphasizing the role of education and public engagement, the chapter advocates for raising AI literacy and fostering informed dialogue. It concludes with a call to action for all stakeholders to collaboratively build a sustainable and ethical AI ecosystem, ensuring that AI technologies benefit society while upholding human values and rights.

Chapter 6

Meenu Chaudhary, Noida Institute of Engineering and Technology,
India
Saloni Tiwari, Noida Institute of Engineering and Technology, India
Ajay Gangele, Noida Institute of Engineering and Technology, India
Loveleen Gaur, Taylor's University, Malaysia

In the dynamic global generation, Generative Artificial Intelligence (GAI) creates wonders beyond human imagination. These models are trained on large datasets and discover ways to generate new content material by figuring out styles and relationships in the statistics. The ethical implications of GAI are complicated and multifaceted. On the one hand, generative AI can revolutionize many industries and enhance our lives in limitless methods. On the other hand, there also are concerns about the potential poor outcomes of generative AI. Additionally, Generative AI could result in task displacement as machines emerge as more and more capable of performing tasks that had been a most effective replacement for human beings. The objectives of this chapter are to explore the moral implications of generative AI and to highlight the ways that technology has been used in many ways. This chapter also highlights the impact of generative AI on society and the economy.

Chapter 7

Bodhibrata Nag, Indian Institute of Management, Calcutta, India

Generative AI, based on neural networks, is transforming the creation of content like text, images, music, and interactive digital experiences. However, it raises ethical and practical questions about the authenticity of AI-generated works, intellectual property rights, and the role of creators. AI is also transforming journalism by integrating algorithms and automation, but raises ethical concerns about potential biases and transparency. To address these issues, the sector should set clear standards, promote diversity, and encourage cooperation among technologists, artists, ethicists, and policymakers.

Gen artificial intelligence (AI) is revolutionizing the higher education sector by employing deep learning models to generate information that closely mimics human content. However, its introduction into educational settings raises questions about factors like academic integrity, moral ethics and potential impacts on critical thinking skills. The emergence of generative artificial intelligence (AI) may seem to be a reason for concern which has led to swift prohibitions by organizations and educational agencies. The chapter proposes to put in detail the benefits of using Generative AI by the stakeholders of the education sector, along with the ethics and values to be taken care of and thereby avoiding the reasons to negate the usage of the AI. This chapter will cover the use of Generative AI in different educational settings, the tools and methods employed, the efficiency of GAI in enhancing teaching and learning, the influence on student outcomes, and the possible drawbacks and moral dilemmas related to its application.

Using Artificial intelligence (AI) in research offers many important benefits for science and society but also creates novel and complex ethical issues. While these ethical issues do not necessitate changing established ethical norms of science, they require the Scientific community to develop new guidance for the appropriate use of AI. In this article, we briefly introduce AI and explain how it can be used in research, examine some of the ethical issues raised when using it, and offer recommendations for responsible use, including: Researchers are responsible for identifying, describing, reducing, and controlling AI-related biases and random errors; Researchers should disclose, describe, and explain their use of AI in research, including its limitations, in language that can be understood by non-experts; Researchers should engage with impacted communities, populations, and other stakeholders concerning the use of AI in research to obtain their advice and assistance and address their interests and concerns, such as issues related to bias.

This chapter examines leaders' perspectives on the ethical use of AI, drawing insights from 17 respondents across various industries. Six open-ended questions were crafted to explore ethical AI usage, with responses analyzed through sentiment and topic analysis. Most leaders expressed positive views on the ethical dimensions of AI and leadership. Findings highlighted the significance of ethical and moral principles in ensuring data privacy and security within AI-augmented systems. Organizational management is actively developing strategies to address the risks and challenges posed by AI. The study identified five key themes: the need for AI systems to be human-centric, the importance of training and development, ethical AI development, data privacy, and the integration of AI into organizational mainframes. Overall, the research underscores the necessity for leaders to cultivate a responsible approach to AI usage.

Chapter 11

This chapter explores the ethical implications of chatbots and AI assistants, focusing on privacy and data security, bias and fairness, transparency and accountability, user autonomy, employment impacts, manipulation and persuasion, user well-being, cultural sensitivity, and sustainability. Privacy and data security are paramount, emphasizing user consent and trust in data handling. Bias in AI systems can perpetuate discrimination, necessitating proactive measures for fairness. Transparency and accountability are critical for user trust, requiring clear communication and regulatory frameworks. User autonomy must be respected, allowing control over AI interactions. The impact on employment highlights the need for strategies to manage job displacement and skill development. AI's potential for manipulation necessitates ethical guidelines to prevent misuse. Ensuring user well-being involves addressing mental health risks and misinformation. Cultural sensitivity in AI design promotes inclusivity. Sustainability considerations address the environmental impact of AI systems.

This chapter depicts how generative AI can redefine the e-commerce industry through enhanced personalised experience, dynamic pricing, and operational efficiency. However, the chapter also illustrates how generative AI can be accompanied by problems in terms of biases and ethical challenges. This chapter discusses ways to uncover, tackle, and prevent bias in AI systems to ensure that no customers are discriminated against. The chapter highlights the necessity of accountability, transparency, and robust privacy standards in AI decision-making to win over users and uphold ethical standards. Generative AI development is expected to become increasingly important when used with decision-making accountability. In this way, Generative AI can promote justice, uphold trust, and achieve sustainable digital economy growth as e-commerce develops. As the e-commerce industry grows, generative AIs can ensure fairness, restore confidence, and drive sustainable development.

Chapter 13

This chapter explores the utilization of generative AI in the criminal justice system, emphasizing the advantages and disadvantages of its applications in risk assessment, evidence analysis, and predictive policing. The chapter focuses on technical challenges such as data bias, algorithm transparency, and the accuracy of AI predictions, all of which can have a significant impact on the fairness and reliability of legal decisions. The preservation of systemic biases, privacy concerns, and striking a balance between personal freedoms and public safety are among the other ethical problems covered. Proposed remedies encompass the adoption of moral protocols, ongoing evaluation and scrutiny of artificial intelligence systems, heightened clarity, and the encouragement of cooperation across many fields. To guarantee that AI in criminal justice enhances fairness and justice, the chapter promotes ongoing, inclusive discussions on its application. It underlines the technology's ability to improve the system while also highlighting the significance of mitigating its hazards.

Chapter 14
Responsible AI Implementation in the Hospitality Sector: Ethical Challenges
and Solutions for VR and AR Applications.. 367
    Milan Sharma, Lovely Professional University, India
    Amrik Singh, Lovely Professional University, India
    Rohit, Faculty of Hotel and Tourism Management, SGT University,
        Gurgaon, India

The hotel sector is leading the charge to harness the power of artificial intelligence
(AI), virtual reality (VR), and augmented reality to improve client experiences and
streamline operations. The moral weight of artificial intelligence (AI) in virtual
reality (VR) and augmented reality (AR) for the hotel industry is the subject of this
study. The effects of artificial intelligence and automation on hotel employees, as
well as the significance of honest and ethical data collecting, are discussed in the
study. This study adds to the continuing conversation on ethical AI deployment
in the hospitality industry by doing a thorough examination of the technological
obstacles presented by AI in virtual reality and augmented reality.

Preface

Generative Artificial Intelligence (AI) represents a profound leap in technological advancement, empowering machines to create content that closely mimics human creativity in various forms, including images, music, and text. As this technology continues to evolve and permeate multiple industries, it is essential to address the accompanying ethical considerations that arise from its use.

In crafting "Responsible Implementations of Generative AI for Multidisciplinary Use," our primary objective has been to educate readers about the complexities of this emerging technology, highlighting both its immense potential and the ethical challenges it presents. We aim to demystify generative AI by breaking down complex concepts into accessible language and offering real-world examples that illustrate the implications of its applications.

A core focus of this book is to promote the importance of ethical considerations in the development and deployment of generative AI. We explore various ethical dilemmas that may arise and discuss possible solutions, emphasizing the critical role that developers, users, and policymakers play in ensuring responsible and ethical practices.

Furthermore, we aspire to stimulate thoughtful discussion among our readers about the future of generative AI. By posing thought-provoking questions and presenting diverse perspectives, we encourage readers to reflect on the societal impact of this technology and to engage in a dialogue about its responsible use.

This book is intended for a wide audience, including AI professionals, researchers, policymakers, students, academics, tech enthusiasts, and business leaders. Each group will find insights relevant to their interests and responsibilities, whether they are directly involved in the creation of AI technologies or in shaping the policies and practices surrounding their use.

Generative AI has the potential to revolutionize industries and foster new forms of human creativity. However, with such power comes the responsibility to ensure that this technology is used ethically, with due consideration for privacy, intellectual property, bias, fairness, and accountability. Through collaboration, dialogue, and the establishment of comprehensive ethical frameworks, we believe it is possible to harness the benefits of generative AI while mitigating the risks.

I hope that this book serves as a valuable resource for anyone interested in the ethical implications of generative AI and inspires a commitment to responsible innovation in this rapidly advancing field.

ORGANIZATION OF THE BOOK

Chapter 1: Generative Insights - Unveiling AI's Evolution and Algorithms

This chapter provides a comprehensive exploration of the evolution of artificial intelligence (AI), focusing on the foundational algorithms and models that have paved the way for today's generative AI technologies. Initially, AI research centered around primitive algorithms with predefined rules, which faced numerous challenges, such as interpretability issues, inadequate data sources, computational burdens, and ethical concerns. As AI evolved, more sophisticated generative algorithms, including probability-based models, Variational Autoencoders (VAEs), Generative Adversarial Networks (GANs), and autoregressive models, emerged. This chapter guides readers through the technical intricacies of these generative models, offering a clear understanding of how they operate and their significance in driving AI's transformative potential. It serves as an essential foundation for grasping the complexities and capabilities of generative AI in various applications.

Chapter 2: The Era of Metaverse and Generative Artificial Intelligence

Generative AI has emerged as a revolutionary force in the evolution of the metaverse, enabling the autonomous creation of new text, images, audio, and video content. This chapter delves into the role of generative AI in shaping the metaverse, offering innovative solutions for content creation and enhancing user experiences. By examining the impact of generative AI on traditional search engines and internet traffic, the chapter highlights the technology's potential to revolutionize information generation and presentation. Additionally, it provides a detailed overview of the technologies underpinning generative AI in the metaverse and offers practical advice on leveraging these tools for creative content production. As the metaverse continues to expand, this chapter underscores the significance of generative AI in driving its future development and the broader digital landscape.

Chapter 3: The Evolution of Ethical Standards and Guidelines in AI

The rapid advancement of AI technologies has necessitated the development of robust ethical standards to address the myriad dilemmas arising from their use. This chapter explores the evolution of AI ethics, focusing on the principles that guide the responsible development and deployment of AI systems. It examines key ethical concerns, such as the disruptive potential of AI-generated content like deepfakes, the devaluation of human work, and the challenges posed to intellectual property rights. By categorizing AI ethics into principles-based, processes-based, and ethical consciousness approaches, the chapter provides a comprehensive framework for understanding the ethical landscape of AI. It emphasizes the need for global collaboration in defining universally applicable ethical guidelines and highlights the importance of transparency, justice, fairness, responsibility, and accountability in AI development. The chapter also stresses the critical role of industry, academia, and public cooperation in building public trust and ensuring that AI technologies benefit society while upholding ethical principles.

Chapter 4: The Pillars of AI Ethics - Transparency, Accountability, and Privacy

As AI technologies continue to advance, the ethical implications of their growing autonomy and complexity have become increasingly significant. This chapter examines the fundamental pillars of AI ethics, with a focus on transparency, accountability, and privacy. It emphasizes the importance of making AI systems comprehensible and explicable to users and stakeholders, ensuring that decision-making processes, data sources, and potential biases are clearly communicated. The chapter also explores the responsibilities of individuals and institutions involved in AI development, highlighting the need for fair and responsible application of AI systems. By addressing these core ethical principles, the chapter provides a foundation for understanding the moral considerations that must guide the creation and use of AI technologies, ultimately fostering trust and accountability in the AI ecosystem.

Chapter 5: Navigating the Ethereal Waters - Establishing Accountability in the Autonomous Age of Generative AI

The increasing autonomy and sophistication of generative AI (GenAI) necessitate a robust framework for accountability in AI-driven decision-making and content creation. This chapter explores the complexities and ethical implications of GenAI, emphasizing the critical need for transparency, fairness, and adaptability in AI governance. It introduces a holistic GenAI accountability framework, providing actionable recommendations for developers, businesses, policymakers, educators, and the public to ensure responsible AI development and deployment. The chapter also highlights the importance of education and public engagement in raising AI literacy and fostering informed dialogue. By advocating for collaborative efforts among all stakeholders, the chapter underscores the necessity of building a sustainable and ethical AI ecosystem that aligns with human values and rights.

Chapter 6: From Code to Conscience - Ethics and Generative Artificial Intelligence

Generative Artificial Intelligence (GAI) has the potential to create content beyond human imagination, offering transformative possibilities across various industries. However, the ethical implications of GAI are complex and multifaceted. This chapter explores the dual nature of GAI, highlighting both its potential to revolutionize industries and enhance our lives, as well as the concerns surrounding its impact on society and the economy. The chapter examines the moral dilemmas associated with GAI, such as job displacement and the erosion of human creativity, and emphasizes the need for responsible and ethical use of this technology. By providing a balanced perspective on the benefits and challenges of GAI, the chapter encourages readers to consider the broader implications of its use and the importance of integrating ethical principles into its development.

Chapter 7: Navigating Ethical Dilemmas in Generative AI - Case Studies and Insights

The transformative power of generative AI, particularly in content creation, presents a range of ethical and practical challenges. This chapter explores these challenges through case studies that illustrate the ethical dilemmas associated with AI-generated works, intellectual property rights, and the role of creators. It also examines the impact of AI on journalism, where algorithms and automation are reshaping content creation but raising concerns about bias and transparency. The chapter calls for the establishment of clear ethical standards, the promotion of diversity, and the fostering of cooperation among technologists, artists, ethicists, and policymakers. By providing insights into the ethical landscape of generative AI, the chapter offers practical solutions for navigating the complex ethical issues that arise in this rapidly evolving field.

Chapter 8: Education in the Era of Generative AI - Understanding the Benefits, Ethics, and Challenges

Generative AI is transforming the higher education sector by enabling the creation of content that closely mimics human-generated material. However, its integration into educational settings raises important ethical questions, such as the impact on academic integrity, moral values, and critical thinking skills. This chapter explores the benefits of generative AI in enhancing teaching and learning, while also addressing the potential drawbacks and moral dilemmas associated with its use. The chapter provides a detailed analysis of the tools and methods employed in educational settings, the efficiency of GAI in improving student outcomes, and the ethical considerations that must be taken into account. By offering a balanced perspective on the use of generative AI in education, the chapter aims to guide educators and policymakers in making informed decisions about the ethical implementation of this technology.

Chapter 9: The Ethical Dilemma of Using Generative AI in Science and Research

The application of AI in scientific research offers significant benefits but also introduces novel and complex ethical issues. This chapter provides an overview of the ethical challenges associated with the use of AI in research, emphasizing the need for new guidance to address these issues. The chapter explores the responsibilities of researchers in identifying, reducing, and controlling AI-related biases and errors, and the importance of transparency in disclosing the use of AI in research. It also highlights the need for researchers to engage with affected communities and stakeholders to address their concerns and interests. By offering practical recommendations for the responsible use of AI in research, the chapter aims to ensure that AI technologies are employed in a way that benefits science and society while upholding ethical standards.

Chapter 10: Ethical Leadership in the Age of AI

Leadership plays a crucial role in ensuring the ethical use of AI technologies across industries. This chapter examines the perspectives of leaders on ethical AI usage, drawing insights from a diverse group of respondents. Through sentiment and topic analysis, the chapter identifies key themes in ethical AI leadership, including the importance of human-centric AI systems, the need for training and development, and the integration of AI into organizational frameworks. The chapter underscores the significance of ethical and moral principles in maintaining data privacy and security within AI-augmented systems. By highlighting the strategies employed by organizational management to address the risks and challenges posed by AI, the chapter provides valuable insights into the role of leadership in cultivating a responsible approach to AI usage.

Chapter 11: Exploring Ethical Dimensions of AI Assistants and Chatbots

As AI assistants and chatbots become increasingly integrated into everyday life, their ethical implications must be carefully considered. This chapter explores the ethical dimensions of these technologies, focusing on issues such as privacy and data security, bias and fairness, transparency and accountability, and the impact on employment. The chapter emphasizes the importance of user autonomy and the need for clear communication and regulatory frameworks to build trust in AI systems. It also addresses the potential for manipulation and the ethical guidelines needed to prevent misuse, as well as the cultural sensitivity and sustainability considerations that must be factored into AI design. By examining the ethical challenges associated with AI assistants and chatbots, the chapter offers guidance on how to navigate these issues in a way that promotes inclusivity, fairness, and user well-being.

Chapter 12: Transformative Potential and Ethical Challenges of Generative AI in E-Commerce

Generative AI has the potential to revolutionize the e-commerce industry by enhancing personalized experiences, dynamic pricing, and operational efficiency. However, this chapter also explores the ethical challenges associated with generative AI, particularly in terms of bias and discrimination. The chapter discusses strategies for uncovering, addressing, and preventing bias in AI systems to ensure that no customers are unfairly treated. It also highlights the importance of accountability, transparency, and robust privacy standards in AI decision-making processes. By offering practical solutions for navigating the ethical challenges of generative AI in e-commerce, the chapter aims to ensure that the technology is used in a way that promotes justice, trust, and sustainable growth in the digital economy.

Chapter 13: Ethical Use of AI in the Criminal Justice System

The application of generative AI in the criminal justice system offers significant advantages, such as improved risk assessment, evidence analysis, and resource management. However, this chapter also examines the ethical concerns associated with the use of AI in this context, including issues related to bias, fairness, and transparency. The chapter provides an overview of the key ethical principles that must guide the use of AI in criminal justice, emphasizing the need for rigorous ethical oversight to ensure that AI technologies do not perpetuate or exacerbate existing biases and inequalities. By offering practical guidance on the ethical use of AI in criminal justice, the chapter aims to promote a more just and equitable legal system that leverages AI in a responsible and ethical manner.

Chapter 14: AI and Intellectual Property - Navigating the Complexities of Ownership and Ethics

As AI systems become increasingly capable of generating original content, questions of intellectual property (IP) rights and ownership have become more pressing. This chapter explores the ethical challenges associated with AI-generated works, focusing on the implications for creators, copyright holders, and the broader legal framework. The chapter discusses the need for updated IP laws that account for the unique nature of AI-generated content, while also addressing the potential for misuse and exploitation. It also examines the role of AI in enhancing creativity and innovation, and the ethical considerations that must be taken into account to ensure that AI-generated works are used in a way that respects the rights of creators and the public. By offering insights into the evolving landscape of IP and AI, the chapter provides practical guidance for navigating the complex ethical and legal issues associated with AI-generated content.

IN CONCLUSION

As we conclude this comprehensive exploration into the responsible implementations of generative AI across multidisciplinary fields, it becomes evident that we are standing at a pivotal moment in the history of technological innovation. The chapters within this volume provide not only a rich tapestry of insights and analyses but also a roadmap for navigating the profound ethical, legal, and societal challenges posed by these advancements. The transformative power of generative AI is undeniable, offering unparalleled opportunities for creativity, efficiency, and problem-solving. However, as with any powerful tool, it also demands a heightened sense of responsibility, transparency, and accountability from all who engage with it.

Through the thoughtful contributions of our esteemed authors, we have aimed to equip researchers, practitioners, policymakers, and educators with the knowledge necessary to harness the full potential of generative AI while adhering to ethical principles that safeguard human values. The diversity of perspectives presented in this book underscores the importance of cross-disciplinary collaboration in addressing the multifaceted challenges of AI and ensuring that its deployment contributes positively to society.

As the editor, I hope this volume serves not only as a scholarly reference but also as a call to action for responsible stewardship in the age of AI. Let us move forward with a commitment to fostering innovation that respects our shared ethical principles, promotes equity, and upholds the dignity of all individuals. By doing so, we can ensure that the benefits of generative AI are realized in ways that enhance, rather than diminish, the human experience.

Chapter 1
Generative Insights Unveiling AI's Evolution and Algorithms

D. Elavarasi

https://orcid.org/0009-0008-4673-4196

Mount Zion College of Engineering and Technology, India

M. S. Ramadevi

Mount Zion College of Engineering and Technology, India

Jayson K. Jayabarathan

Mount Zion College of Engineering and Technology, India

S. Robinson

Mount Zion College of Engineering and Technology, India

ABSTRACT

Artificial intelligence (AI) has the capability for machines to learn from experience, change their inputs and perform actions as if they were human. To begin with, AI researchers focused on primitive algorithms that have predefined rules. There are also a number of shortcomings when it comes to these algorithms like interpretability problems, inadequate data sources, computation resource, data description and quality, ethical consideration, overfitting and underfitting computational burden, data collection and bias, high probability of error, lack of enough trained instances no causality as well as reproducibility issues among others. This chapter will guide you through understanding of Generative AI by discussing fundamental algorithms and models used in powering this game-changing technology. In this, it investigates into the basics by using some generative algorithms like probability-based models,

DOI: 10.4018/979-8-3693-9173-0.ch001

VAEs, GANs and autoregressive models.

1 INTRODUCTION

Generative AI has emerging as one of the most prominent areas of study, greatly influencing not only computer vision and natural language processing but also creative arts. In this paper, the authors dwell into the basics of generative AI by looking at its needs, models, types, and metrics of evaluation to have a better understanding of this relatively new area of study. Cao, Y et al (2023) generative AI is an attempt to apply algorithms and models to the generation of synthetic data closer to reality. From entertainment to healthcare and finance, many industries rely on the generation of realistic and novel data. Zhang et al (2023), generative AI has opened completely new frontiers of opportunities in the creation of synthesized images, text generation, musical composition, and chatbot development in human likeness.

It has created a rapid development in generative AI with the increase in large-scale data and deep learning techniques availability. This includes data that is generated to have characteristics similar to real-world data; this kind of generation finds applications in data augmentation, anomaly detection, and some very creative domains such as music composition. Informed decisions from the design and implementation of generative systems can now be drawn by researchers and practitioners when given the knowledge of requirements, models, types of data generation, and criteria for evaluation in generative AI.

"Generative AI Market Size to Hit around USD 118.06 Bn by 2032, (2023)", discussed the growth of the market related to it, one can base the impact that generative AI drives. As per Precedence Research, in 2022, the global generative AI market was valued at $10.79 billion. Now it is expected to reach almost $118.06 billion by 2032, growing at a compound annual growth rate of 27.02% from 2023 to 2032. The fact that it is now in demand has underscored the recognition of the potential for generative AI to empower a multitude of special capabilities across various sectors.

This chapter aims to analyze deep prerequisites, models, generative types, and evaluation criteria utilized in generative AI. This chapter aims to synthesize contributions from the existing literature along with summarized research findings in this field as part of a contribution and guiding direction for the researcher, practitioner, and enthusiast. In essence, this chapter looks into basic requirements for development, commonly used models, formats of inputs and outputs, and performance metrics of a generative AI system. It promotes practical implementation and evaluation of various generative AI applications by providing for the effective realization of the same.

Bandi, A et al, (2023), the basic knowledge and terminology is essential for understanding the foundation of generative AI. This encompasses understanding contents in generative models where the algorithms provide new data similar to the provided dataset. The likelihood function is the probability of observing data on some given parameters and "latent space" which is the abstract space for representing data points.

Generative models operate on the two basic principles of learning representations and probability distributions. They aim at understanding probability distributions that originate from the underlying structure of data. This primarily entails reducing the gap between generated data distribution and actual data distribution which can be measured using adversarial training techniques or maximum likelihood estimation among others. Understanding these concepts facilitates the efficient design and assessment of generative models.

Figure 1. A timeline illustrating key milestones in the development of generative AI, from Turing Machines to GPT-4

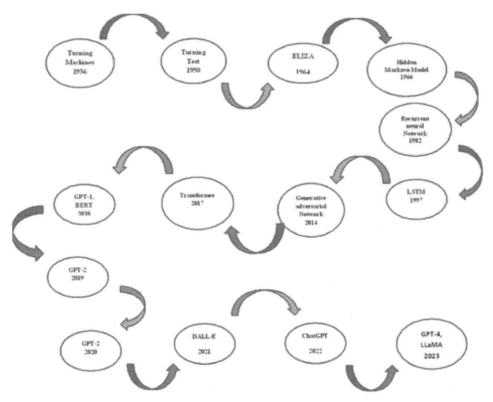

Figure 1 A flowchart in chronological order illustrating significant developments in artificial intelligence from Turing Machines in 1936 to GPT-4 in 2023 with arrows indicating progression.

Zhang, E. Y., Cheok,, (2023), a few markers probably indicate how long the journey has been in the evolution of artificial intelligence and machine learning. In 1936, Alan Turing introduced Turing machines, laying a theoretical framework for understanding computation and algorithmic processes—the laying of the foundation of modern computer science. In 1950, he set out with the Turing Test, the first practical measure for machine intelligence: a measure of whether a machine's behavior can be indistinguishable from that of a human. This gave rise to ELIZA, the first chatbot created in 1964 that was geared toward imitating human conversations. This was where the development of natural language processing began.

This was followed by the creation of Recurrent Neural Networks in 1982, an artificial neural network designed mainly for handling sequential data input with the ability to retain memory, and then the Hidden Markov Model in 1996, which was the very first statistical modeling algorithm to predict sequential data. In 1997, Long Short-Term Memory (LSTM) neural networks were designed to overcome the vanishing gradient problem of RNNs and allow them to process even longer data sequences. In 2014, a new scheme for generating data was introduced by the Generative Adversarial Network (GAN), which generates new data based on the training datasets.

In 2017, transformers, coming from the attention mechanism, supplied a scalable, efficient architecture for large language models and made great improvements in performance and training efficiency. In 2018, OpenAI and Google developed far-reaching GPT-1, with 117 million parameters, and BERT for increasing the language models' understanding. In 2019, the breakthrough in text generation with 1.5 billion parameters was achieved by GPT-2. Then, in 2020, GPT-3 was developed with 175 billion parameters, showing its translational, essay writing, and code generation potential.

In 2021, OpenAI published DALL-E, able to generate high-quality images from textual descriptions at a quality really pushing the boundaries of AI creativity. ChatGPT, released in 2022 and 2023, set a new standard for natural, coherent, and context-aware interactions in generative models, showing just how fast and continuing the development of artificial intelligence.

Generative Artificial Intelligence comes in various models, each designed for specific tasks or types of media creation. Some of the better-known types include (Kumar, L, 2023) Generative Adversarial Networks (GANs) Probability-based Models, (Khamparia, A, 2021) Variational Autoencoders (VAEs) and Autoregressive models, among others. We'll dive into each of these in more depth in the following sections.

A GAN is a special approach in the field of generative AI (Kumar, L, 2023). In sharp contrast to traditional, annotated data-driven methods, a GAN suggests a new framework that typically involves two major components: a Generator and a Discriminator. It is the Generator's job to produce, constantly, synthetic data from noise inputs in a manner such that the outputs created resemble real data. On the other hand, a Discriminator acts like a critic, who looks at whether a given sample is sourced from the real data distribution or from the Generator.

VAEs, proposed by Kingma et al. (2013), provide a different angle of view towards generative AI. The basic constituent parts of VAE are an encoder and a decoder. The former compresses the input data into a lower-dimensional space, called the latent space, and the latter reconstructs back such latent representation into its original form.

In the process, VAEs inject variability in the latent space with the help of a standard Gaussian distribution. This variability allows the model to learn diversified and meaningful representations of the input data. This will basically lead to the end result of getting outputs supposed to not only resemble the input data in the statistical properties—mean and variance—but also offer a structured way of coming up with new samples from the learned data distribution.

This leaves complex ethical issues that call for careful attention. First and foremost is bias and discrimination: the way AI systems are likely to perpetuate unconscious prejudices flowing within society. Privacy is another critical concern, mainly in the case when generative AI crosses borders with personal data. Deep fakes, for example, can be used to create credible images or videos without consent, bringing about the risk of defamation or misinformation. Churning out transparency and accountability of the AI operations breeds trust. Transparency on how AI makes its decisions is important to the creators and the users, making it clear and able to justify the output. This would go a long way to bring clarity between genuine and synthetic content, hence warding off legal and ethical challenges like intellectual property cases. Another ethical dimension encompasses fair use, balancing innovation with creators' rights in a digital environment where AI can create either derivative or wholly new works. AI-generated content can run amok in the spread of disinformation; hence, it is of serious concern to the public discourse and trust in digital media. Another issue is the environmental impact of AI development. Computational demands alone in the training of AI models make a large carbon footprint. This, therefore, creates the need for sustainability in AI research and deployment. Paul, R. K et al., (2023) these ethical challenges are not about strategy but a moral imperative. Alajaji, S et al., (2023) assurance that generative AI will be of service to society, upholding our values and doing no harm to people and the environment, needs to be made certain.

The current chapter is a survey of the very latest research in generative artificial intelligence. It delivers profound and systematic analysis on cutting-edge developments and offers an overview that gives insight into present works on the landscape in the realm concerned and also discussed about the ethical consideration of GenAI various models. To this end, we have filtered and analyzed relevant research papers according to specific criteria to make sure that we will capture the most relevant and newest information available in generative AI.

After this introductory section, Sections 2 review the basic early architecture of Generative adversarial Networks and their variants, Variational Autoencoders (VAE), Probability Based Models, Autoregressive Models. Section 3 deeply explores the recent applications and the advancements in Generative AI application-specific techniques. Section 4 provides the Challenges and opportunities of Generative AI. Lastly, Section 5 concludes the work and highlights the future directions of generative AI.

2. TYPES OF GENERATIVE AI MODEL

In the previous chapter we discussed about the emergence of Generative AI Model. This chapter will delve into the types of Generative AI Algorithms, their applications and ethical considerations associated with each.

Figure 2. Types of generative AI models

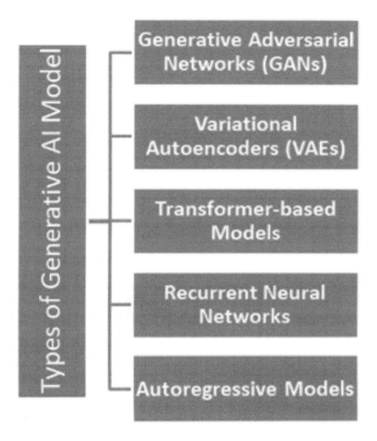

Figure 2 shows the different types of Generative AI models that are widely applied in many domains.

2.1 Generative Adversarial Networks

In 2014, Ian Good fellow proposed Generative Adversarial Networks (GAN) as a new technique in generative AI. GAN is a term referring to the two neural networks that serve as generator and discriminator respectfully. Simply put, when we say generative models, we mean they create new examples, while discriminative models differentiate between various types of data examples.

Discriminative models are like supervised learning which is able to classify the data as like classification model. The generator seeks out realistic data samples, while the discriminator distinguishes between actual and fraudulent data. Making use of this adversarial training process, GAN can produce realistic, high-quality data in a various domains like images, text and music.

Architecture

Figure 3. Architecture of generative adversarial networks

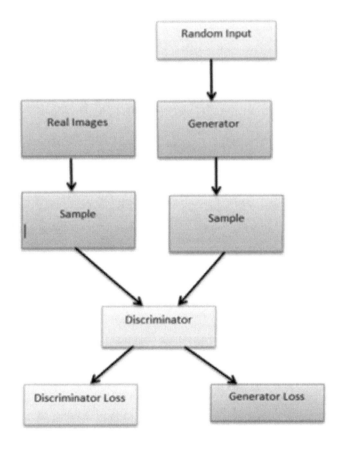

GAN architecture diagram illustrating the sample generation process.

The discriminator and the generator are the two primary parts of the GAN architecture.

Generator: The generator network creates data samples related to the training set using random noise as input fed to the generator. The capability to map points from a low dimensional latent space to the data space using gradient descent and backpropagation. Through this, producing samples that make more realistic over time.

Discriminator: The discriminator receives fake samples generated by the generator and actual data samples from the training dataset. Liu, Y et al., (2024) the discriminator can distinguish between fake and actual datasets by assigning a low probability to the fake datasets and high probability to the actual datasets. The discriminator goal is to get better discrimination, which might help the generator create a realistic sample.

In a minimax game structure, the generator and discriminator are taught concurrently during the training process. While the discriminator seeks to optimize its accuracy in differentiating between real and fake samples, the generator seeks to decrease the likelihood that the discriminator would correctly categorize its generated samples as fake.

Applications of GAN

The versatility and efficiency of Generative Adversarial Networks have been demonstrated by their application in multiple domains:

Image Synthesis: GANs ability to produce, more accurate photorealistic images in a variety of prospective such as artistic works, landscape and portraits images. Few examples are image-to-image translation, image in painting and style transfer. Face app also uses GAN network for face editing options and photograph editing possible.

Text Generation: GANs capable to generate text that is similar and related to the context by effective learning of the fundamental structures of textual inputs. Natural language processing tasks including text generation, dialogue generation and language translation have been handled by GANs. Also do text to image translation.

Drug Discovery: To create new chemical compounds with desired features, GANs are used in molecular design and drug discovery. They are able to suggest potential chemicals for therapeutic development and conduct effective chemical space exploration.

Video Generation: By simulating temporal connections between frames, GANs have been expanded to produce realistic video sequences. Video synthesis, video editing, and video prediction are examples of applications.

2.2 Variational Autoencoders (VAE)

Algorithms require a large number of instructions that manipulate datasets during execution with the aim of automated problem-solving. Automatically analyzing, filtering, and comprehending different kinds of data to interpret various content and types of media is another solution that people implement. None the less, even though standards demand such actions following the instructions properly can lead to making unnecessary errors.

Let us discuss how Variational Autoencoders (VAEs) overcome the drawbacks of Autoencoders. Variational autoencoders, are an advancement in traditional auto-encoders that bring in a sense of uncertainty about the data formation process. The incorporation makes them better learners of latent factors because they make use of many other types of models which in turn makes them more versatile.

Variational Autoencoders (VAEs) are generative models that produce new samples from the underlying probability distribution of the dataset; in order to prevent overfitting in the course of training of a VAE, regularized constraints are imposed on the model's parameters such that the latent space has good properties for generation. Liu, Y et al., (2024) architecture is made up of an encoder-decoder which is given to the input data. The encoder takes the data into a latent form and the decoder uses this form to generate the initial data as it appears in between steps. To do that, we have programmed the maximizing of the information of a variational encoder (VE) through minimizing discrepancy between source and reconstruction words.

Architecture

An autoencoder, a neural network pattern, compacts input information such as pictures, images and photos by using an encoder to record essential characteristics on lower dimensions which we refer to as dimensionality reduction. This bottleneck is known as compressed form that minimizes data complexity. Therefore, from this compressed form decoder turns back the input. Thus, the latency of the reconstruction fidelity during training by autoencoder is improved using the model original input such as measuring pixel-wise differences in image data usually. Making use of autoencoders for optimizing bandwidth on mobile devices calls for reducing the amount of information that moves over the network upon dispatch and then increasing it after reception, like when processing images. They are also very good for removing noise from corrupted images so that it becomes easier to recognize them.

Figure 4. Architecture of VAE

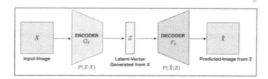

Figure 4 shows the flow from an input image (X) through an encoder to produce a latent vector (Z), which is then decoded to generate a predicted image (X).

Kullback-Leibler (KL) divergence

KL stands for Kullback-Leibler (KL) divergence, named after Solomon Kullback and Richard Leibler. Kullback-Leibler divergence measures the difference between a probability distribution and its expected value compared to the other one. In variational autoencoders' peculiar realm, it quantifies how the learned latent distribution differs from a predefined desired distribution normally taken as a standard Gaussian distribution.

Mathematically, for two probability distributions P and Q, the KL divergence from P to Q is defined as:

$$D_{KL}(P\|Q) = \sum_i P(i)\log\left(\frac{P(i)}{Q(i)}\right)$$

In the network, the trick of reparameterization solves the problem of backpropagation through stochastic nodes by dividing it into mu, sigma and a trainable parameter epsilon as a constant random element. When writing code in TensorFlow, the mean and standard deviation parameters are learned by the encoder network. Then we sample a latent vector to use in the sampling operation which follows this time. To compute how far away the learned distribution is from the standard Gaussian one, one would need to calculate KL divergence which adds up in the end with other costs while backpropagation takes place.

Disentangled variational autoencoders aim to ensure that each neuron in the latent distribution learns distinct features of the input data. By introducing a hyperparameter in the loss function to regulate the presence of KL divergence, the network emphasizes uncorrelated learning of latent variables. This ensures that each variable contributes meaningfully to the compression process. In visualizing the results of a disentangled representation, a simple dataset of images generated from four latent factors (x position, y position, object size, and rotation) demonstrates the effectiveness of the approach in capturing distinct features.

When training a disentangled variational autoencoder, the objective is to ensure that the model learns to encode input information into specific latent variables. By examining visual results, we observe that increasing the beta factor in the autoencoder encourages the utilization of only a subset of latent variables for encoding information. These latent variables correspond to distinct features of the input data, such as position, scale, and rotation. Interestingly, other latent variables remain fixed at a standard distribution, indicating their redundancy in encoding input information.

Variational autoencoders are used as feature extractors, compressing high-dimensional input spaces into compact representations in cases like reinforcement learning. Nevertheless, disentangling latent spaces requires some bargaining: oversimplification causes overfitting, whereas over-disentanglement results in poor outcomes because missing complexity could degrade performance.

2.3 Probability-based Models

Probability-based models, which are a subset of generative AI, are able to create new data instances that are similar in appearance to the dataset given using probability theory. These models use probability distributions so as make authentic samples and this helps them hold secret any underlying patterns therein. They can produce a wide range of data in different domains including text, music or images with an ability to adjust the level of complexity depending on what they learn in statistical training data.

Architecture

Probability-based models use different algorithms, each with its distinctive method of modeling and generation in some way or another. The Autoregressive model is an example of such an algorithm that uses preceding elements to predict probabilities distributions over sequences. Autoregressive architectures are employed in models like PixelRNN or PixelCNN when generating detailed pictures of high resolution one pixel at the time. Another way of creating a probability distribution is the method called Variational Autoencoder (VAE). VAE works by sampling from a latent space in order to maximize the likelihood of observed data. In turn, VAE creates a latent representation for the data and then produces new samples from it. Markov models and Bayesian networks are also utilized, particularly in applications such as natural language processing and audio recognition.

Strengths and Limitations

Probability based models offers several strengths are as follows:

Flexibility: They are able to produce a variety of samples across different fields and model complex data distributions.

Interpretability: Probabilistic models are good at handling unforeseen circumstances or missing data and generating samples help uncover hidden uncertainties.

Versatility: Through the use of perfect models and teaching methods that are incorporated in their design to fit many different types of tasks and databases.

However, these models also have limitations:

Computational Complexity: For large-scale and high-dimensional datasets training and inference can be computationally difficult.

Mode Collapse: When the generator generates just a small number of sample variations, a model, especially generative adversarial networks (GANs), may experience mode collapse.

Comparative Evaluation

Autoregressive models have numerous of pro and cons when compared to other generative models.

2.4 Autoregressive Models

A generative model is good at capturing a sequential dependency in data is the autoregressive model. This model generates data in a sequential way, with each element being trained on the previous elements. Autoregressive models are generative models they are models can train data set and then once trained the model can generate new observations that the model is successful could have come from the same data set look like samples generated from the same distribution from which gather the data. This method provides data with high precision in a variety of domains such as natural language, picture creation, time series with complex dependencies.

Modelling the conditional probability distribution of each element in a sequence given the preceding elements is a key component of the autoregressive model approach. Neural networks are usually used for this, with each element's output distribution determined by the network's hidden state as well as the elements that were generated earlier.

Model Architecture: To model sequential data, autoregressive models frequently use transformer or recurrent neural network (RNN) architectures. Sequences are processed sequentially by RNN-based models, like LSTM (Long Short-Term Memory) and GRU (Gated Recurrent Unit), which update their hidden states at each time step. Conversely, transformer-based models can effectively use self-attention techniques to capture long-range relationships.

Training: Training data to increase chance data is as seeing it uses the maximum likelihood estimation to train autoregressive models. The model predicts each element of the sequence during training with parameters being tuned to minimize the gap between anticipated and observed distribution.

Sampling: Sampling is the process of repeatedly extracting a sequence of newly developed such that once they have been trained autoregressive models can generate samples from the learned conditional distributions. The model generates elements one at a time, conditioning each new prediction on the components it has already created, starting with an initial seed.

In the section 3 comparison of various AI Generative model is explained.

Comparison of GenAI Models

Table 1. Comparison of algorithms with various parameters

Parameter	Probability-based Models	Variational Autoencoders (VAEs)	Generative Adversarial Networks (GANs)	Autoregressive Models
Architecture	Bayesian Networks, Markov Models	Encoder-Decoder Architecture	Generator-Discriminator Architecture	RNNs, Transformers
Learning Method	Maximum Likelihood Estimation	Variational Inference	Adversarial Training	Maximum Likelihood Estimation
Latent Space	Explicit Latent Space	Continuous Latent Space	Implicit Latent Space	N/A (Sequential)
Sample Generation	Direct Sampling from Explicit Distribution	Sampling from Latent Space	Direct Sampling from Generator Network	Sequential Generation
Mode of Operation	Modeling Data Distribution	Learning Latent Representations	Minimax Game Between Generator and Discriminator	Sequential Conditional Generation
Strengths	Flexibility, Interpretability	Scalability, Continuous Latent Space	High-Quality Samples, Diverse Outputs	Captures Sequential Dependencies
Limitations	Computational Complexity, Limited Scalability	Mode Collapse, Discrete Latent Space	Mode Collapse, Training Instability	Sequential Generation, Limited Parallelism

Table 1 provides a concise comparison of these generative models across different parameters, highlighting their respective strengths and limitations.

3. APPLICATIONS OF GENERATIVE AI

In previous section we discussed about the GenAI models and comparative study of those models. In this section we will discuss about the applications of Generative AI in various domain.

Figure 5. Applications of generative AI

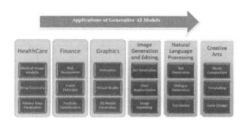

Figure 5 shows the application of GenAI in various domains like healthcare, Finance, Graphics, NLP, Image Augmentation and Creative Arts

Health Care

Generative intelligence is transforming the way in which medical imaging is conducted. At the moment, generative intelligence has found applications in image creation, enhancement, pathological image analysis, recognition and auxiliary diagnostics in the field of medical imaging. These implementations combine deep learning with generative models to help healthcare personnel analyze images and diagnose diseases in a more accurate and efficient manner. The use of generative intelligence in medical imaging has achieved some remarkable results. For instance, photorealistic medical image data can be synthesized using generative models like Generative Adversarial Networks (GANs) to increase the variety and quantity of data sets and thus improve the generalization power and performance of the model. Also, by using the generative model, it is possible to enhance medical images for example denoising, enhancing contrast and sharpness hence improving their quality and visualization effect. The segmentation and feature extraction from medical imagery is another application area where doctors can accurately identify and locate sick areas with the help of generative intelligence.

Finance

Generative AI has a lot of uses in finance, especially in the generation of models such as GANs and VAEs. The following are some examples worth mentioning

Portfolio Optimization - AI can generate synthetic data derived from historical market data to resemble several market scenarios. This helps one achieve an optimized investment portfolio by evaluating different risk and return profiles.

Algorithmic Trading - Generative models can create trading signals or forecast market trends based on patterns recognized in prior data. Furthermore, they could also be used to expand training data sets with synthetic data for improved model performance

Risk Management- Financial institutions' ability to evaluate and control risks is significantly enhanced through simulated scenarios generated via AI. For example, generation of synthetic datasets with extreme values serves as an avenue for model stress testing.

Fraud Detection - Anomalies could be detected by generative models through generation of expected patterns associated with financial transactions so as to unearth irregularities possibly related to frauds

Customer Service & Chatbots - In banking/finance industry, AI-based chatbots that rely on generative models provide more human-like responses improving client interactions.

Credit Scoring- Using vast amounts of data, generative models can predict creditworthiness.

Image Generation and Editing

Generative AI in graphics has significantly advanced various fields, including computer graphics, art, and design. Here are some key applications and examples

3D Shape Creation

Research is currently underway in this area to create realistic 3D representations of objects.

Better shapes can be created using GAN-based shape generation regarding source similarity.

In addition, precise shapes can be manufactured and adjusted to achieve the required shape.

Create Music

Generative AI can help you create original songs for commercials and other creative projects. Please note that using copyrighted material in your workout data may constitute copyright infringement. But there's nothing you can't fix with a bit of legal advice. Let generative AI be your muse and fuel your creativity.

Generate Videos

Generative AI can create videos ranging from short clips to feature films. This can be done through image generation to create visual content, text generation to create scripts or storyboards, and music generation to create soundtracks.

Synthetic Data

Generative AI is a way of making fake data. This type of data isn't collected from real-life situations but is created by a computer program. It's handy for keeping where the original data came from a secret. For instance, instead of using real patient information, like medical records, to study health trends, researchers can use generated data to protect people's privacy.

- Amazon is using synthetic data to train Alexa's language system
- Google's Waymo uses synthetic data to train its self-driving cars
- Health insurance company Anthem works with Google Cloud to generate synthetic Data
- American Express & J.P. Morgan are using synthetic financial data to improve fraud detection
- Roche is using synthetic medical data for clinical research
- German insurance company Provincial tests synthetic data for predictive analytics

Increase Image Resolution

While creating new documents from existing content, generative AI employs various techniques. One such technique is called a Generic Adversarial Network (GAN). The generator and the discriminator form a GAN, which generates new data and ensures that it is factual. High-resolution image renderings can be generated with GAN-based techniques using super-resolution GANs. This technique can create high-quality copies of medical documents and archives that are too expensive to store in a high-resolution format.

Creation of an Instance Image

Generative AI is best known for its ability to create real photos of people. When the input data is an image of someone's face, the model is trained on it and then generates an image with the same face.

Image-to-Image Conversion

It involves changing the external components of an image while maintaining its internal components, such as colour, media, or shape. Such a transformation may involve changing the daytime image into the night-time image. Essential characteristics of an image can also be changed, such as its colour or style, using this transformation.

Natural Language Processing

One of the key roles of Natural Language Processing (NLP) in Generative AI (GenAI) is to allow machines to comprehend, analyze, and produce human language, leading to various applications and examples in this field.

Text-To-Speech Generator

Text-to-speech Generation has several commercial uses, including marketing, education, podcasting, and advertising. To make class notes more engaging, teachers can turn them into audio files. This technique can also provide educational material to the blind or visually impaired. Text-to-speech provides companies with multiple voice and language repertoire capabilities and cost savings on voice actors and equipment.

Conversational Agents

Chatbots and Virtual Assistants like siri, Alexa and Google Assistant use NLP to understand and respond queries.

4. CHALLENGES AND ETHICAL CONSIDERATIONS

Now it's time to consider the technical and ethical challenges of GenAI to the society and personal life. In the beginning of any new technology, the development process needs to start taking into account how it will be ethically used in future instances.

Past occurrences show instances where technology was developed thoughtlessly that later made it difficult to unwrap the moral and ethical issues involved. When it comes to GANs, there is still much debate surrounding their implications with regards being used to generate texts or other artifacts because there are conflicting views on whether they are right or wrong.

Tad Roselund, Managing Director and Senior Partner at BCG consultancy, highlights that the risks associated with generative AI vary in their severity and impact. In order to effectively counter these difficulties, organizations must adopt an all-encompassing perspective that includes a clearly defined plan, strong governance structures, and a strong dedication to ethical AI practices. The establishment of a corporate culture focused on ethics of generative AI should address eight important issues.

Harmful Content Distribution

Artificial intelligence systems can produce the data from a given text prompt. Despite saving time, there is a possible danger associated with their wrong use, knowingly or unknowingly. For example, in-house AI-email might display potentially unwelcome phrases to staff. According to Greenstein, leveraging generative AI alongside human labor and prior systems is essential if the content is to be produced adherently to ethics of the organisation while reinforcing the brand attributes.

Copyright and Legal Exposure

Companies leveraging popular generative AI tools face considerable copyright and legal risks, usually because the tools rely heavily on broad image and text databases obtained from varied online sources. There is difficulty in determining where the data comes from, often requiring labels on its origin. Due to the private formulas pharmaceutical companies rely on when processing medicinal substances or financial organizations participating in confidential transactions, this often presents challenges to them. Besides that, it may so happen that.

Data Privacy Violations

According to the Montreal AI Ethics Institute's founder and principal researcher Abhishek Gupta, Generative AI large language models (LLMs) are often trained on datasets that have personally identifiable information (PII) about people highlighted by Abhishek Gupta, founder and principal researcher at the Montreal AI Ethics Institute. One of the best things about it is that this information is very easy to pull out with just a small amount of text. People can have trouble recognizing or asking for these kinds of data be deleted when using traditional search engines unlike consumers who cannot easily locate them on typical search engines. Therefore, companies involved in constructing or refining LLMs must prioritize measures to prevent the embedding of PII within these models. Additionally, they should establish protocols to facilitate the easy removal of PII from LLMs in compliance with privacy regulations.

Sensitive Information Disclosure

Generative AI is democratizing AI capabilities and making them more accessible. This combination of democratization and accessibility, Roselund said, could potentially lead to a medical researcher inadvertently disclosing sensitive patient information or a consumer brand unwittingly exposing its product strategy to a third party. The consequences of unintended incidents like these could irrevocably breach patient or customer trust and carry legal ramifications. Roselund recommended that companies institute clear guidelines, governance and effective communication from the top down, emphasizing shared responsibility for safeguarding sensitive information, protected data and IP.

The challenges in training generative AI models are energy-consuming and leave behind large carbon footprints. Large models, such as those of OpenAI's GPT-3, have consumed about 1,287 MWh for training, which is associated with massive emissions of CO_2. Many data centres that run the AI models are still running off fossil fuels, meaning their carbon footprint is higher than usual, even with offsetting in the form of renewable energy credits Berthelot, A et al., (2024). Techniques like early stopping during training could reduce energy usage by 80%, while optimizing hardware configuration for AI inference reduced it by 10-20%. Improvements in this area are better coupled with a transition to renewable energies to minimize AI's impact on climate change.

To overcome the ethical consideration the regulatory frame work was designed by various countries in the paper "Balancing Innovation and Regulation in the Age of Generative Artificial Intelligence," Xukang Wang and Ying Cheng Wu have emphasized the requirement for regulatory frameworks that are flexible and adaptable to ensure control over the rapid development of generative AI technologies. The authors indicate that traditional and rigid regulation styles no longer suit the fast pace at which AI develops and evolves. They propose some solutions in their article about how to ensure relevant and effective regulations—first, the risk-based regulation—modulating regulations based on the risk level of the AI application. Stricter oversight would have to be placed on high-risk systems, with more flexibility accorded to lower-risk ones. In any case, continuous monitoring and updating are essential, so regulatory bodies will have to reassess developments in AI from time to time and change the rules accordingly. International collaboration is also needed to harmonize AI regulation through global cooperation, sharing best practices, and common standards.

Moreover, developing and promoting ethical guidelines and good practices covering transparency, accountability, data privacy, and fairness would be instrumental. Finally, the engagement of different stakeholders—including developers, policymakers, experts from the industry, and the public—ensures that other perspectives are considered and regulations address broad concerns. On these lines, the authors believe that adaptive regulatory frameworks can balance innovation with public interest and ethical standards in the age of generative AI, Wu, Y et al., (2024).

Table 2. Ethical challenges

Challenges	Issues	References
Trust and lack of interpretability	Content produced by generative AI could be violent, offensive or erotic	(Zhuo et al., 2023)
Lack of transparency, lineage, and trustworthiness	Using AI Models fake and misleading content or images or videos could be created which would impact serious social and political consequences and sometimes it personally affects the reputation of a person	Donald Metzler et al., 2021
Bias and discrimination	Gender stereotypes, religious stereotypes	Zhuo et al., 2023)
Privacy and copyright implications	Generative AI sometime may reveal sensitive and private data	Fang et al., 2017; Siau & Wang, 2020)
Uncertainty in Model robustness and security	Uncertainty in the models output will results in dilemma for the end users to trust the result.	Karthik Abinav Sankararaman et al., 2022

Table 2 Weisz, J et al., 2024 addressed the ethical challenges are addressed in detailed. The table gives some of the major challenges of generative AI. In case it generates hazardous or inappropriate, it means trust and interpretability issues. Due to transparency that might not be, fake or misleading content will be created; this will have serious social and political consequences, which means damaging the reputation of people. Weisz, J et al., (2024) bias and discrimination mean perpetuating stereotypes in gender and religion. Privacy and copyright occur in cases where generative AI inadvertently reveals sensitive data. Eventually, model robustness and security issues create uncertainty in the reliability of AI-generated outputs.

5. FUTURE TRENDS IN GENERATIVE AI

Generative AI has marked a decisive moment in the year 2023 with the introduction of chatGPT. Toward the end of 2024, the generative AI landscape is likely to undergo rapid evolution, introducing a handful of trends that promise to transform technology and its applications. Shama, M. (2024). addressed with advances in multimodal AI models and the rise of small language models, these trends will shape the technological landscape as well as redefine interactions, creativity, and understanding of AI.

Emergence of Multimodal AI Models

Large language models, such as OpenAI's GPT4, Meta's Llama 2 and Mistral, indicate advances. They allow incorporation of text, audio, images, and videos during content creation and improvement unlike earlier text-based LLMs since they are now multi-modal AI models. Prediction making and generation of outcomes are done using this algorithm which combines speech along with text as well as images or even other categories of data such as graphs and tables. AI experts are expecting significant evolution in the future with multimodal approaches that use hybrid datasets like pictures, text, and recordings. In the transition to multimodal models, AI will achieve more common sense and dynamism compared to current models.

Capable and Powerful Small Language Models

LLMs are trained using large datasets of size terabytes which are extracted from the billions of available free resources. LLMs are trained on massive datasets such as Common Crawl and the Pile. The terabytes of data comprising these datasets were extracted from billions of publicly accessible websites. On the other hand, Small language models are trained on limited datasets that are composed of esteemed sources such as textbooks, journals, and authoritative content. These models require smaller terms of parameter count as well as less memory and storage. Also it runs on less powerful and quite less expensive hardware components. SLMs produce content of comparable quality to some of their larger counterparts, despite being a fraction of the size of LLMs.

The Rise of Autonomous Agents

Autonomous agents are software programs that can operate on their own. Without human intervention these agents can build generative AI models which prevail over the prompt engineering methods used in generative AI. Autonomous agents are implemented using advanced algorithms and machine learning techniques to learn and adapt to new situations and make predictions and decisions with less human intervention. For example, tools like custom GPTs created by OpenAI make effective use of autonomous agents that propelled drastic changes in the field of AI.

Autonomous agents use natural language processing, computer vision, and machine learning to make predictions, actions and analyze different data types by applying the current context.

Frameworks such as Lang Chain and Llama Index are some of the popular tools used to build agents based on the LLMs.

Through intelligent and responsive interaction, Autonomous agents are highly efficacious in improving the customer experience. Overall human intervention is reduced in the field like travel, hospitality, retail and education.

Open Models Will Become Comparable With Proprietary Models

In 2024, open, generative AI models are expected to evolve significantly, with some predictions suggesting that they will be comparable to proprietary models. The comparison between open and proprietary models, on the other hand, is complex and depends on a variety of factors, including the specific use cases, development resources, and data used to train the models.

Meta's Llama 2 70B, Falcon 180B and Mistral AI's Mixtral-8x7B became extremely popular in 2023, with comparable performance to proprietary models such as GPT 3.5, Claude 2 and Jurassic-2.

In the future, the gap between open models and proprietary models will be narrowed, providing enterprises with a great option for hosting generative AI models in hybrid or on premises environments.

In 2024, the next iteration of models from Meta, Mistral, and possibly new entrants will be released as viable alternatives to proprietary models available as APIs.

Cloud Native Becomes Key to On-Prem GenAI

Kubernetes is already the preferred environment for hosting generative AI models. Key players such as Hugging Face, OpenAI, and Google are expected to leverage cloud native infrastructure powered by Kubernetes to deliver generative AI platforms.

Tools such as Text Generation Inference from Hugging Face, Ray Serve from AnyScale, and vLLM already support running model inference in containers. In 2024, we will see the maturity of frameworks, tools, and platforms running on Kubernetes to manage the entire lifecycle of foundation models. Users will be able to pre-train, fine-tune, deploy, and scale generative models efficiently.

Key cloud native ecosystem players will provide reference architectures, best practices, and optimizations for running generative AI on cloud native infrastructure. LLMOps are extended to support integrated cloud native workflows.

Generative AI can combine with other technological advances, including the Internet of Things, blockchain, and edge computing, to increase their potential. On this note, generative AI analyzes and predicts data from these connected devices to help in decision-making and automate IoT devices. Generative AI in blockchain helps in the development of smart contracts and makes them more secure by code generation and validation. Generative AI benefits edge computing in this way: real-time data processing and model updates are possible directly on local devices. It will reduce latency and bandwidth usage at the same time. Together, such technologies can actuate a more coherent and efficient technology ecosystem, pushing innovation and optimization across diverse applications

7. CONCLUSION

We covered generative AI democratization while pointing out accessibility together with prospective risks like data leaks. It is under the impact of being guidelines and communication to protect sensitive information. AI that can generate content is connected to intellectual property and legal predicaments stemming from incorporation of information from myriad websites. If you provide AI in common, it might lead to leaking of private data. For instance, commercial applications include selling products or services by means of voice overs, tele-learnings among others but not limited thereto. Also, it is capable of composing melodies, altering characteristics of pictures as well as enhancing pixels in photos.

Interpretability, multi-modal generation, and anomaly detection should be what people in the future focus on. More has to be done to talk about ethical and societal aspects of generative models. Gen AI has the capability to create videos, come up with images based on a certain text, and has future trends like chatGPT.

Models that are meant for the development of sequence data can be applied in diverse areas such as time series analysis and processing of natural language. The quality of data generated can be assessed using FID scores, perceptual metrics, and human judging of generative models. Large language models trained on datasets containing personally identifiable information have spurred concerns over privacy infringements. Inception Score is used prior to assessing the quality of the generated images under diversity. The former are utilized to establish the latter's diversity and quality. Computational challenges exist in training and inferring generative AI models for large-scale datasets.

REFERENCES

Alajaji, S. A., Khoury, Z. H., Elgharib, M., Saeed, M., Ahmed, A. R., Khan, M. B., & Sultan, A. S. (2023). Generative Adversarial Networks In Digital Histopathology: Current Applications, Limitations, Ethical Considerations, and Future Directions. *Modern Pathology*, ●●●, 100369. PMID:37890670

Bandi, A., Adapa, P. V. S. R., & Kuchi, Y. E. V. P. K. (2023). The Power of Generative Ai: A Review of Requirements, Models, Input–Output Formats, Evaluation Metrics, And Challenges. *Future Internet*, 15(8), 260. DOI:10.3390/fi15080260

Berthelot, A., Caron, E., Jay, M., & Lefèvre, L. (2024). Estimating The Environmental Impact of Generative-AI Services Using An LCA-Based Methodology. [Energy consumption]. *Procedia CIRP*, 122, 707–712. DOI:10.1016/j.procir.2024.01.098

Cao, Y., Li, S., Liu, Y., Yan, Z., Dai, Y., Yu, P. S., & Sun, L. (2023). A Comprehensive Survey of Ai-Generated Content (Aigc): A History of Generative AI From GAN To Chatgpt. *arXiv preprint arXiv:2303.04226*.

Feuerriegel, S., Hartmann, J., Janiesch, C., & Zschech, P. (2024). Generative AI. *Business & Information Systems Engineering*, 66(1), 111–126. DOI:10.1007/s12599-023-00834-7

Fui-Hoon Nah, F., Zheng, R., Cai, J., Siau, K., & Chen, L. (2023). Generative AI and ChatGPT: Applications, challenges, and AI-human collaboration. *Journal of Information Technology Case and Application Research*, 25(3), 277–304. DOI:10.1080/15228053.2023.2233814

Generative AI Market Size to Hit around USD 118.06 Bn by 2032. 2023. Available online: https://www.globenewswire.com/en/news- release/2023/05/15/2668369/0/en/Generative-AI-Market-Size-to-Hit-Around-USD-118-06-Bn-By-2032.html/ (accessed on 29 June 2023).

Kenthapadi, K., Lakkaraju, H., & Rajani, N. (2023, August). Generative AI Meets Responsible Ai: Practical Challenges And Opportunities. In *Proceedings of the 29th ACM SIGKDD Conference on Knowledge Discovery and Data Mining* (pp. 5805-5806). DOI:10.1145/3580305.3599557

Khamparia, A., Gupta, D., Rodrigues, J. J., & de Albuquerque, V. H. C. (2021). Dcavn: Cervical Cancer Prediction and Classification Using Deep Convolutional And Variational Autoencoder Network. *Multimedia Tools and Applications*, 80(20), 30399–30415. DOI:10.1007/s11042-020-09607-w

Kingma, D. P., & Welling, M. (2013b). Auto-encoding variational bayes. arXiv preprint arXiv:1312.6114.

Kumar, L., & Singh, D. K. (2023). A Comprehensive Survey on Generative Adversarial Networks Used For Synthesizing Multimedia Content. *Multimedia Tools and Applications*, 82(26), 40585–40624. DOI:10.1007/s11042-023-15138-x

Linkon, A. A., Shaima, M., Sarker, M. S. U., Nabi, N., Rana, M. N. U., Ghosh, S. K., & Chowdhury, F. R. (2024). Advancements and Applications of Generative Artificial Intelligence And Large Language Models on Business Management: A Comprehensive Review. *Journal of Computer Science and Technology Studies*, 6(1), 225–232. DOI:10.32996/jcsts.2024.6.1.26

Liu, Y., Du, H., Niyato, D., Kang, J., Xiong, Z., Kim, D. I., & Jamalipour, A. (2024). Deep Generative Model and Its Applications In Efficient Wireless Network Management: A Tutorial And Case Study. *IEEE Wireless Communications*, 31(4), 199–207. DOI:10.1109/MWC.009.2300165

Metzler, D., Tay, Y., Bahri, D., & Najork, M. 2021. Rethinking Search: Making Domain Experts Out of Dilettantes. In ACM SIGIR Forum, Vol. 55.

Moulaei, K., Yadegari, A., Baharestani, M., Farzanbakhsh, S., Sabet, B., & Afrash, M. R. (2024). Generative artificial intelligence in healthcare: A Scoping Review on Benefits, Challenges And Applications. *International Journal of Medical Informatics*, 188, 105474. DOI:10.1016/j.ijmedinf.2024.105474 PMID:38733640

Paul, R. K., & Sarkar, B. (2023) Generative AI and Ethical Considerations For Trustworthy Ai Implementation. *Journal ID, 2157*, 0178.

Sankararaman, K. A., Wang, S., & Fang, H. 2022. BayesFormer: Transformer with Uncertainty Estimation. arXiv preprint arXiv:2206.00826 (2022).

Sengar, S. S., Hasan, A. B., Kumar, S., & Carroll, F. (2024). Generative Artificial Intelligence: A Systematic Review and Applications. *arXiv preprint arXiv:2405.11029.*

Shama, M. (2024). Generative Artificial Intelligence in Finance. Artificial Intelligence (Ai) and Business, 70.

Weisz, J. D., He, J., Muller, M., Hoefer, G., Miles, R., & Geyer, W. (2024, May). Design Principles for Generative AI Applications. In *Proceedings of the CHI Conference on Human Factors in Computing Systems* (pp. 1-22).

Wu, Y. C., & Wang, X. (2024). Balancing Innovation and Regulation in the Age of Generative Artificial Intelligence. Journal of Information Policy, 14. Regulatory Framework.

Xu, J., Wu, B., Huang, J., Gong, Y., Zhang, Y., & Liu, B. (2024). Practical Applications of Advanced Cloud Services and Generative AI Systems in Medical Image Analysis. arXiv preprint arXiv:2403.17549.

Zhang, C., Zhang, C., Zheng, S., Qiao, Y., Li, C., Zhang, M., & Hong, C. S. (2023). A Complete Survey On Generative Ai (Aigc): is Chatgpt From Gpt-4 To Gpt-5 All You Need? *arXiv preprint arXiv:2303.11717.*

Zhang, E. Y., Cheok, A. D., Pan, Z., Cai, J., & Yan, Y. (2023). From Turing To Transformers: A Comprehensive Review and Tutorial On The Evolution And Applications of Generative Transformer Models. *Sci*, 5(4), 46. DOI:10.3390/sci5040046

Chapter 2
The Era of Metaverse and Generative Artificial Intelligence

Syed Ibad Ali
https://orcid.org/0000-0001-6312-6768
Parul Institute of Engineering and Technology, Parul University, Vadodara, India

Mohammad Shahnawaz Shaikh
https://orcid.org/0000-0002-1763-8989
Parul Institute of Engineering and Technology, Parul University, Vadodara, India

Smita Shahane
https://orcid.org/0009-0009-8204-7729
Ajeenkya D.Y. Patil University, Pune, India

Kamini Sharma
https://orcid.org/0009-0000-7033-5429
Indus University, India

Kiran Macwan
https://orcid.org/0000-0003-2610-2593
Parul Institute of Engineering and Technology, Parul University, Vadodara, India

ABSTRACT

A type of artificial intelligence known as "generative AI" is capable of creating new text, pictures, audio, and video material on its own. Filling in the gaps in the metaverse's evolution, generative AI offers creative methods for content creation in the metaverse. Products like ChatGPT have the ability to improve search results, change how information is generated and presented, and open up new markets. points

DOI: 10.4018/979-8-3693-9173-0.ch002

for the traffic on the internet. This is anticipated to have a substantial influence on conventional search engine offerings, speeding up industry innovation and modernization. In addition to providing an overview of the technologies and potential uses of generative AI in the development of metaverse technology, this article provides advice on how to make generative AI more useful for producing creative material.

INTRODUCTION

Artificial intelligence (AI) has considerable promise for greatly improving the metaverse by automating intelligent decision-making and producing highly customized user experiences.. Web3 offers customers improved security and privacy for online financial transactions thanks to its distributed network design. Furthermore, data security and integrity are ensured by the immutable data storage and transfer methods made available by block chain technology. In the era of Web3, generative AI solutions such as Chat Generative Pre-trained Transformer (ChatGPT) may solve issues with digital assets and content production and bridge critical gaps in the development of Web3 to become productivity tools. It is anticipated that generative AI technologies will hasten the emergence of the Web3 era by providing Web3 contributors and producers with more dependable and practical productivity tools. The industry has been paying close attention to ChatGPT and other generative AI technologies because of its inventiveness and adaptability. ChatGPT, which uses deep learning models to produce content in a wide range of settings and meet a wide range of demands, has the potential to significantly increase the efficiency and quality of content generation and distribution. Apart from these advantages, ChatGPT may help remove barriers, improve human comprehension and creativity, and produce invaluable discoveries and breakthroughs.

ChatGPT may leverage multi-modal AI technologies to evaluate, understand, and generate information more thoroughly by employing several perceptual modes. This will enable adaptive feedback and real-time perception and response to material, which will ultimately result in the production of richer and more varied types of content. Reconstructed content production will use technologies like speech synthesis, picture generation, and virtual characters. The process of producing high-quality content for the metaverse has been made much easier by the development of vital technologies like ChatGPT as parts of the metaverse engine layer as a result of technological developments in AI for Generative Content (AIGC). The amount of material in the metaverse has not yet kept up with user demand, and only a small number of businesses can afford the high expense of creating metaverse locations. Furthermore, expensively constructed virtual environments sometimes lack enthusiasm, transparency, and sophistication. Yet, if AI can help designers lower the barriers—for example, by offering consistent sceneries with simple descriptions—the cost of building metaverse settings can be significantly reduced. The amount of content in the metaverse will increase significantly thanks to generative AI, which will also fuel the growth of sectors like augmented reality and virtual reality (VR and AR). The aforementioned statement holds particular significance in the era of AIGC. It will be imperative for all platforms creating or preparing to create metaverse spaces to ascertain the metaverse's ability to leverage AI to attain material richness and hence draw users even in the absence of users. This paper reviews the technical foundations of generative AI, as well as its possible applications and uses in the metaverse. concerning the subject of technical advancements in the metaverse. In addition, the generative AI-based ChatGPT is investigated as a means of reducing the technological barrier to the realization of creativity in the metaverse age.

Figure 1 shows a comparison of the workflows for generative AI using deep learning techniques and conventional computer vision approaches.

Figure 1. Comparison of the workflow

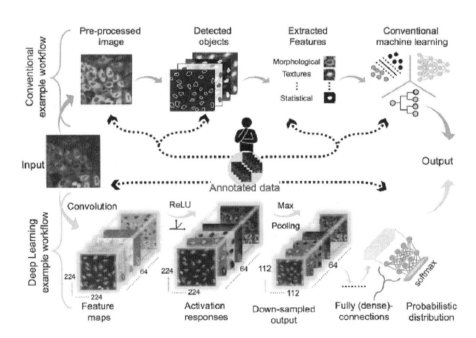

Application of generative AI to the implementation of different metaverse components

Theoretical Foundation of Generative AI

Studying generative AI has become crucial as a result of the advancement of AI technology. One special kind of AI is called "generative AI," which uses input data to create new content on its own (Lim et al., 2023). Machine learning, computer vision, image processing, and natural language processing (NLP) make up the theoretical foundation of generative AI. Andriulli et. al. (2022) contend that machine learning is a fundamental building block of generative artificial intelligence. The subject of machine learning focuses on creating efficient algorithms with data so that computers may learn new things from it. It can help generative AI produce varied content based on various datasets and learn new material from enormous volumes of data. AI relies heavily on machine learning, which includes both generating and discriminative models. While generative models explicitly forecast a distribution and produce fresh data, discriminative models use data to estimate a conditional probability for classification and decision-making. As a result, two categories of AI systems exist: generative AI and discriminative AI. The technology behind discriminative AI is rather advanced, and it has been essential in the previous ten years of the AI age.

Samant et al. (2022) stressed the critical relevance of NLP as a theoretical foundation for generative AI in their study. NLP is a field that focuses on human language and employs generative AI to make it easier to understand human language. This allows for the generation of a variety of content using different language data. Another essential building block for generative AI is image processing, which is the study of picture data to produce new knowledge and a variety of content from various image collections. Computer vision serves as a theoretical foundation for generative AI in that it deals with the processing of picture data to produce a variety of content and learn new things from various image collections. The processes for deep learning and conventional computer vision approaches are shown in Figure 1. Furthermore, by facilitating smooth communication between the actual and virtual worlds, a few basic computer vision methods may improve users' experiences in virtual environments. In conclusion, the theoretical underpinnings of generative AI include machine learning, natural language processing, computer vision, and image processing. These theories can assist generative AI in producing diverse content depending on different datasets and in learning new content from enormous volumes of data. Furthermore, these ideas can facilitate generative AI's use across a range of domains and contribute to its ongoing progress.

PROPOSED METHODOLOGY

Construction of Metaverse Buildings Based on Generative AI

According to Castelli & Manzoni's study from 2022, generative AI is a kind of machine learning system that can create new output data on its own based on input data that is given. By using this technology to create metaverse buildings, architects will be able to quickly create intricate structures, increasing the effectiveness of building design. Ghannad & Lee (2022) claim that generative AI may assist architects in swiftly creating complex architectural forms. Architects may design intricate architectural structures with generative AI, including intricate external designs like arches, circles, triangles, quadrilaterals, and more. Furthermore, generative AI can help architects swiftly design complex interior environments, including finishes, layouts, and decorations. According to Haleem & Javaid (2022), generative AI can help architects create complicated materials more rapidly. Using generative AI, architects may build complex materials like titanium, copper, titanium alloy, wood, metal, cement, ceramics, steel, rubber, oak, and bamboo.

Intricate wall decorations like murals, sculptures, tapestries, and other embellishments may be quickly and effectively created by architects with the help of generative AI. Furthermore, sophisticated functional qualities like sound absorption, robustness, wind resistance, water resistance, cold resistance, moisture resistance, and corrosion resistance may be designed with the use of generative AI by architects. dampness, dust, and water. Finally, complex energy-efficient features like groundwater efficiency, geothermal efficiency, wind energy efficiency, surface water efficiency, and photovoltaic energy pool efficiency may be swiftly created using generative AI. To put it simply, the efficacy of building design may be greatly increased by including generative AI into the development of metaverse buildings. The work by Lu et al. (2022) shows that generative AI may help with complicated form creation and interior decorating, as well as with the design of functional features and sophisticated materials, energy management techniques, and energy pool efficiency.

Figure 2. The evolution of interaction methods based on generative AI

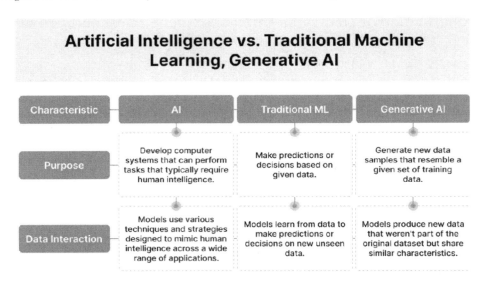

Player avatar and dialog with non-player characters based on generative AI.

It makes sense for metaverse participants to employ generative AI., according to research by Machado et al. (2021), since it allows developers to create a greater variety of game material. Developers may generate unique monsters, maps, and equipment by utilizing generative AI. Additionally, it may be utilized to create unique metaverse games with custom incentives, rules, and sanctions. Consequently, it is possible to improve the gameplay experience and raise player engagement. Huang et al. (2021) discovered that generative AI can improve players' comprehension of game characters and settings, leading to a more engaging gaming experience. With the use of generative AI, players may design unique tactics and smoothly combine various components. Generative AI can function as an intelligent agent within the game. Players may exercise more influence over the scripts and characters in the game, alter the look of the game, comprehend the characters' emotional states, and conduct more effective game analysis by using generative AI. maximize the player experience overall and the game's difficulty. In order to improve the game experience, developers may also use generative AI to mimic player behavior and generate new material. The usage of generative AI in player avatars, according to Ramirez Gomez and Lankes (2021), offers game creators tremendous chances to create more immersive game worlds and provide players an even more exhilarating gaming experience.

In their work, Dobre et al. (2022) pointed out that generative AI might improve the creation of more lifelike non-player characters (NPCs) for video games. By utilizing machine learning techniques, developers may give non-player characters (NPCs) a wider variety of behavioral traits, which will allow them to react differently to different in-game scenarios. Scripted reactions from NPCs are restricted, in contrast to this. By allowing the incorporation of a greater variety of personality characteristics and behaviors, generative AI may also help create more diversified NPCs. These developments in NPC design allow for more natural and customized player-NPC interactions as NPCs react instantly to player actions and words. Players may anticipate increasingly realistic and varied interactions between in-game characters as developers employ machine learning to produce an increasingly diversified and realistic cast of NPCs. Furthermore, developments in brain-computer interfaces may completely transform how people engage with virtual environments by enabling them to communicate with in-game entities directly through brainwaves. Users would be liberated from temporal and spatial limitations as a result, as virtual worlds would appear to them directly in their minds. The development of interactive techniques based on generative AI is seen in Figure 2.

According to Eysenbach (2023), generative AI in dialogue is a technique that uses artificial intelligence to produce dialogue in natural language. It may comprehend players' intents by taking the context into account and provide matching replies based on the information they offer. Thanks to this technology, NPCs may react differently to player actions and show real emotions during interactions. By allowing players to engage with virtual characters, both generative AI in dialogue and NPCs are useful technologies that may improve the immersive experience of games. These technologies might be used to improve everyday communication in other domains like robots and virtual assistants.

Multilingual Translation Based on Generative AI in the Metaverse

A virtual environment known as the "metaverse" is becoming more and more well-liked as a venue for amusement, cooperation, and communication. The demand for multilingual translation technology is growing as it develops. Because generative AI technology allows users to converse in several languages, it provides a varied and inclusive experience. Razumovskaia et al. (2022) described in their study how multilingual translation technology functions by utilizing natural language processing (NLP) algorithms to identify the user's input language and convert it into the language of other users. There are several applications for this technology, including multilingual communication and text-based conversational automated translation. Additionally, it enables audio discussions, enabling users to converse despite language barriers. An example of a generative AI-based one-stop content production application is shown in Figure 3 to aid with automated translation.

Figure 3. One-stop content generation application based on generative AI

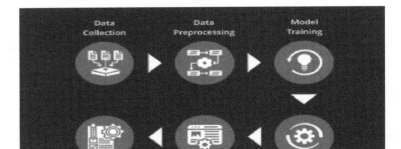

Research by Natarajan et al. (2022) claims that there are several advantages to multilingual translation technology. Businesses may swiftly launch their goods and services into foreign markets by using it to drastically cut down on the time and expense of human translation. It also has the ability to translate words effectively between languages, which aids in understanding people from various cultural backgrounds.

Additionally, by using this technology, organizations may improve their comprehension of foreign markets and create effective marketing plans. Numerous industries, including international travel, education, business, medical, law, journalism, and media, have made extensive use of multilingual translation technology. Moreover, it may be applied to social media platform content sharing, enabling users to access material in several languages. Liu and colleagues (2020) examined the nascent field of generative AI-driven multilingual translation technology, which facilitates automated language translation between several languages. In order to achieve autonomous language translation, neural networks are used to learn the mapping connections between various languages. Multilingual translation technology powered by generative artificial intelligence (AI) offers several benefits over conventional machine translation techniques great precision. It can provide translation results that are more accurate and translate documents quickly because it employs neural networks to understand the mapping relationships between various languages. It can produce translation results more quickly since it learns the mapping relationships between several languages using neural networks; 3. increased scalability. It can achieve higher scalability in terms of translation outcomes since it learns the mapping relationships between multiple languages using neural networks.

Chakravarthi et al. (2021) claim that text may be automatically translated across languages without the need for human involvement thanks to generative AI, a multilingual machine translation system built on deep learning. This is not like typical machine translation techniques since deep learning is used by generative AI to create multilingual translation. Text generation, which converts the source text into the target language, and text comprehension, which comprehends the information in the source text, make up the two primary components of the process. After analyzing the source text and creating the target text appropriately, the neural network understands the target text and correctly interprets the data in the source text. The multilingual translation technology of generative AI still faces some difficulties, notwithstanding its benefits. Practical application may encounter performance problems due to its dependence on enormous volumes of data and processing power. Furthermore, bias and noise in the data may impair its accuracy.

With its ability to translate across languages quickly, accurately, and fluently, generative AI's multilingual translation technology holds the promise of increasing the effectiveness of international communication. Furthermore, it may help companies grow their operations in international markets by making it easier for them to introduce their goods and services to other nations quickly and accurately.

GENERATIVE AI IN METAVERSE CONTENT CREATION

Implementation of Generative AI

In accordance with Barnes (2022), generative AI generates new samples in the form of $F(\bullet)$ by learning a probability distribution $p(x)$. For example, machine algorithms for facial synthesis take into account the constraints of facial models, features of the face, and the physical principles of biomechanics. They also learn from a large quantity of data, including images, text, and language. This enables the computer to sample and render from a subspace that it has learnt to be associated with human faces. Ye & Wang (2022) claim that the core of generative AI is deep neural networks, which are trained on enormous datasets to discover their underlying patterns and probability distributions. These networks then produce new data by using generative models. The two methods below can be used to develop generative AI.

(1) Conditional probability-based generative models, or autoregressive models, are able to produce material that is connected to previously created content. Transformers and recurrent neural networks are examples of popular autoregressive models.

(2) Generative models based on adversarial learning, known as Generative Adversarial Networks (GANs), are capable of producing realistic data, including audio and picture. To increase a GAN's ability to produce realistic data, it is necessary to engage in a friendly competition between its discriminator and generator during training. However, GAN training is challenging because it has to prevent local generator optima while upholding a fine balance between the discriminator's and generator's performance.

According to Talan & Kalinkara (2023), the transformer algorithm, which is based on a self-attention mechanism and is shown in Fig. 4, is one of the primary algorithms employed in generative AI. Transformers can concurrently take into account global information, eliminating the problem of local incoherence, in contrast to recurrent neural networks. Two primary encoders and a decoder make up the transformer. The input sequence data is converted by the encoder into a set of feature vectors, which are then used by the decoder to create the output sequence. Elements: An In modern generative AI models, deep learning neural networks are frequently employed in addition to conventional techniques. According to Kshetri & Dwivedi (2023), deep learning makes predictions by learning from data using enormous neural networks. When external stimuli trigger connected neurons in a neural network, the neurons transmit messages to each other.

Using these techniques, generative AI models with NLP and image recognition skills may be created. Music, painting, and other creative applications are also produced using generative AI models. According to Chow et al. (2022), GPT-3 is a sophisticated language model that, given prompts, can produce writing that resembles that of a person. It is based on the Transformer architecture, which manages massive amounts of linguistic data effectively. GPT-3 has attracted a lot of interest because of its capacity to produce text that is cohesive and contextually relevant in a variety of applications.

Park & Kim (2022) state that OpenAI's DALL-E is a generative model that can create original images from written descriptions. By combining GPT-3's language creation capabilities with picture generating techniques, DALL-E is able to generate visually appealing images that complement the supplied text. Artificial agents may learn to make decisions by interacting with their environment and experiencing rewards or penalties for their behaviors. This process is known as reinforcement learning. This technique may be used to optimize generative models to generate output of superior quality.

Metaverse Content Creation Based on Generative AI

According to Weking et al. (2023), the two key components of the metaverse—content and interactivity—are what are propelling its expansion. The metaverse has been given fresh life by recent developments in application scenarios and the maturing of underlying interactive technologies. Immersive, low latency, diverse, ubiquitous, and clearly identifiable multimodal material is the cornerstone of the metaverse and its primary route. AI will permeate every aspect of the metaverse ecosystem, contributing to improved distribution, quicker content creation, enhanced content display, and increased terminal application performance.

Zhang et al. (2022) point out that generational AI is predicated on deep learning techniques and artificial neural networks that replicate the composition and operations of the human brain. These systems use networked nodes or neurons that function at different levels of a hierarchical structure to store, process, and transfer data. The use of generative AI has tremendously enhanced NLP tasks such as dialogue systems, machine translation, and text summarization. One of the most popular and effective Transformer models for machine translation tasks is the one based on the attention mechanism.

In computer vision, generative AI may also be applied to picture synthesis and restoration, according to Dimcea et al. (2023). One of the most typical models is the Deep Convolution Generative Adversarial Network model, which is based on generative adversarial networks and may produce realistic pictures. Generative AI may be used to audio applications like music production and voice synthesis. One of the most well-known models, the Wavenet model, can create realistic artificial speech and music. As seen in Fig. 5, there are two primary types of generative AI models: single-modal and multi-modal. While multi-modal models can accept input from several sources and produce a variety of output forms, single-modal models only receive instructions from input types that correspond to their output.

Notaro (2022) pointed out that the fast advancement of several technologies, including generative artificial intelligence, is contributing to the establishment of the metaverse. With extremely few textual instructions, this deep learning neural network-based method may produce unique conceptual art and other material. By combining generative AI with other interactive technologies like digital twins and extended reality (XR), one may increase accessibility to AI-generated content inter-action, which has been made feasible by advancements in algorithms and computing capacity. Thanks to developments in computer power, modelling techniques for artificial intelligence, and algorithms, multi-modal material production has grown more diversified and efficient.

CONCLUSION

The metaverse is going to be a complex, massive system with a massive amount of data being processed at once. The digital world will witness an exponential growth in information complexity, beyond the human brain's processing capacity, as a result of the growing dimensionality of information. This article explores the important role of the metaverse in a number of fields, including industry, government, and scientific research. It does this by providing a thorough assessment of major technologies in the metaverse and the application of generative AI. With the integration of key technologies such as blockchain, AR/VR, Internet of Things, and artificial intelligence, the metaverse facilitates the high-quality growth of the intelligent economy.

Future advances in computational power are expected to be primarily driven by these technologies, creating a significant demand for computing resources and altering the way that computation is deployed. When paired with the specialized platform of the metaverse, artificial intelligence (AI), which is a key element of future technological advancement and is demonstrated by apps like ChatGPT, has the potential to establish a new social and economic space where socioeconomic activities may flourish inside a closed-loop environment. Nonetheless, much thought and deliberation are required for the creation and use of the metaverse. Achieving a balance between privacy protection and technical advancement, maintaining the openness and variety of the metaverse, and resolving any possible security problems are important challenges to take up.

REFERENCES

Ai, D., Jiang, G., Lam, S. K., & Li, C. (2023). Computer vision framework for crack detection of civil infrastructure—A review. *Engineering Applications of Artificial Intelligence*, 117, 105478. DOI:10.1016/j.engappai.2022.105478

Ali, S. I.. (2023). Marketing policy in service enterprises using deep learning model. *International Journal of Intelligent Systems and Applications in Engineering*, 12, 239–243. Retrieved January 4, 2024, from https://ijisae.org/index.php/IJISAE/article/view/4066

Andriulli, F., Chen, P. Y., Erricolo, D., & Jin, J. M. (2022). Guest editorial machine learning in antenna design, modeling, and measurements. *IEEE Transactions on Antennas and Propagation*, 70(7), 4948–4952. DOI:10.1109/TAP.2022.3189963

Baía Reis, A., & Ashmore, M. (2022). From video streaming to virtual reality worlds: An academic, reflective, and creative study on live theatre and performance in the metaverse. *International Journal of Performance Arts and Digital Media*, 18(1), 7–28. DOI:10.1080/14794713.2021.2024398

Cai, Q., Wang, H., Li, Z., & Liu, X. (2019). A survey on multi-modal data-driven smart healthcare systems: Approaches and applications. *IEEE Access : Practical Innovations, Open Solutions*, 7, 133583–133599. DOI:10.1109/ACCESS.2019.2941419

Castelli, M., & Manzoni, L. (2022). Generative models in artificial intelligence and their applications. *Applied Sciences (Basel, Switzerland)*, 12(9), 4127. DOI:10.3390/app12094127

Dr Ali, S. I.. (2023). *Causal convolution employing Almeida–pineda recurrent backpropagation for mobile network design*. ICTACT Journals., DOI:10.21917/ijct.2023.0460

Dr Ali, S. I.. (2023) An Innovation of Algebraic Mathematical based statistical Model for complex number Theory. In: *ICDT 2023 IEEE Explorer Conference Proceedings* DOI: DOI:10.1109/ICDT57929.2023.10151169

Gaafar, A. A. (2021). Metaverse in architectural heritage documentation & education. *Adv. Ecol. and Environ. Res.*, 6(10), 66–86.

Gao, F., Wang, C., Li, L., & Zhang, D. (2020). Altitude information acquisition of uav based on monocular vision and mems. *Journal of Intelligent & Robotic Systems*, 98(3-4), 807–818. DOI:10.1007/s10846-019-01018-w

Godwin-Jones, R. (2023). Emerging spaces for language learning: AI bots, ambient intelligence, and the metaverse. *Language Learning & Technology*, 27(2), 6–27.

Gordon, S. (2022). Virtual navigation and geospatial mapping tools, customer data analytics, and computer vision and simulation optimization algorithms in the blockchain-based metaverse. *Rev. Contemp. Philos.*, (21), 89–104.

Guo, X., Wang, Z., Zhu, W., He, G., Deng, H. B., Lv, C. X., & Zhang, Z. H. (2022). Research on DSO vision positioning technology based on binocular stereo panoramic vision system. *Defence Technol.*, 18(4), 593–603. DOI:10.1016/j.dt.2021.12.010

Guo, Y., Yu, T., Wu, J., Wang, Y., Wan, S., Zheng, J., & Dai, Q. (2022). Artificial Intelligence for Metaverse: A Framework, CAAI. *Artificial Intelligence Research*, 1(1), 54–67.

Hawkins, M. (2022). Virtual employee training and skill development, workplace technologies, and deep learning computer vision algorithms in the immersive metaverse environment. *Psychosociological Issues in Human Resource Management*, 10(1), 106–120. DOI:10.22381/pihrm10120228

Hawkins, M. (2023). Metaverse live shopping analytics: Retail data measurement tools, computer vision and deep learning algorithms, and decision intelligence and modeling, J. Self-Governance Manage. *Econ.*, 10(2), 22–36.

Jovanovic, M., & Campbell, M. (2022). Generative artificial intelligence: Trends and prospects. *Computer (Long Beach Calif)*, 55(10), 107–112.

Lim, W. M., Gunasekara, A., Pallant, J. L., Pallant, J. I., & Pechenkina, E. (2023). Generative AI and the future of education: Ragnarök or reformation? A paradoxical perspective from management educators. *International Journal of Management Education*, 21(2), 100790. DOI:10.1016/j.ijme.2023.100790

Mondal, S., Das, S., & Vrana, V. G. (2023). How to bell the cat? A theoretical review of generative artificial intelligence towards digital disruption in all walks of life. *Technologies*, 11(2), 44. DOI:10.3390/technologies11020044

Pal, S., Rabehaja, T., Hitchens, M., & Hill, A. (2019). On the design of a flexible delegation model for the Internet of Things using blockchain. *IEEE Transactions on Industrial Informatics*, 16(5), 3521–3530. DOI:10.1109/TII.2019.2925898

Perkins, J. (2022). Immersive metaverse experiences in decentralized 3d virtual clinical spaces: Artificial intelligence-driven diagnostic algorithms, wearable internet of medical things sensor devices, and healthcare modeling and simulation tools. *American Journal of Medical Research (New York, N.Y.)*, 9(2), 89–104. DOI:10.22381/ajmr9220226

Poggi, M., Tosi, F., Batsos, K., Mordohai, P., & Mattoccia, S. (2021). On the synergies between machine learning and binocular stereo for depth estimation from images: A survey. *IEEE Transactions on Pattern Analysis and Machine Intelligence*, 44(9), 5314–5334. DOI:10.1109/TPAMI.2021.3070917 PMID:33819150

Popescu, G. H. (2022). K. Valaskova, J. Horak, Augmented reality shopping experiences, retail business analytics, and machine vision algorithms in the virtual economy of the metaverse, J. Self-Governance Manage. *Econ.*, 10(2), 67–81.

Ray, P. P. (2023). *ChatGPT: a comprehensive review on background, applications, key challenges, bias, ethics, limitations and future scope, Internet Things Cyber-Phys.* Syst.

Samant, R. M., Bachute, M. R., Gite, S., & Kotecha, K. (2022). Framework for deep learning-based language models using multi-task learning in natural language understand-ing: A systematic literature review and future directions. *IEEE Access : Practical Innovations, Open Solutions*, 10, 17078–17097. DOI:10.1109/ ACCESS.2022.3149798

Shaikh, M. S., Ali, S. I., Deshmukh, A. R., Chandankhede, P. H., Titarmare, A. S., & Nagrale, N. K. (2024). AI business boost approach for small business and shopkeepers: Advanced approach for business. In Ponnusamy, S., Assaf, M., Antari, J., Singh, S., & Kalyanaraman, S. (Eds.), *Digital Twin Technology and AI Implementations in Future-Focused Businesses* (pp. 27–48). IGI Global., DOI:10.4018/979-8-3693-1818-8.ch003

Sleaman, W. K., Hameed, A. A., & Jamil, A. (2023). Monocular vision with deep neural networks for autonomous mobile robots navigation. *Optik (Stuttgart)*, 272, 170162. DOI:10.1016/j.ijleo.2022.170162

Watson, R. (2022). The virtual economy of the metaverse: Computer vision and deep learning algorithms, customer engagement tools, and behavioral predictive analytics. *Linguistic Philos. Investig.*, (21), 41–56.

Wu, X., Guan, F., & Xu, A. (2020). Passive ranging based on planar homography in a monocular vision system. *Journal of Information Processing Systems*, 16(1), 155–170.

Chapter 3
The Evolution of Ethical Standards and Guidelines in AI

Bodhibrata Nag

https://orcid.org/0000-0002-6621-359X

Indian Institute of Management Calcutta, Kolkata, India

ABSTRACT

AI ethics focuses on the systematic study of what is right and wrong, with the aim of providing principles to resolve ethical dilemmas. AI products, such as deepfakes, have raised concerns about their potential to disrupt traditional industries, devalue human work, and threaten intellectual property rights. AI ethics are intertwined with the need for an understanding of potential risks and benefits. We can categorize AI ethics into principles-based, processes-based, and ethical consciousness approaches. Key themes emerging from AI ethics include transparency, justice, fairness, non-maleficence, responsibility and accountability, privacy, beneficence, freedom and autonomy, trust in AI, dignity, sustainability, and solidarity. The development of AI ethics requires defining universally applicable guidelines and promoting global collaboration. Collaboration between industry, academia, and the public is critical for detecting and evaluating AI-generated content, addressing the social and economic impacts of AI displacements, and building public trust in AI technologies.

DOI: 10.4018/979-8-3693-9173-0.ch003

INTRODUCTION TO AI AND ETHICAL CONSIDERATIONS

Artificial Intelligence (AI) systems, particularly those based on machine learning (ML) and deep learning (DL), operate through complex algorithms that learn patterns from large datasets (Goodfellow, Bengio, & Courville, 2016) (LeCun, Bengio, & Hinton, 2015) (Schmidhuber, 2015).

Machine learning is the process of teaching computers to use data to make predictions or choices without being explicitly programmed to do so. Deep learning is a type of machine learning that uses neural networks with many levels (hence the name "deep") to model high-level concepts in data, like speech and image recognition. Neural networks (Hecht-Nielsen, 1992), which are based on the structure of the human brain, are made up of nodes (neurons) that are connected to each other and process input in layers. Each layer takes in info and changes it into a slightly less concrete form. As the model learns, the weights of these links are changed based on how wrong its predictions were. This is usually done using a method called backpropagation. This process keeps going over and over until the model meets a certain level of accuracy.

Artificial intelligence (AI) has swiftly progressed from a theoretical notion to a revolutionary technology that profoundly affects multiple facets of society. This chapter examines the historical progression of AI ethics, current obstacles, and forthcoming paths to offer a thorough comprehension of the ethical deliberations in AI.

Concluding Summary Points:

- AI's rapid development from theoretical concepts to practical applications.
- The necessity of ethical considerations to navigate the complexities of AI.
- Importance of understanding AI's impact on society and the need for robust ethical frameworks.

The foundational concepts of AI and its profound impact on various sectors lay the groundwork for understanding the evolution of ethical standards in this rapidly advancing field.

Historical Developments in AI and Ethics

Early Foundations (1940s-1950s): The foundation for AI was established in the 1940s with notable contributions by pioneers like Warren McCulloch and Walter Pitts, who invented neural networks. (McCulloch & Pitts, 1943). Alan Turing, a pioneer of modern computing, investigated the philosophical and ethical aspects of computing, including machine awareness and the societal influence of machines, in his seminal work "Computing Machinery and Intelligence" in 1950. He contemplated the possibility of machines exhibiting consciousness or the ability to think, which raised concerns regarding the predictability of computer behavior and the concept of free will. (Turing, 1950). Turing pondered on the nature of human cognitive processes and how they may be mimicked or copied by machines, implying that the brain could be regarded a type of machine. He also questioned machine intelligence's societal influence, taking into account the ethical implications of computers outperforming people in cognitive activities and the potential for them to modify societal structures and human interactions. Turing's philosophical inquiries went beyond mere technical functionality to consider how machines might influence concepts of identity, autonomy, and ethics. (Copeland, 2004). The 1956 Dartmouth Summer Research Project Proposal first coined the term "artificial intelligence (AI)." It was spearheaded by John McCarthy, Marvin Minsky, Nathaniel Rochester, and Claude Shannon (Press, 2016). The project's core hypothesis was that a machine could simulate all aspects of learning and intelligence with precision. The proposal aimed to explore how machines could use language, form abstractions and concepts, solve problems typically reserved for humans, and improve themselves. The project aimed to: (a) investigate the programming of computers to understand and manipulate language, a fundamental aspect of human thought. (b) build on existing theoretical and experimental work by exploring neuron networks and their potential to form concepts. (c) concentrate on the potential of intelligent machines to engage in self-improvement activities, implying an abstract study of this aspect of intelligence, and (d) categorize and simulate various abstractions, an essential role in intelligent reasoning and decision-making. (McCarthy, Minsky, Rochester, & Shannon, 2006).

1960s: The first industrial robot, Unimate, is introduced in 1961, raising ethical questions about job displacement and the role of robots in human environments. Joseph Weizenbaum developed ELIZA from 1946 to 1967 which was an early natural language processing program. He later questions the ethical implications of AI (Weizenbaum, 1976). The first expert system, DENDRAL, is developed in 1965, leading to discussions on the ethical use of AI in decision-making processes.

1970s: Isaac Asimov's "Three Laws of Robotics" formalized in his works in 1976, addressed ethical guidelines for robots (Asimov, 1976). The Lighthill Report (Lighthill, 1972) criticized AI research, leading to reduced funding and raising questions about the societal impact and feasibility of AI.

1980s: There was a resurgence of AI with the development of expert systems like MYCIN prompting discussions on ethical decision-making in medical applications (Buchanan & Shortliffe, 1984). James Moor's paper "What is Computer Ethics?" published in 1985 lays the foundation for the field of computer ethics (Moor J. H., 1985).

1990s: IBM's Deep Blue defeated chess champion Garry Kasparov in 1997, raising ethical questions about AI's role in competitive environments and human-AI interactions (Campbell, Hoane, & Hsu, 2002). The launch of Sony's AIBO robot dog in 1999 brought AI's integration into consumer products, prompting discussions about the ethics of AI in everyday life.

2000s: The introduction of the Roomba vacuum cleaner in 2002 raised privacy concerns and the ethical implications of AI in household appliances. STANLEY, the autonomous vehicle developed by Stanford, won the DARPA Grand Challenge in 2005, highlighting ethical issues related to autonomous systems and safety (Thrun, 2006).

2010s: IBM's Watson wins Jeopardy! In 2011, leading to ethical debates on AI in information retrieval and decision-making (Ferrucci, 2010). The acquisition of DeepMind by Google and the development of AlphaGo in 2014, which defeated human champions, raises ethical concerns about AI surpassing human abilities (Silver, 2016). The Asilomar AI Principles were established in 2017, providing ethical guidelines for AI research and applications (Future of Life Institute, 2017). The AI Now Institute released a report in 2018 highlighting the social and ethical implications of AI, particularly in surveillance and bias (AI Now Institute, 2018).

2020s: The release of OpenAI's GPT-3 in 2020 raises ethical issues about the misuse of language models and the potential for generating misleading or harmful content (Brown, 2020). DeepMind's AlphaFold achieved significant breakthroughs in protein folding in 2021, showcasing AI's potential in scientific research and raising ethical considerations about the impact on the field of biology (Jumper, 2021). The launch of DALL-E 2 and other advanced generative models in 2022 highlighted the ethical implications of AI in creative fields, including issues of copyright, authenticity, and the potential for misuse (Ramesh, 2022). Continued advancements in AI technology over 2023-24 have prompted ongoing ethical debates, including the environmental costs of training large models, the impact on job markets, and the need for robust regulatory frameworks (Bender, Gebru, McMillan-Major, & Shmitchell, 2021) (Floridi L., 2018).

In summary, the evolution of AI ethics has been marked by significant milestones, from early theoretical foundations to the complex ethical issues that emerged alongside technological advancements. This timeline demonstrates the continuous evolution of ethical considerations in AI, highlighting key developments and the corresponding ethical challenges they pose. Understanding this history is essential for tackling the ethical challenges we face today.

Concluding Summary Points:

• Early foundational work by pioneers like Turing and McCulloch.

- Key milestones such as the Dartmouth Conference and the development of expert systems.
- The emergence of ethical considerations as AI technologies advanced.

Understanding the historical milestones in AI provides a backdrop for exploring the development of AI ethics, which has evolved alongside technological advancements.

Development of AI Ethics

Ethics has been defined by (Kazim & Koshiyama, 2021) as the "rational and systematic study of the standards of what is right and wrong, and morality as the commonly used term for notions of good and bad in more common use of the English language". (Moor, 2006) defined computer ethics as the analysis of the nature and social impact of computer technology and the corresponding formulation and justification of policies for the ethical use of such technology. (Anderson, 2011) defined machine ethics as being concerned with giving machines ethical principles or a procedure for discovering a way to resolve the ethical dilemmas they might encounter, enabling them to function in an ethically responsible manner through their own ethical decision making. It may also be noted that law and ethics are closely intertwined to ensure accountability and enforcement.

Concerns about AI ethics predate the term "artificial intelligence," with early discussions in literature (Goel, 2022). Isaac Asimov had proposed the famous three laws of robotics in 1942: "(i) First Law- A Robot may not injure a human being, or through inaction allow a human being to come to harm, (ii) Second Law- A Robot must obey the orders given it by human beings except where such orders would conflict with the First Law and (iii) Third Law-A Robot must protect its own existence as long as such protection does not conflict with the First or Second Laws." (Ellison & Asimov, 1978)

Computer ethics has been a topic of conversation since the 1970s, with the emergence of AI raising concerns. Early pioneers like Norbert Wiener and Joseph Weizenbaum understood how digital computing could impact aspects of life such as warfare and labor. By the 1980s and 1990s, computer ethics had developed into a field of applied ethics, with organizations like ACM and BCS creating guidelines for professionals.

As AI technology progressed and the pervasive diffusion of computing technology progressed into various aspects of life, so did the complexity of dilemmas concerning privacy, surveillance, autonomy, and ownership increase in magnitude (Stahl, Timmermans, & Mittelstadt, 2016). The idea of AI systems making decisions independently sparked discussions on responsibility and accountability.

Three main schools of thought, namely utilitarianism, rights-based ethics, and virtue ethics, play roles in AI ethics discussions (Kazim & Koshiyama, 2021). Utilitarianism looks at the consequences of actions taken by AI systems aiming to maximize happiness while minimizing suffering. Rights-based ethics require AI systems to uphold individual rights such as privacy and freedom of expression. However, challenges arise in ensuring these rights are respected when they clash with objectives or others' rights. Virtue ethics emphasizes the importance of character and moral virtues, which both individuals and AI systems should demonstrate. It emphasizes the development of virtues such as honesty, justice, and empathy in AI systems.

AI ethics draws inspiration from engineering ethics, technological philosophy, and science and technology studies (STS) (Kazim & Koshiyama, 2021). Engineering ethics deals with the values of engineering practice, taking into account the environmental impacts of engineering decisions. Philosophers such as Heidegger and Marcuse have contributed to discussions about the philosophy of technology by highlighting the risks associated with technological advancements. STS investigates how societal, cultural, and political factors influence research and technological innovation, as well as how these innovations impact society. By using methods to study the relationship between society and technology, STS challenges assumptions about technology. Having an understanding of these areas is essential for navigating the complex ethical dilemmas brought about by AI technologies. This comprehensive approach underscores the nature of AI ethics by integrating exploration, engineering principles, and empirical research to guide ethical AI development.

(Coeckelbergh, 2020) raises important questions regarding the moral implications of advanced artificial intelligence technologies. This is in regard to dilemmas about the moral status of artificial intelligence. The discussions predominantly revolve around two aspects: moral agency and moral patiency. Moral agency questions the moral significance of AI, regardless of its independence or sentience. On the other hand, moral patiency raises concerns about the permissibility of causing harm to or dismantling AI systems, prompting deeper contemplation on the essence of intellect and consciousness. These dialogues underscore the importance of establishing frameworks to navigate the changing terrain of AI technology, grappling with practical ethical challenges while pondering aspects like intelligence, consciousness, and the interplay between humans and machines.

Concluding Summary Points:

- Definition and scope of AI ethics.
- Influence of early discussions on current ethical frameworks.
- The integration of various ethical theories and interdisciplinary insights into AI ethics.

As AI technology advanced, so did the complexity of ethical dilemmas, leading to the formalization of AI ethics as a distinct field of study.

Contemporary Ethical Challenges

Privacy, Bias, and Transparency in AI

Artificial Intelligence (AI) systems, particularly those based on machine learning (ML) and deep learning (DL), operate through complex algorithms that learn patterns from large datasets. This data dependency raises significant ethical issues related to privacy, bias, and transparency (Pant, 2023).

Privacy Concerns

AI systems often require vast amounts of personal and sensitive information to function effectively. This data dependency raises significant privacy issues. For instance, the use of AI in facial recognition technology has led to widespread surveillance concerns. In 2019, it was revealed that the facial recognition startup Clearview AI had scraped billions of images from social media platforms without user consent to build its database, sparking a significant outcry over privacy violations and the misuse of personal data. Ensuring compliance with data protection regulations like the General Data Protection Regulation (GDPR) is essential to safeguard individuals' privacy (Voigt & Von dem Bussche, 2017). Stricter data governance principles are driving the development of more effective consent mechanisms, transparency in data usage, and privacy-preserving technologies such as differential privacy and federated learning.

Intellectual Property Issues

Intellectual property issues in AI are equally complex. Generative AI models like OpenAI's GPT-3 and DALL-E 2 can create content that closely resembles human work, including text, images, and music. This capability blurs the lines of authorship and ownership. In 2021, a legal case involving the AI-generated artwork "Edmond de Belamy" brought these issues to the forefront. The artwork, created by the Paris-based collective Obvious using a generative adversarial network (GAN), sold for $432,500 at Christie's auction house. This raised questions about the ownership of AI-generated works and whether AI systems or their developers should hold copyrights. Furthermore, the use of AI in content creation has led to disputes over the originality and authenticity of works, with concerns that AI might infringe on existing copyrights by producing derivative works.

Bias in AI

Bias in AI models can enter at various stages: during data collection, model training, or through the algorithms themselves (Bolukbasi, Chang, Zou, Saligrama, & Kalai, 2016) (Sambasivan, 2021). If the training data lacks diversity, the model may perform poorly or unfairly when applied to different demographic groups, exacerbating social inequalities. The COMPAS algorithm used in the United States for assessing the risk of recidivism among offenders faced significant scrutiny when a 2016 ProPublica investigation revealed that it exhibited racial bias, disproportionately labeling African American defendants as high-risk compared to their white counterparts. Addressing bias requires rigorous testing, the use of diverse datasets, and continuous monitoring to ensure fairness and accountability in AI systems.

Transparency and Explainability

Many AI systems, especially deep learning models, function as "black boxes" with decision-making processes that are not easily interpretable by humans (Doshi-Velez & Kim, 2017). This lack of transparency can be problematic in critical applications like healthcare and criminal justice, where understanding the rationale behind decisions is crucial for accountability and trust. For instance, IBM's Watson for Oncology, an AI system designed to recommend cancer treatments, faced criticism for providing unsafe and incorrect treatment advice in some instances, highlighting the importance of transparency in AI decision-making processes (Strickland, 2019) (Ross & Swetlitz, 2017). The principles-based approach to AI ethics emphasizes transparency, requiring AI systems to be explainable and understandable. The European Union's AI Act, for example, proposes strict requirements for high-risk AI systems, including transparency and accountability measures to safeguard privacy and IP rights.

Integrated Ethical Approach

Addressing these ethical challenges requires an integrated approach that prioritizes fairness, transparency, and accountability. Policymakers, developers, and researchers must collaborate to ensure that AI technologies are designed and implemented in ways that respect the rights and dignity of all individuals, particularly the most vulnerable. This includes incorporating diverse datasets, conducting thorough impact assessments, and engaging with affected communities throughout the AI development process. By tackling these issues head-on, stakeholders can promote ethical AI development that respects individual privacy, ensures fairness, and fosters trust in AI systems.

Concluding Summary Points:

- Privacy concerns data usage in AI.
- Bias in AI models and its societal implications.
- The importance of transparency and explainability in AI systems.

With the rise of AI systems that rely on large datasets, significant ethical challenges related to privacy, bias, and transparency have come to the forefront.

AI's Effect on Job Displacement

Artificial intelligence (AI) and automation technologies are rapidly transforming industries, with significant implications for job displacement and economic inequality. As AI systems become more capable, tasks traditionally performed by humans are increasingly automated, leading to concerns about widespread job losses. Current data indicates that sectors such as manufacturing, transportation, and retail are particularly vulnerable. For instance, a report by McKinsey Global Institute (McKinsey Global Institute, 2024) estimates that up to 800 million jobs could be displaced by automation by 2030, affecting approximately one-fifth of the global workforce. However, the impact of AI is not uniformly distributed; it disproportionately affects low- and middle-skill jobs, exacerbating economic inequality (Brynjolfsson & McAfee, 2014). High-skill jobs, particularly those requiring complex problem-solving and creativity, are less likely to be automated and may even see increased demand (McKinsey Global Institute, 2017). Further, it might result in critical areas such as aviation and health care being affected in a disastrous way in the event of malfunctioning or subversion of automated AI systems. (Floridi, 2018)

The rise of AI also creates new job opportunities, particularly in tech and AI-related fields (Autor, 2015). However, these new roles often require advanced skills and education, leading to a skills gap. Individuals without access to higher education or retraining programs are at risk of being left behind, further widening the economic divide. The World Economic Forum (World Economic Forum, 2023) emphasizes the importance of reskilling and upskilling initiatives to mitigate these impacts, suggesting that proactive policies and investment in education are critical to ensuring a more equitable future.

Future predictions suggest that while AI will continue to drive economic growth and productivity, it will also necessitate significant societal adjustments. Policymakers are urged to consider measures such as universal basic income, progressive taxation, and robust social safety nets to address the potential economic disruptions caused by AI. Additionally, fostering public-private partnerships to create inclusive growth strategies can help manage the transition to an AI-driven economy.

In conclusion, while AI and automation promise substantial economic benefits, they also pose significant challenges related to job displacement and economic inequality. Addressing these issues requires a multifaceted approach, including policy interventions, educational reforms, and inclusive economic strategies to ensure that the benefits of AI are broadly shared across society.

AI and Vulnerable Populations

Artificial Intelligence (AI) has significant implications for vulnerable populations, including individuals with disabilities, the elderly, and marginalized groups. AI technologies have the potential to empower these populations by enhancing accessibility and providing tailored support. For example, AI-powered assistive technologies, such as speech recognition and natural language processing, can facilitate communication for individuals with speech or hearing impairments. Similarly, AI-driven mobility aids, like autonomous wheelchairs and navigation systems, can enhance independence for those with physical disabilities (Reddy, Fox, & Purohit, 2021). For the elderly, AI can offer solutions to improve quality of life and safety. AI-based health monitoring systems can track vital signs and predict potential health issues, enabling timely interventions and reducing the burden on healthcare providers (Topol, 2019). Additionally, social robots equipped with AI can provide companionship and support, addressing loneliness and improving mental health among the elderly (Broadbent, Stafford, & MacDonald, 2018).

However, the deployment of AI also poses significant risks to these vulnerable groups. Bias in AI algorithms can lead to discriminatory outcomes, exacerbating existing inequalities. For instance, facial recognition systems have been shown to perform less accurately for individuals with darker skin tones, which can lead to false identifications and reinforce racial biases (Buolamwini & Gebru, 2018). In 2019, San Francisco became the first major city to ban the use of facial recognition technology by government agencies due to its potential for misuse and the risk of infringing on citizens' privacy rights. Marginalized groups may also face greater risks of surveillance and privacy violations, as AI technologies are increasingly used for monitoring and data collection without adequate consent or oversight (Eubanks, 2018). Furthermore, the automation of jobs through AI threatens employment opportunities for low-income individuals and those with lower educational levels, potentially widening socioeconomic disparities (West, 2018). Malicious use of AI could occur through instances of cyber attacks, automated hacking, data poisoning of existing AI infrastructure, and deployment of autonomous weapon systems (Brundage, 2018).

To mitigate these risks, it is crucial to develop inclusive AI systems that prioritize fairness, transparency, and accountability. Policymakers, developers, and researchers must collaborate to ensure that AI technologies are designed and implemented in ways that respect the rights and dignity of all individuals, particularly the most vulnerable (Noble, 2018). This includes incorporating diverse datasets, conducting thorough impact assessments, and engaging with affected communities throughout the AI development process.

Environmental Impact

The environmental costs of training large AI models have become a significant concern in the field of AI development. Training state-of-the-art models like GPT-3 and similar large-scale neural networks requires substantial computational power, leading to high energy consumption. For instance, training a single deep learning model can emit as much carbon dioxide as five cars in their lifetimes (Strubell, Ganesh, & McCallum, Energy and Policy Considerations for Deep Learning in NLP, 2019). This immense energy use not only contributes to greenhouse gas emissions but also strains local power grids, particularly in regions where renewable energy sources are scarce. The carbon footprint of AI is amplified by the need for extensive data storage and cooling systems for data centers, further exacerbating environmental impacts (Bender, Gebru, McMillan-Major, & Shmitchell, 2021).

To address these challenges, the AI community is increasingly focusing on sustainable practices. One approach is optimizing algorithms to be more energy-efficient. Techniques such as model pruning, quantization, and knowledge distillation can reduce the computational requirements of AI models without significantly sacrificing performance (Han, Mao, & Dally, 2016). Additionally, researchers are exploring the use of more efficient hardware, such as application-specific integrated circuits (ASICs) and field-programmable gate arrays (FPGAs), which can perform AI computations more efficiently than general-purpose processors.

Another critical strategy is the deployment of AI models in regions with abundant renewable energy resources. By locating data centers in areas with access to wind, solar, or hydroelectric power, companies can significantly reduce the carbon footprint associated with AI operations. Furthermore, there is a growing emphasis on transparency and reporting, with initiatives encouraging organizations to disclose the environmental impact of their AI projects and adopt green certifications (Schwartz, Dodge, Smith, & Etzioni, 2020).

Collaboration between industry, academia, and policymakers is essential to develop and enforce regulations that promote sustainable AI development. Encouraging the adoption of energy-efficient practices and investing in renewable energy infrastructure will be crucial in mitigating the environmental costs associated with the rapid advancement of AI technologies.

Concluding Summary Points:

- AI's impact on job markets and economic inequality.
- Risks and benefits of AI for vulnerable populations.
- Environmental costs associated with large-scale AI models.

Contemporary applications of AI bring forth new ethical challenges, particularly concerning job displacement, vulnerable populations, and environmental impact.

Interdisciplinary Perspectives of AI's Societal Impacts

To fully comprehend the ethical implications of AI, it is essential to integrate perspectives from law, sociology, anthropology and philosophy, which collectively enrich our understanding of how AI impacts broader societal structures.

From a legal standpoint, AI poses significant challenges regarding privacy, accountability, and regulation. Legal scholars emphasize the need for robust frameworks that ensure AI systems operate within ethical and legal boundaries, highlighting issues such as data protection and the responsibility of AI creators and users (Gasser & Almeida, 2017).

Sociological perspectives bring attention to the societal implications of AI, including how it reinforces or mitigates social inequalities and affects human interactions. Sociologists argue that AI technologies can perpetuate existing biases and power imbalances, necessitating critical examination of the social contexts in which AI is developed and deployed (Eubanks, 2018). Philosophical insights, particularly from ethics, provide foundational principles for evaluating the moral dimensions of AI, such as fairness, autonomy, and justice. Philosophers like (Floridi L., 2013; Floridi L., 2018) advocate for a comprehensive ethical framework that addresses the potential harms and benefits of AI, ensuring that these technologies align with human values and societal good. By incorporating these interdisciplinary views, we gain a holistic understanding of the ethical considerations surrounding AI and can develop more nuanced and effective guidelines for its responsible use.

Sociological perspectives also help explore how AI technologies influence social structures, power dynamics, and inequalities. For instance, studies reveal that AI systems can reinforce existing social biases and inequalities, especially when trained on biased data sets. This understanding is critical for developing AI systems that promote social equity and justice (Noble, 2018). Psychologists contribute by examining the cognitive and emotional impacts of AI on individuals (Turkle, 2011) (Sherry, 2019). AI technologies, such as social robots and virtual assistants, affect human behavior, emotional well-being, and mental health. For example, prolonged interaction with AI companions can impact social skills and emotional development, particularly in children. Additionally, the increasing use of AI in decision-making processes raises concerns about autonomy and trust, which are central themes in psychological research.

Anthropological studies offer insights into how different cultures perceive and interact with AI (Suchman, 2007) (Dourish & Bell, 2011). Cultural values and norms shape the acceptance and integration of AI technologies across societies. For example, anthropologists have documented how varying cultural contexts influence the ethical considerations and regulatory approaches to AI, highlighting the importance of culturally sensitive AI development. Understanding these cultural nuances is essential for creating AI systems that are ethically and socially acceptable in diverse settings.

Incorporating interdisciplinary perspectives ensures that AI development is not solely driven by technical feasibility but also considers the broader social, psychological, and cultural implications. This holistic approach fosters the creation of AI systems that are not only innovative but also ethically responsible and socially beneficial. Addressing these interdisciplinary insights can lead to more equitable, trustworthy, and culturally aware AI technologies, ultimately enhancing their positive impact on society .

Cultural Context

The application of ethical AI standards varies significantly across different cultural and social contexts, particularly in non-Western countries. In Western contexts, ethical AI discussions often emphasize individual privacy, transparency, and accountability, reflecting the values enshrined in liberal democratic traditions. Conversely, in many non-Western cultures, communal values and collective well-being often take precedence over individual rights. For instance, in many Asian countries, there is a greater emphasis on harmony, social stability, and collective benefit, which can lead to different approaches to data privacy and AI deployment. In China, the government's use of AI for social credit systems and surveillance illustrates a utilitarian approach where the perceived benefits for societal control and security outweigh individual privacy concerns. Japan looks at AI ethics as a tool to enhance its "reputation, influence and international competitiveness" (Wright, 2024).

Similarly, in Africa, AI ethics must consider the continent's unique socio-economic challenges, such as digital divide, access to technology, and the need for inclusive development. Initiatives like the African Union's Convention on Cyber Security and Personal Data Protection highlight the importance of balancing technological advancement with safeguarding citizens' rights and promoting digital equity. Latin

American countries, too, grapple with issues of inequality and governance, influencing their ethical AI frameworks to prioritize transparency and anti-corruption measures. The collaborative framework proposed by the Inter-American Development Bank emphasizes the need for ethical AI that addresses regional socio-economic disparities and fosters inclusive growth.

The diversity in ethical AI standards across cultures underscores the importance of context-sensitive approaches that respect and integrate local values and needs. International organizations and AI developers must engage with local stakeholders to ensure that ethical guidelines are adaptable and relevant to different cultural settings. By acknowledging and incorporating diverse cultural perspectives, the global discourse on AI ethics can become more inclusive and effective, promoting equitable and just outcomes in AI applications worldwide.

Ethics Perspectives From Non-Western Cultures

Enriching the discussion of AI ethics by including perspectives from non-Western cultures, particularly from Africa, Asia, and Latin America, offers a more comprehensive and inclusive understanding of the ethical challenges and solutions in AI development and deployment. In Africa, the concept of Ubuntu, which emphasizes community, mutual care, and interconnectedness, provides a unique ethical framework for AI that prioritizes communal well-being over individual benefits (Eze, 2010). This perspective can lead to AI applications that foster social cohesion and collective progress rather than exacerbating inequalities. Similarly, in Asia, Confucian ethics, which stress harmony, respect for authority, and moral rectitude, can guide AI development towards enhancing societal harmony and respecting cultural traditions (Lai, 2008). For instance, AI systems designed with these principles can be tailored to support social stability and respect for hierarchical relationships, which are deeply rooted in many Asian cultures.

Latin American perspectives often highlight the importance of social justice and equity. The ethics of AI in this context are deeply intertwined with historical and socio-political dynamics, such as colonialism and economic disparity. Latin American scholars and practitioners advocate for AI that addresses systemic inequities and empowers marginalized communities. This includes ensuring fair access to AI technologies and using AI to combat social injustices, such as improving healthcare and education for underserved populations.

Incorporating these non-Western ethical frameworks into the global discourse on AI ethics can address the dominance of Western-centric views and promote a more balanced approach to AI governance. It encourages the development of AI systems that are culturally sensitive and globally equitable, fostering international collaboration and understanding. By integrating these diverse perspectives, the

AI community can better anticipate and mitigate ethical issues, ensuring that AI benefits are distributed more fairly and ethically across different cultural contexts.
Concluding Summary Points:

- Legal, sociological, and philosophical insights into AI's societal impact.
- The need for culturally sensitive AI development.
- Holistic understanding of ethical considerations through interdisciplinary collaboration.

To address AI's societal impacts comprehensively, it is crucial to incorporate perspectives from various disciplines, including law, sociology, anthropology, and philosophy.

Real-World Examples and Case Studies

Real-world applications of AI have demonstrated both positive and negative impacts, underscoring the importance of ethical considerations in practice. The following examples illustrate how AI systems can both adhere to and violate ethical principles, highlighting contemporary challenges (Kazim & Koshiyama, 2021).

Social Media: Cambridge Analytica Scandal

The 2018 scandal involving Cambridge Analytica exposed how AI algorithms could be manipulated to influence public opinion and elections. AI algorithms used data harvested from millions of Facebook profiles without user consent to create targeted political advertisements (Cadwalladr & Graham-Harrison, 2018). This case raised ethical issues about data privacy, consent, and the impact of AI on democratic processes, highlighting the need for stringent data protection regulations and ethical AI practices.

Financial Sector: JPMorgan Chase's COIN

In the financial sector, AI algorithms are used for fraud detection, risk assessment, and personalized financial advice. JPMorgan Chase's COIN (Contract Intelligence) platform uses AI to review legal documents and has reportedly saved the company 360,000 hours of work annually. Despite these benefits, ethical issues such as transparency, accountability, and fairness remain crucial, particularly when algorithms make decisions that significantly impact individuals' financial lives.

Healthcare: Predicting Protein Structures

AI-driven diagnostic tools have significantly improved the accuracy and efficiency of disease detection. For instance, AI systems like Google's DeepMind have achieved remarkable success in predicting protein structures, which has profound implications for understanding diseases and developing new treatments (Jumper, 2021). However, the deployment of AI in healthcare also raises ethical concerns, such as patient privacy, data security, and the potential for biased outcomes if the training data lacks diversity (Ossa, 2024) (Mennella, 2024).

Creative Industries: AI-Generated Content

Generative AI models like OpenAI's GPT-3 and DALL-E 2 have opened new possibilities for content creation, from writing to visual arts. These technologies enable the rapid generation of high-quality content but also raise concerns about authorship, copyright, and the potential for misuse in creating misleading or harmful materials. For example, deepfake videos can generate realistic but fake images and videos, leading to misinformation and reputational damage.

In 2018 an edited video of Nancy Pelosi, the U.S. Speaker of the House was shared online making her speech seem garbled hinting at intoxication or health issues (Mervosh, 2019). There was another incident involving actor Tom Cruise, where fake videos portraying him in activities gained popularity on social media (Metz, 2021). These fake videos were so lifelike that many viewers mistook them for real. In 2019 the CEO of a UK energy company fell victim to a scam where €220,000 was transferred to a supplier after receiving a phone call from someone impersonating the CEO's boss using AI generated voice mimicry (Damiani, 2019).

AI in Law Enforcement: Predictive Policing

Predictive policing algorithms, such as those used by the Los Angeles Police Department, aim to allocate resources more efficiently by predicting crime hotspots (Singer, 2019) (Mittelstadt, 2016). However, these systems have been criticized for perpetuating racial biases and disproportionately targeting minority communities (Angwin, Larson, Mattu, & Kirchner, 2016) (Chouldechova, 2017). The potential for AI to reinforce existing prejudices necessitates rigorous ethical scrutiny and the implementation of fairness and transparency standards (O'Neil, 2016).

These examples demonstrate that while AI has the potential to bring about significant benefits, it also poses serious ethical challenges. They also illustrate the dual-edged nature of AI applications, underscoring the need for robust ethical frameworks to guide the responsible development and deployment of AI technologies.

Addressing these requires a comprehensive approach that includes robust ethical guidelines, continuous monitoring, and active involvement of diverse stakeholders to ensure that AI technologies are developed and deployed responsibly. Ensuring that AI systems are transparent, accountable, and fair is essential to maximize their benefits while mitigating potential harms.

Concluding Summary Points:

- Case studies like COMPAS, facial recognition, and Watson for Oncology illustrate ethical challenges.
- The dual-edged nature of AI technologies.
- Importance of continuous ethical oversight and accountability.

Real-world examples of AI applications highlight both the potential benefits and ethical pitfalls, underscoring the need for robust ethical frameworks.

Development of Ethical Principles and Frameworks

Growing applications of artificial intelligence (AI) in a variety of sectors are creating moral conundrums and upending accepted professional ethics (Boddington, 2017). Being able to make decisions on its own, artificial intelligence (AI) can make it challenging to hold someone accountable for any damage these choices may cause. This phenomenon may progressively reduce human control, bring up moral and autonomy issues, and maintain biases in the training data. The capacity of AI systems to analyze large volumes of personal data poses significant privacy risks, challenging existing legal and ethical frameworks meant to protect individual rights. They might, also automate jobs, which could raise moral questions about worker replacement, as discussed earlier. AI systems can be used maliciously, for example, in cyber-attacks or autonomous weapons. This necessitates a rethinking of security practices and ethical guidelines in technology development.

While IEEE and ACM first developed professional codes of conduct, they lacked specific guidance for AI development. Nevertheless, they laid the groundwork for subsequent, more detailed ethical frameworks. Professions such as law, engineering, and medicine adhere to traditional codes of ethics that emphasize responsibility, integrity, and client confidentiality. Reevaluating these ideas, though, is required as AI develops to take into consideration the participation of AI entities and the wider effects on society.

The healthcare industry is advancing through various methods such as utilizing AI diagnostic tools. This also raises concerns about safeguarding patient privacy and well-being, ensuring transparency in AI decision-making processes, and effectively communicating findings to patients and their families. The use of AI in the legal profession may necessitate new laws governing the use and security of data. This is caused by possible changes in the techniques for gathering and assessing evidence.

Engineers creating AI systems must consider not only how well their systems work, but also how they might affect society in the long run, such as possible changes in power dynamics and job displacement. As AI systems increasingly entered important decision-making procedures in the early 2000s, transparency, accountability, and fairness became crucial. Accountability is the responsibility of AI developers and operators; transparency is the degree of clarity in AI systems' operation; and fairness is the guarantee that AI systems do not exacerbate or prolong current social injustices and prejudices. (Boddington, 2017)

Ethical Principles

(Kazim & Koshiyama, 2021) propose three approaches to AI ethics: principles-based, processes-based, and ethical consciousness. Principle-based design entails creating ethical guidelines for AI system design and use. Few examples of principle-based approach are given below. Process-based integration incorporates ethical considerations into AI technology development, advocating for 'ethical by design' practices. Examples include IEEE Ethically Aligned Design (IEEE Global Initiative on Ethics of Autonomous and Intelligent Systems, 2019), which emphasizes human rights, well-being, accountability, transparency, and awareness of misuse. Ethical consciousness advocates for moral awareness and a culture shift among AI developers, promoting an ethical mindset that goes beyond compliance with guidelines and laws.

The principles-based approach to AI is a framework that emphasizes the alignment of AI with human values, fairness, transparency, accountability, and safety. An example of such principles is the Google's AI Principles (Google, 2023), which aims to be socially beneficial, avoid unfair bias, build and test for safety, be accountable to people, incorporate privacy design principles, uphold high standards of scientific excellence, and make AI available for uses that accord with these principles.

The Asilomar AI Principles (Future of Life Institute, 2017), published by the Future of Life Institute in 2017, is the second example of a principles-based approach to AI. It also emphasizes the importance of AI systems being safe and secure throughout their operational lifetime. They emphasize failure transparency(if an AI system causes harm, it should be possible to ascertain why), judicial transparency(any involvement by an autonomous system in judicial decision-making should provide a satisfactory explanation auditable by a competent human authority), responsibility(designers and builders of advanced AI systems are stakeholders in the moral implications of their use, misuse, and actions, with a responsibility), value alignment(highly autonomous AI systems should be designed so that their goals and behaviors can be assured to align with human values throughout their operation), human values(AI systems should be designed and operated so as to be compatible with ideals of human dignity, rights, freedoms, and cultural diversity), personal privacy(people should have the right to access, manage and control the data they generate, given AI systems' power to analyze and utilize that data), liberty and privacy(application of AI to personal data must not unreasonably curtail people's real or perceived liberty), shared benefit(AI technologies should benefit and empower as many people as possible), shared prosperity(economic prosperity created by AI should be shared broadly, to benefit all of humanity), human control(humans should choose how and whether to delegate decisions to AI systems, to accomplish human-chosen objectives), non-subversion(power conferred by control of highly advanced AI systems should respect and improve, rather than subvert, the social and civic processes on which the health of society depends), risks(risks posed by AI systems, especially catastrophic or existential risks, must be subject to planning and mitigation efforts commensurate with their expected impact), and recursive self-improvement(AI systems designed to recursively self-improve or self-replicate in a manner that could lead to rapidly increasing quality or quantity must be subject to strict safety and control measures).

The Montreal Declaration for Responsible AI (Morandín-Ahuerma, 2023) is the third example of a principles-based approach to AI. It aims to permit the growth of the well-being of all sentient beings, respect people's autonomy, protect privacy and intimacy from intrusion and data acquisition and archiving systems (DAAS), maintain bonds of solidarity among people and generations, meet intelligibility, justifiability, and accessibility criteria, contribute to a just and equitable society, maintain social and cultural diversity, exercise caution in AI development, and ensure strong environmental sustainability.

The European Commission's High-Level Expert Group on Artificial Intelligence published the Ethics Guidelines for Trustworthy AI in 2019 as the fourth example of a principles-based approach to AI. It introduced the concept of "trustworthy AI," which encompasses lawful, ethical, and robust AI systems. These guidelines propose seven key requirements for achieving trustworthy AI, including human agency and oversight, privacy, and accountability (Stamboliev & Christiaens, 2024).

The IEEE Global Initiative on Ethics of Autonomous and Intelligent Systems launched the Ethically Aligned Design document, which is the fifth example of a principles-based approach to AI. It is a comprehensive framework that discusses ethical considerations and offers recommendations for value-based design. This framework covers principles like human rights, well-being, data agency, and effectiveness, offering a holistic approach to ethical AI.

The development of AI ethics guidelines by different stakeholders—public sector organizations, research institutes, and private companies—has increased dramatically. But most of these recommendations come from industrialized nations, which raises the possibility of bias (Gupta & Heath, 2020) in the framing of global AI ethics and as well as the possibility of advancing inequality in the Global South (Astobiza, 2022).

The themes emerging in the field of AI ethics guidelines include transparency, justice, fairness, non-maleficence, responsibility and accountability, privacy, beneficence, freedom and autonomy, trust in AI, dignity, sustainability, and solidarity (Jobin, Ienca, & Vayena, 2019) (Ryan & Stahl, 2021). Transparency means increasing explainability and understanding of machine mechanics, which is critical for validating and improving AI systems. Justice and fairness involve preventing bias, respecting diversity, inclusion, equality, and fair access to AI. It may note here that there are differing interpretations of these concepts among different communities (Gupta & Heath, 2020). The principle of non-maleficence dictates that AI must not inflict harm through discrimination, privacy violations, or bodily harm, and it must also avoid misuse. Responsibility and accountability involve ensuring AI systems operate ethically, safely, and within legal frameworks, while accountability involves assigning blame or praise to individuals or organizations for their outcomes. AI complicates traditional accountability concepts due to its ability to learn and adapt independently. Determining liability for actions taken by AI systems, particularly in autonomous vehicles, is therefore complex and unresolved. AI products may be treated under product liability laws, but these are not fully adapted to unique issues. Debates continue around granting legal personhood to AI entities to take care of these issues. (Frankish & Ramsey, 2014) (Wallach & Allen, 2009).

Privacy in AI involves ensuring responsibility, security, and compliance with data protection laws, such as the European Union's GDPR (General Data Protection Regulation).Beneficence refers to the obligation to act for the benefit of others, promoting good and preventing harm. AI technologies should promote human and societal well-being while minimizing potential harms and risks. Freedom and autonomy should promote individuals' rights to act independently and make choices without external coercion. Transparent operations and clear consent mechanisms are crucial for supporting users' freedom and autonomy. Trust in AI is crucial for its reliability, safety, and alignment with human values. Trust is built through rigorous testing, validation, and continuous monitoring. Dignity is linked to human rights and is seen as something that AI should not only preserve but also promote (Santos, 2002). To uphold dignity, it is suggested that AI developers should respect it in their processes, and governance and new legislation should support its maintenance in the context of AI technologies. Sustainability in AI technology involves promoting long-term environmental responsibility and societal well-being, enhancing ecological balance, contributing to social equity, and ensuring economic viability for future generations. Designing AI solutions that minimize environmental impact, promote fair benefits, and support economic development without depleting natural resources is essential. The impact of AI on the labor market primarily affects social solidarity, highlighting the need for a robust social safety net and equitable distribution of AI's benefits to uphold social cohesion, prevent negative effects from escalating social divisions, and prevent radical individualism.

(Kazim & Koshiyama, 2021) summarized the above themes in six buckets: human agency and oversight, safety, privacy, transparency, fairness, and accountability. AI should enhance human autonomy and decision-making capabilities, ensuring users understand interactions. It should be secure, reliable, and robust against manipulations and errors, with fallback plans and risk assessments. Privacy is crucial, and systems should use only the necessary data for their function. Transparency is essential, with clear explanations available for decision-making processes. AI should operate without bias, offering equal opportunity and treatment across different user demographics. Systems should be accessible to all and designed to avoid reinforcing existing inequalities. Finally, clear accountability mechanisms must be in place to track decisions and address any negative impacts or failures.

Ethics washing and ethics shopping are terms used to misuse ethical guidelines and discussions for ulterior motives. Ethics washing involves using ethical language and symbolism to superficially demonstrate commitment to ethical standards in AI development, deflecting criticism, mitigating regulation, and maintaining public trust without accountability. Ethics shopping involves selectively endorsing ethical standards or frameworks that are least disruptive to business operations, often choosing those that are the least stringent. Companies aim to argue compliance while continuing practices that might be ethically dubious or controversial. For example, a corporation might align with an ethics framework that imposes minimal constraints on data usage, allowing more freedom to exploit user data for profit while claiming ethical compliance. (Bayamlıoğlu, Baraliuc, Janssens, & Hildebrandt, 2018).

(Mittelstadt B., 2019) has warned some AI ethics initiatives, particularly from industry, are seen as virtue-signaling to delay regulation (Hagendorf, 2020), producing vague principles that lack specific recommendations. Though many AI ethics initiatives mirror medical ethics, AI development lacks common aims and fiduciary duties, professional history and norms, proven methods to translate principles into practice, and robust accountability mechanisms, unlike medicine. AI developers often face pressure to prioritize company interests, unlike medical practitioners who align with patient welfare. AI lacks a shared professional culture or ethical standards, unlike medicine with its longstanding codes and moral obligations. AI lacks proven methods to translate high-level principles into actionable requirements, unlike medicine with its established frameworks. Concerns regarding self-regulation arise because AI lacks professional and legal accountability frameworks akin to those found in medicine.

Balancing ethics principles that apply universally with the cultural, social, and political variations seen across different countries presents a significant challenge. It involves defining what ethical behavior for AI entails and establishing guidelines that go beyond national borders. The lack of agreement on ethical principles stems from varying cultural values, resulting in differing opinions on the ideal form of ethical AI. There is a discussion regarding whether international organizations or individual governments should take the lead in regulating AI, with some advocating for robust global regulations and others favoring customized rules. Building trust and promoting global collaboration through the establishment of AI ethics standards is crucial, to ensure the fair development and utilization of AI technologies for the benefit of all humanity. This necessitates continuous dialogue, interdisciplinary cooperation, and a dedication to harmonizing ethical viewpoints. (Coeckelbergh, 2020)

Ethical Frameworks

Ethical guidelines serve as principles or statements that outline the expected behaviors and outcomes in the development and deployment of AI technology. They provide guidance on essential considerations and priorities, including ensuring AI safety, promoting transparency, and upholding fairness. Distinct from frameworks, which offer a structured method for analyzing and deciding on ethical matters (Coeckelbergh, 2020), ethical guidelines are more about providing practical guidance for navigating complex ethical dilemmas faced by individuals or organizations. Ethical frameworks often include questions, criteria, or considerations to assist users in assessing aspects of their actions or policies. Two prominent ethical frameworks include (a) the Ken Blanchard and Norman Vincent Peale frameworks, which emphasize legality, fairness, and user experience impact, and (b)the Markkula Center Framework which outlines five approaches to addressing ethical issues based on utilitarianism, rights, fairness, common good principles, and virtue ethics (Siau & Wang,, 2020).

Concluding Summary Points:

- Various principles-based approaches to AI ethics.
- Themes such as transparency, fairness, and accountability.
- The need for balancing universal principles with cultural variations.

The development of ethical principles and frameworks provides a structured approach to ensuring responsible AI development and deployment.

The Role of Public Perception and Media in Shaping AI Ethics

Public perception and media representation plays a critical role in shaping the ethical landscape of AI. Media portrayals can influence societal attitudes, which in turn impact policy and ethical discourse. The popular news stories and popular movies and other kind of media storytelling affects how people understand and accept AI technology, especially creative AI. (Cave & Dihal, 2019) surveyed 300 fictional and non-fictional works to categorize the fundamental fears about artificial intelligence into the following four dichotomies, each paired with a corresponding hope. These fears are deeply intertwined with their counterpart hopes, often reflecting a tension between the benefits AI could bring and the potential worst-case scenarios if those technologies go awry or are misused.

1. Inhumanity: As a result of the drastic modifications required to achieve immortality through AI, a person may literally lose their identity, along with their human values and feelings.
2. Obsolescence: This fear is linked to the desire for ease and the release from taxing labor. Many people fear they will have no purpose or place in society as a result of AI rendering human labor unnecessary. It can take two forms: voluntary obsolescence, where people choose to live a dull and meaningless life due to the ease that comes with using an overly capable AI system, and involuntary obsolescence, where people are left idle without their consent.
3. Alienation: There is a fear of alienation that AI-driven interactions could take the place of real human connections, resulting in loneliness and the loss of meaningful relationships. This includes worries that AI will become too perfect or dominant in meeting human needs and negating the need for human interaction.
4. Uprising: There is a concern that AI could turn against humanity. This includes situations in which artificial intelligence (AI), particularly in the form of autonomous weapons or systems, acquires excessive control or becomes autonomous, which could spark a rebellion against human rulers.
5. There is concern that AI might surpass human intelligence and create situations in which AI could endanger human existence. Because of this, the term "superintelligence" was created to describe intellect that greatly surpasses human cognitive ability. This might occur if AI starts to develop on its own without human intervention. If such a superintelligent AI gets out of control and starts acting against human interests, it could endanger humankind. A superintelligent AI might decide to focus on self-preservation or other instrumental objectives. The "control problem," or how to create a superintelligence that will behave in a way consistent with human values and interests, is a central theme. (Bostrom, 2014)

6. One of the bigger effects of AI on society is its potential to exacerbate inequalities, as those with access to AI technology could rapidly outpace those without. Additionally, the role of AI in decision-making processes in sectors like finance, healthcare, and law enforcement raise ethical questions about fairness, bias, and the accountability of AI systems.

The way media has shown creative AI, like deepfake technology, made people aware of how it can be abused through misinformation, fraud and loss of privacy. The sensationalism that is typical to these kinds of stories may make people even more worried and create an atmosphere of fear, which then may lead to calls for tighter moral rules and governments to do something. On the other hand, if good stories about generative AI are shown (for example, how it can be used in art, education, or healthcare), it may make people more open to those advances in technology. Public perception, as shaped by media narratives, plays a pivotal role in the prioritization of ethical issues. Problems that are important to most people, like how AI affects their work, or their privacy tend to gain more attention from policymakers and ethical standard-setting bodies.

Public advocacy groups, researchers, and citizens use media to raise awareness and push for ethical AI development. Social media platforms like Twitter, Facebook, and Reddit play a crucial role in shaping public perception on AI ethics. These platforms enable widespread engagement and debate, raising ethical concerns globally. The media's role in shaping ethical standards in AI is complex, as it can oversimplify complex issues or propagate misinformation, raising concerns about its ethical responsibility in reporting on AI.

Concluding Summary Points:

- Media's role in highlighting ethical issues and shaping public opinion.
- Impact of media portrayals on policy and ethical guidelines.
- Importance of responsible media reporting on AI technologies.

Public perception and media representation significantly influence the ethical discourse on AI, shaping societal attitudes and policy responses.

Regulatory Responses to AI Ethics

The rise of generative AI has led to the necessity of developing and enacting regulatory frameworks to tackle its unique ethical issues. AI's hypersonic rate and agile nature of development make regulation and control through traditional legal frameworks almost impossible. The debate about ethical versus legal regulation of

AI systems is about the most suitable way to regulate such a rapidly developing technology (Elliott, 2022) (Ebers, 2019) (Chesterman, 2023).

Ethical regulation, also known as ethical self-regulation, is a softer, more flexible approach in which the organization producing or using AI voluntarily follows ethical rules and principles. Ethical regulation allows more room for maneuver, which leads to more innovation and best practices. The issue with ethical regulation stems from its lack of enforceability and standardization, which allows different organizations to adhere to different ethical principles, resulting in nearly uncontrollable and unpredictable AI-related ethical behavior.

Legal regulation, on the other hand, is about governmental or sometimes intergovernmental bodies enacting specific laws and regulations that control the behavior of AI or behavior in relation to AI. Compared to ethical regulation, legal regulation is more standard and enforceable. A legal framework can aid in fostering trust in AI within society, and ensuring the public is confident that AI regulations are in place to prevent any potentially harmful outcomes. The problem with legal regulation is its rigidity and control orientation. It is challenging to adapt or implement legal regulations to the new, rapidly developing AI technology. Legal regulations can cause regulatory competition due to different jurisdictions having different legal environments. Considering the contrast between ethical and legal regulation of AI, a balanced approach is required where the legal aspect guides and the ethical aspect helps innovate. Dynamic regulation, which incorporates ethical principles, is more appropriate for the rapidly evolving world of AI development today. International cooperation in ethical and legal AI regulation can help reduce conflicts between various jurisdiction standards.

The rapid advancement of artificial intelligence (AI) and its increasing impact on society have necessitated the development and enactment of regulatory frameworks to address unique ethical challenges. These regulatory responses vary globally, reflecting different cultural, political, and legal contexts. This section explores various global approaches and their effectiveness in governing AI.

European Union

The European Union (EU) has taken a proactive stance in regulating AI through comprehensive frameworks. The General Data Protection Regulation (GDPR), implemented in 2018, addresses data privacy and security, emphasizing transparency, consent, and the right to explanation (Voigt & Von dem Bussche, 2017). The proposed Artificial Intelligence Act aims to classify AI applications by risk level and enforce stringent requirements for high-risk AI systems, such as those used in critical infrastructure and law enforcement (European Commission, 2021) (Laux,

2024). These regulations underscore the importance of transparency, accountability, and human oversight in AI applications.

United States

In the United States, regulatory responses to AI are more fragmented, with various state-level initiatives complementing federal guidelines. The California Consumer Privacy Act (CCPA) of 2018 provides protections similar to GDPR, focusing on consumer rights regarding personal data. Additionally, California's Bot Disclosure Law mandates that automated accounts on social media disclose their non-human nature to prevent deceptive practices (California Legislative Assembly, 2018). Despite these efforts, the U.S. lacks a comprehensive federal AI regulation, resulting in a patchwork of state laws and initiatives. The Algorithmic Accountability Act proposes that companies assess and mitigate biases in their automated systems, but it remains a legislative proposal without nationwide implementation.

China

China's regulatory approach emphasizes state control and surveillance capabilities. The New Generation Artificial Intelligence Development Plan highlights the need for AI ethics and governance, but it is often criticized for prioritizing governmental oversight over individual privacy and freedoms (Roberts, et al., 2021). This approach reflects China's broader socio-political landscape, where AI is leveraged to enhance state control and societal stability.

Canada

Canada's Directive on Automated Decision-Making, implemented in 2019, mandates transparency, accountability, and fairness in AI systems used by the federal government (Treasury Board of Canada Secretariat., 2023). This directive sets a precedent for responsible AI deployment in public services, ensuring that automated decisions are explainable and equitable. The directive emphasizes the need for algorithmic impact assessments and provides guidelines for ethical AI use in government operations.

Singapore

Singapore has adopted a forward-thinking approach with its Model AI Governance Framework, which promotes transparency and explainability in AI systems. The framework encourages organizations to adopt ethical practices and align with international standards. The testing framework consists of 11 AI ethics principles of transparency, explainability, repeatability/reproducibility, safety, security, robustness, fairness, data governance, accountability, human agency and oversight, inclusive growth, societal and environmental well-being. (Singapore Personal Data Protection Commission, 2020)

Effectiveness of Global Approaches

The effectiveness of these regulatory frameworks varies based on their scope, enforceability, and adaptability. The EU's comprehensive approach provides a robust model for balancing innovation with ethical considerations, setting high standards for AI governance. The U.S. approach, while innovative, suffers from a lack of uniformity and comprehensive federal oversight. China's regulatory framework, though effective in achieving state goals, raises significant ethical concerns regarding individual rights. Canada's directive offers a balanced approach to transparency and fairness in AI, particularly in public services, while Singapore's framework serves as a model for integrating ethical principles into AI governance.

Challenges and Future Directions

Regulatory responses to AI ethics face several challenges, including defining AI, addressing ethical concerns, addressing global disparities, integrating AI across various sectors, addressing data protection, addressing employment impacts, building public trust in AI technologies, addressing cross-sectoral impacts, addressing rapid technological advancements, and enforcing regulations effectively (Owczarczuk, 2023). Global disparities in AI governance and the potential for regulatory competition further complicate efforts to establish uniform standards. To address these challenges, a balanced approach that incorporates both ethical and legal regulations is necessary. Dynamic regulation, which evolves alongside technological advancements, can help mitigate the risks associated with AI while promoting innovation.

Defining AI is challenging due to its broad range of applications, including virtual assistants, automatic news aggregation, image and speech recognition, translation, financial trading, self-driving cars, and automated weapon systems (Ebers, 2019). Ethical concerns include potential biases in decision-making processes, privacy concerns, and the risk of enhanced surveillance capabilities. To ensure responsible AI technology development and use, (McLennan, 2020) suggests embedding ethicists in the AI development team.

Regulating competition among regions and countries can lead to a "race to the bottom" where jurisdictions compromise on ethical standards to attract AI businesses. Global disparities in AI governance can create challenges in international cooperation and standardization. Enforcement of regulations requires resources and expertise that many regulatory bodies are still developing.

Rapid technological advancements pose significant challenges in ensuring new AI applications are safe and ethical. Tailored approaches are needed to address sector-specific issues. Data protection is crucial as AI systems rely heavily on data, raising concerns about privacy and security.

Regulatory frameworks must address the social and economic impacts of AI displacements and build public trust in AI technologies. Cross-sectoral impacts require a coordinated approach across traditional regulatory domains.

International cooperation is crucial for developing cohesive AI regulations that transcend national borders. Collaboration between regulators, AI developers, researchers, ethicists, and the broader public is essential to ensure that AI technologies are developed and deployed responsibly. By fostering global dialogue and harmonizing ethical standards, stakeholders can build trust in AI systems and ensure their benefits are broadly shared across society.

In conclusion, regulatory responses to AI ethics aim to prevent harm and foster an environment where ethical AI can thrive. Effective regulations can build public trust in AI systems and promote innovation while ensuring that generative AI serves the public interest. Collaboration between regulators, AI developers, researchers, ethicists, and the broader public is essential for this goal.

Concluding Summary Points:

- Overview of regulatory frameworks in the EU, US, China, Canada, and Singapore.
- Effectiveness and challenges of these regulatory approaches.
- Necessity of dynamic regulation and international cooperation.

Global regulatory responses to AI ethics reflect diverse approaches and effectiveness, highlighting the need for adaptive and comprehensive frameworks.

Ethical Standards in AI Research and Development

Ethical norms in AI research and development are critical to the integrity of the AI research projects and promotion of society's well being. It assumes importance due to AI's capability for potential harm to privacy, autonomy, and even the societal framework, and the ethical dilemmas posed go beyond conventional academic considerations. Responsible Research and Innovation (RRI) principles are vital for ethical AI research. These principles require researchers to anticipate wider consequences of their study and actively engage with all stakeholders. Ethical Review Boards of academic institutions also have crucial supervisory function in AI research (Knight,

2024). They evaluate proposed experiments, analyze potential risks and benefits, and ensure adherence to ethical rules. Academia bears a unique obligation to establish ethical benchmarks for AI research, integrate ethics into the educational program, prepare prospective AI professionals, and promote cross-disciplinary research. Partnerships among computer scientists, ethicists, social scientists, and legal specialists have the potential to provide more thorough and pragmatic ethical principles.

Concluding Summary Points:

- Importance of Responsible Research and Innovation (RRI) principles.
- Role of Ethical Review Boards in supervising AI research.
- Academia's responsibility in promoting ethical AI development.

Ethical standards in AI research and development are critical for ensuring the integrity and societal benefits of AI technologies.

CONCLUSION

The evolution of ethical standards and guidelines in artificial intelligence (AI) is a testament to the rapid advancements in technology and the growing recognition of its profound impact on society. From the early theoretical foundations laid by pioneers like Turing and McCarthy to contemporary discussions on bias, transparency, and privacy, the journey of AI ethics reflects the dynamic interplay between innovation and ethical responsibility.

AI's integration into various sectors, including healthcare, criminal justice, finance, and social media, underscores the need for robust ethical frameworks that address unique challenges. Examples of the COMPAS algorithm, facial recognition technology, IBM's Watson for Oncology, and the Cambridge Analytica scandal highlight both the potential and pitfalls of AI applications. These cases emphasize the importance of fairness, transparency, accountability, and the need for continuous monitoring to ensure ethical AI deployment.

The interdisciplinary perspectives from law, sociology, anthropology, and philosophy enrich our understanding of AI's societal impacts. These perspectives reveal how AI technologies can reinforce or mitigate social inequalities, affect human interactions, and influence cultural norms. By integrating these insights, we can develop more nuanced and effective ethical guidelines that consider the broader social, psychological, and cultural implications of AI.

Global regulatory responses to AI ethics, such as the EU's GDPR, the US's CCPA, and Singapore's Model AI Governance Framework, demonstrate varying approaches to addressing these challenges. The effectiveness of these frameworks depends on their scope, enforceability, and adaptability to rapid technological advancements. A balanced approach that incorporates both ethical and legal regulations, along with international cooperation, is essential for fostering responsible AI development.

As AI continues to advance, new ethical challenges will emerge, including the environmental impact of large-scale AI models, job displacement, and the integration of AI with other advanced technologies. Addressing these challenges requires a proactive and adaptive regulatory approach that evolves alongside technological progress. Collaboration between policymakers, developers, researchers, and the public is crucial for ensuring that AI technologies are designed and implemented in ways that respect human values and rights.

Future directions in AI ethics must focus on fostering public trust, ensuring fairness and accountability, and promoting sustainable AI practices. Ethical AI development should prioritize transparency, explainability, and inclusivity, ensuring that AI benefits are broadly shared and do not exacerbate existing inequalities. By embedding ethical considerations into the AI lifecycle—from design and development to deployment and regulation—we can create AI systems that enhance human well-being and societal good.

In conclusion, the evolution of ethical standards and guidelines in AI is an ongoing process that requires continuous reflection, adaptation, and collaboration. By learning from past experiences, embracing interdisciplinary insights, and fostering a culture of ethical awareness, we can navigate the complex landscape of AI ethics and ensure that AI technologies contribute positively to society.

REFERENCES

AI Now Institute. (2018, December). *AI Now Report2018*. Retrieved from https://ainowinstitute.org/AI_Now_2018_Report.pdf

Anderson, M. (2011). *Machine ethics*. Cambridge University Press. DOI:10.1017/CBO9780511978036

Angwin, J., Larson, J., Mattu, S., & Kirchner, L. (2016, May 23). *ProPublica*. Retrieved from Machine Bias: https://www.propublica.org/article/machine-bias-risk-assessments-in-criminal-sentencing

Asimov, I. (1976). *Robot Visions*. Penguin Books.

Astobiza, A. M. (2022). Ethical Governance of AI in the Global South: A Human Rights Approach to Responsible Use of AI. *MDPI Proceedings*, 1-5.

Autor, D. H. (2015). Why Are There Still So Many Jobs? The History and Future of Workplace Automation. *The Journal of Economic Perspectives*, 29(3), 3–30. DOI:10.1257/jep.29.3.3

Bayamlıoğlu, I. E., Baraliuc, I., Janssens, L., & Hildebrandt, M. (2018). *Being Profiled: Cogitas Ergo Sum: 10 years of "Profiling the European Citizen"*. Amsterdam University Press. DOI:10.5117/9789463722124

Bender, E. M., Gebru, T., McMillan-Major, A., & Shmitchell, S. (2021). On the Dangers of Stochastic Parrots: Can Language Models Be Too Big? *Proceedings of the 2021 ACM Conference on Fairness, Accountability, and Transparency* (pp. 610-623.). New York City: Association for Computing Machinery. DOI:10.1145/3442188.3445922

Boddington, P. (2017). *Towards a Code of Ethics for Artificial Intelligence*. Springer. DOI:10.1007/978-3-319-60648-4

Bolukbasi, T., Chang, K. W., Zou, J. Y., Saligrama, V., & Kalai, A. T. (2016). Man is to Computer Programmer as Woman is to Homemaker? Debiasing Word Embeddings. *Advances in Neural Information Processing Systems*, ●●●, 4349–4357.

Bostrom, N. (2014). *Superintelligence- Paths, Dangers and Strategies*. Oxford University Press.

Broadbent, E., Stafford, R., & MacDonald, B. (2018). Acceptance of Healthcare Robots for the Older Population: Review and Future Directions. *International Journal of Social Robotics*, ●●●, 257–271.

Brown, T. e. (2020, July 22). *Language Models are Few-Shot Learners.* Retrieved from ARXIV: https://arxiv.org/abs/2005.14165

Brundage, M. e. (2018). *The Malicious Use of Artificial Intelligence: Forecasting, Prevention, and Mitigation.* Oxford: Future of Humanity Institute.

Brynjolfsson, E., & McAfee, A. (2014). *The Second Machine Age: Work, Progress, and Prosperity in a Time of Brilliant Technologies.* W.W. Norton & Company.

Buchanan, B., & Shortliffe, E. (1984). *Rule-based expert systems: The MYCIN experiments of the Stanford heuristic programming project.* Addison-Wesley.

Buolamwini, J., & Gebru, T. (2018). Gender Shades: Intersectional Accuracy Disparities in Commercial Gender Classification. *Proceedings of the 1st Conference on Fairness, Accountability and Transparency*, (pp. 77-91.).

Cadwalladr, C., & Graham-Harrison, E. (2018, March 17). *Revealed: 50 million Facebook profiles harvested for Cambridge Analytica in major data breach.* Retrieved from The Guardian: https://www.theguardian.com/news/2018/mar/17/cambridge-analytica-facebook-influence-us-election

California Legislative Assembly. (2018, July 1). *California Civil Code, Section 17940 (2018). Bot Disclosure Law.* Retrieved from Casetext-Thomson Reuters: https://casetext.com/statute/california-codes/california-business-and-professions-code/division-7-general-business-regulations/part-3-representations-to-the-public/chapter-6-bots/section-17941-unlawful-use-of-bots

Campbell, M., Hoane, A. J.Jr, & Hsu, F. H. (2002). Deep Blue. *Artificial Intelligence*, 134(1-2), 57–83. DOI:10.1016/S0004-3702(01)00129-1

Cave, S., & Dihal, K. (2019). Hopes and fears for intelligent machines in fiction and reality. *Nature Machine Intelligence*, 1(2), 74–78. DOI:10.1038/s42256-019-0020-9

Chesterman, S. (2023, May). *From Ethics to Law: Why, When, and How to Regulate AI.* Retrieved from NUS Law Working Paper: https://law.nus.edu.sg/wp-content/uploads/2023/05/014_SimonChesterman.pdf

Chouldechova, A. (2017). Fair prediction with disparate impact: A study of bias in recidivism prediction instruments. *Big Data*, 5(2), 153–163. DOI:10.1089/big.2016.0047 PMID:28632438

Coeckelbergh, M. (2020). *AI Ethics.* The MIT Press. DOI:10.7551/mitpress/12549.001.0001

Copeland, J. B. (2004). *The Essential Turing-Seminal Writings in Computing, Logic, Philosophy, Artificial Intelligence, and Artificial Life plus The Secrets of Enigma.* Oxford University Press.

Damiani, J. (2019, September 3). *A Voice Deepfake Was Used To Scam A CEO Out Of $243,000.* Retrieved April 30, 2024, from Forbes: https://www.forbes.com/sites/jessedamiani/2019/09/03/a-voice-deepfake-was-used-to-scam-a-ceo-out-of-243000/?sh=2354d82f2241

Data & Society. (2018, April 18). *Algorithmic Accountability: A Primer.* Retrieved from Data & Society: https://datasociety.net/library/algorithmic-accountability-a-primer/

Doshi-Velez, F., & Kim, B. (2017, March 2). *Towards a Rigorous Science of Interpretable Machine Learning.* Retrieved from ARXIV: https://arxiv.org/abs/1702.08608

Dourish, P., & Bell, G. (2011). *Divining a Digital Future: Mess and Mythology in Ubiquitous Computing.* MIT Press. DOI:10.7551/mitpress/9780262015554.001.0001

Ebers, M. (2019). *Algorithms and Law (Regulating AI and Robotics: Ethical and Legal Challenges).* Cambridge University Press.

Elliott, A. (2022). *The Routledge Social Science Handbook of AI.* Routledge.

Ellison, H., & Asimov, I. (1978). *I, Robot: The Illustrated Screenplay.* Warner Books.

Eubanks, V. (2018). *Automating Inequality: How High-Tech Tools Profile, Police, and Punish the Poor.* St. Martin's Press.

European Commission. (2021, April 4). *Proposal for a Regulation laying down harmonised rules on artificial intelligence (Artificial Intelligence Act).* Retrieved from EUR-LEX: https://eur-lex.europa.eu/legal-content/EN/TXT/?uri=CELEX%3A52021PC0206

Eze, M. O. (2010). *Intellectual History in Contemporary South Africa.* Palgrave Macmillan. DOI:10.1057/9780230109698

Ferrucci, D. e., Brown, E., Chu-Carroll, J., Fan, J., Gondek, D., Kalyanpur, A. A., Lally, A., Murdock, J. W., Nyberg, E., Prager, J., Schlaefer, N., & Welty, C. (2010). Building Watson: An Overview of the DeepQA Project. *AI Magazine*, 31(3), 59–79. DOI:10.1609/aimag.v31i3.2303

Floridi, L. (2013). *The Ethics of Information.* Oxford University Press. DOI:10.1093/acprof:oso/9780199641321.001.0001

Floridi, L., Cowls, J., Beltrametti, M., Chatila, R., Chazerand, P., Dignum, V., Luetge, C., Madelin, R., Pagallo, U., Rossi, F., Schafer, B., Valcke, P., & Vayena, E. (2018). AI4People—An Ethical Framework for a Good AI Society: Opportunities, Risks, Principles, and Recommendations. *Minds and Machines*, 28(4), 689–707. DOI:10.1007/s11023-018-9482-5 PMID:30930541

Frankish, K., & Ramsey, W. M. (2014). *The Cambridge Handbook of Artificial Intelligence*. Cambridge University Press. DOI:10.1017/CBO9781139046855

Future of Life Institute. (2017, August 11). *Asilomar AI Principles*. Retrieved from Future of Life Institute: https://futureoflife.org/open-letter/ai-principles/

Gasser, U., & Almeida, V. A. (2017). A Layered Model for AI Governance. *IEEE Internet Computing*, 21(6), 58–62. DOI:10.1109/MIC.2017.4180835

Goel, A. K. (2022). Looking back, looking ahead: Humans, ethics, and AI. *AI Magazine*, 43(2), 267–269. DOI:10.1002/aaai.12052

Goodfellow, I., Bengio, Y., & Courville, A. (2016). *Deep Learning*. MIT Press.

Google. (2023). *RESPONSIBILITY: Our Principles*. Retrieved from Google AI: https://ai.google/responsibility/principles/

Gupta, A., & Heath, V. (2020, September 14). AI ethics groups are repeating one of society's classic mistakes. *MIT Technology Review*. Retrieved from .

Hagendorf, T. (2020). The Ethics of AI Ethics: An Evaluation of Guidelines. *Minds and Machines*, 30(1), 99–120. DOI:10.1007/s11023-020-09517-8

Han, S., Mao, H., & Dally, W. J. (2016, February 15). *Deep Compression: Compressing Deep Neural Networks with Pruning, Trained Quantization and Huffman Coding*. Retrieved from ARXIV: https://arxiv.org/abs/1510.00149

Hecht-Nielsen, R. (1992). Theory of the Backpropagation Neural Network. In Wechsler, H. (Ed.), *Neural Networks for Perception* (Vol. 2, pp. 65–93). Academic Press. DOI:10.1016/B978-0-12-741252-8.50010-8

IEEE Global Initiative on Ethics of Autonomous and Intelligent Systems. (2019, March 25). *Ethically Aligned Design: A Vision for Prioritizing Human Well-being with Autonomous and Intelligent Systems (A/IS)*. Retrieved from IEEE Standards Association: https://standards.ieee.org/news/ieee-ead1e/

Jobin, A., Ienca, M., & Vayena, E. (2019). The global landscape of AI ethics guidelines. *Nature Machine Intelligence*, 1(9), 389–399. DOI:10.1038/s42256-019-0088-2

Jumper, J. e., Evans, R., Pritzel, A., Green, T., Figurnov, M., Ronneberger, O., Tunyasuvunakool, K., Bates, R., Žídek, A., Potapenko, A., Bridgland, A., Meyer, C., Kohl, S. A. A., Ballard, A. J., Cowie, A., Romera-Paredes, B., Nikolov, S., Jain, R., Adler, J., & Hassabis, D. (2021). Highly accurate protein structure prediction with AlphaFold. *Nature*, 596(7873), 583–589. DOI:10.1038/s41586-021-03819-2 PMID:34265844

Kazim, E., & Koshiyama, A. S. (2021). A high-level overview of AI ethics. *Patterns (New York, N.Y.)*, ●●●, 1–12. PMID:34553166

Knight, S., Shibani, A., & Vincent, N. (2024). Ethical AI governance: Mapping a research ecosystem. *AI and Ethics*. Advance online publication. DOI:10.1007/s43681-023-00416-z

Lai, K. L. (2008). *An Introduction to Chinese Philosophy*. Cambridge University Press. DOI:10.1017/CBO9780511800832

Laux, J. (2024). Three pathways for standardisation and ethical disclosure by default under the European union artificial intelligence act. *Computer Law & Security Review: The International Journal of Technology Law and Practice*.

LeCun, Y., Bengio, Y., & Hinton, G. (2015). Deep Learning. *Nature*, 521(7553), 436–444. DOI:10.1038/nature14539 PMID:26017442

Lighthill, J. (1972, July). *Artificial Intelligence: A General Survey*. Retrieved from CHILTON: https://www.chilton-computing.org.uk/inf/literature/reports/lighthill_report/p001.htm

McCarthy, J., Minsky, M. L., Rochester, N., & Shannon, C. E. (2006). A Proposal for the Dartmouth Summer Research Project on Artificial Intelligence, August 31, 1955. *AI Magazine*, ●●●, 12–14.

McCulloch, W. S., & Pitts, W. (1943). A Logical Calculus of Ideas Immanent in Nervous Activity. *The Bulletin of Mathematical Biophysics*, 5(4), 115–133. DOI:10.1007/BF02478259

McKinsey Global Institute. (2017, December 1). *Jobs Lost, Jobs Gained: Workforce Transitions in a Time of Automation*. Retrieved from McKinsey & Company: https://www.mckinsey.com/~/media/mckinsey/industries/public%20and%20social%20sector/our%20insights/what%20the%20future%20of%20work%20will%20mean%20for%20jobs%20skills%20and%20wages/mgi-jobs-lost-jobs-gained-executive-summary-december-6-2017.pdf

McKinsey Global Institute. (2024, July 16). *Jobs lost, jobs gained: What the future of work will mean for jobs, skills, and wages.* Retrieved from McKinsey & Company: https://www.mckinsey.com/featured-insights/future-of-work/jobs-lost-jobs-gained -what-the-future-of-work-will-mean-for-jobs-skills-and-wages

McLennan, S. e., Fiske, A., Celi, L. A., Müller, R., Harder, J., Ritt, K., Haddadin, S., & Buyx, A. (2020). An embedded ethics approach for AI development. *Nature Machine Intelligence*, 2(9), 488–490. DOI:10.1038/s42256-020-0214-1

Mennella, C., Maniscalco, U., De Pietro, G., & Esposito, M. (2024). Ethical and regulatory challenges of AI technologies in healthcare: A narrative review. *Heliyon*, 10(4), e26297. DOI:10.1016/j.heliyon.2024.e26297 PMID:38384518

Mervosh, S. (2019, May 24). *Distorted Videos of Nancy Pelosi Spread on Facebook and Twitter, Helped by Trump.* Retrieved April 30, 2024, from The New York Times: https://www.nytimes.com/2019/05/24/us/politics/pelosi-doctored-video.html

Metz, R. (2021, August 6). *How a deepfake Tom Cruise on TikTok turned into a very real AI company.* Retrieved April 30, 2024, from CNN Business: https://edition.cnn .com/2021/08/06/tech/tom-cruise-deepfake-tiktok-company/index.html

Mittelstadt, B. (2019). Principles alone cannot guarantee ethical AI. *Nature Machine Intelligence*, 1(11), 501–507. DOI:10.1038/s42256-019-0114-4

Mittelstadt, B. D., Allo, P., Taddeo, M., Wachter, S., & Floridi, L. (2016). The ethics of algorithms: Mapping the debate. *Big Data & Society*, 3(2), 1–21. DOI:10.1177/2053951716679679

Moor, J. (2006). The Nature, Importance, and Difficulty of Machine Ethics. *IEEE Intelligent Systems*, 21(4), 18–21. DOI:10.1109/MIS.2006.80

Moor, J. H. (1985). What is Computer Ethics? *Metaphilosophy*, 16(4), 266–275. DOI:10.1111/j.1467-9973.1985.tb00173.x

Morandín-Ahuerma, F. (2023, September 20). Montreal Declaration for Responsible AI: 10 Principles and 59 Recommendations. Retrieved from *OSF Preprints*: https:// osf.io/preprints/osf/sj2z5DOI:10.31219/osf.io/sj2z5

Noble, S. U. (2018). *Algorithms of Oppression: How Search Engines Reinforce Racism.* NYU Press. DOI:10.18574/nyu/9781479833641.001.0001

O'Neil, C. (2016). *Weapons of Math Destruction: How Big Data Increases Inequality and Threatens Democracy.* Crown Publishing Group.

Ossa, L. e. (2024). Integrating ethics in AI development: A qualitative study. *BMC Medical Ethics.* PMID:38262986

Owczarczuk, M. (2023). Ethical and regulatory challenges amid artificial intelligence development. *Ekonomia i Prawo. Economics and Law*, 295-310.

Pant, A. e. (2023). Ethics in the Age of AI: An Analysis of AI Practitioners' Awareness and Challenges. *Journal of the Association for Computing Machinery*.

Press, G. (2016, December 30). *A Very Short History Of Artificial Intelligence (AI)*. Retrieved May 4, 2024, from Forbes: https://www.forbes.com/sites/gilpress/2016/12/30/a-very-short-history-of-artificial-intelligence-ai/?sh=68b5d5776fba

Ramesh, A. e. (2022, April 13). *Hierarchical Text-Conditional Image Generation with CLIP Latents*. Retrieved from ARXIV: https://arxiv.org/abs/2204.06125

Reddy, S., Fox, J., & Purohit, M. P. (2021). Artificial intelligence-enabled healthcare delivery. *Journal of the Royal Society of Medicine*, ●●●, 372–376. PMID:30507284

Roberts, H., Cowls, J., Morley, J., Taddeo, M., Wang, V., & Floridi, L. (2021). The Chinese Approach to Artificial Intelligence: An Analysis of Policy, Ethics, and Regulation. In Floridi, L. (Ed.), *Ethics, Governance, and Policies in Artificial Intelligence* (pp. 47–79). Springer. DOI:10.1007/978-3-030-81907-1_5

Ross, C., & Swetlitz, I. (2017, September 5). *IBM pitched its Watson supercomputer as a revolution in cancer care. It's nowhere close*. Retrieved from STAT News: https://www.statnews.com/2017/09/05/watson-ibm-cancer/

Ryan, M., & Stahl, B. C. (2021). Artificial intelligence ethics guidelines for developers and users: clarifying their content and normative implications. *Journal of Information, Communication and Ethics in Society*, 61-86.

Sambasivan, N. (2021). Re-imagining Algorithmic Fairness in India and Beyond. *FAccT '21: Proceedings of the 2021 ACM Conference on Fairness, Accountability, and Transparency* (pp. 315-328). New York City: ACM. DOI:10.1145/3442188.3445896

Santos, B. S. (2002). *Towards a New Legal Common Sense: Law, Globalization, and Emancipation*. Butterworths.

Schmidhuber, J. (2015). Deep Learning in Neural Networks: An Overview. *Neural Networks*, 61, 85–117. DOI:10.1016/j.neunet.2014.09.003 PMID:25462637

Schwartz, R., Dodge, J., Smith, N. A., & Etzioni, O. (2020). Green AI. *Communications of the ACM*, 63(12), 54–63. DOI:10.1145/3381831

Sherry, T. L. (2019). AI and Emotional Intelligence: Understanding the Impact on Human Relationships. *Journal of Human-Computer Interaction*, 303-320.

Siau, K., & Wang, W. (2020). Artificial Intelligence (AI) Ethics: Ethics of AI and Ethical AI. *Journal of Database Management*, 31(2), 74–87. DOI:10.4018/JDM.2020040105

Silver, D. e., Huang, A., Maddison, C. J., Guez, A., Sifre, L., van den Driessche, G., Schrittwieser, J., Antonoglou, I., Panneershelvam, V., Lanctot, M., Dieleman, S., Grewe, D., Nham, J., Kalchbrenner, N., Sutskever, I., Lillicrap, T., Leach, M., Kavukcuoglu, K., Graepel, T., & Hassabis, D. (2016). Mastering the Game of Go with Deep Neural Networks and Tree Search. *Nature*, 529(7587), 484–489. DOI:10.1038/nature16961 PMID:26819042

Singapore Personal Data Protection Commission. (2020). *Singapore's Approach to AI Governance*. Retrieved from Singapore Personal Data Protection Commission: https://www.pdpc.gov.sg/help-and-resources/2020/01/model-ai-governance-framework

Singer, N. (2019, May 14). *San Francisco Bans Facial Recognition Technology*. Retrieved from NY Times: https://www.nytimes.com/2019/05/14/us/facial-recognition-ban-san-francisco.html

Stahl, B. C., Timmermans, J., & Mittelstadt, B. D. (2016). The Ethics of Computing: A Survey of the Computing-Oriented Literature. *ACM Computing Surveys*, 48(4), 1–38. DOI:10.1145/2871196

Stamboliev, E., & Christiaens, T. (2024). How empty is Trustworthy AI? A discourse analysis of the Ethics Guidelines of Trustworthy AI. *Critical Policy Studies*, 1–18. DOI:10.1080/19460171.2024.2315431

Strickland, E. (2019, April 1). *IBM Watson, Heal Thyself: How IBM Overpromised and Underdelivered on AI Health Care*. Retrieved from IEEE Spectrum: https://read.nxtbook.com/ieee/spectrum/spectrum_na_april_2019/ibm_watson_heal_thyself.html

Strubell, E., Ganesh, A., & McCallum, A. (2019). Energy and Policy Considerations for Deep Learning in NLP. *Proceedings of the 57th Annual Meeting of the Association for Computational Linguistics* (pp. 3645-3650). PKP Publishing Services Network. DOI:10.18653/v1/P19-1355

Strubell, E., Ganesh, A., & McCallum, A. (2020). Energy and Policy Considerations for Deep Learning in NLP. *Proceedings of the AAAI Conference on Artificial Intelligence* (pp. 13693-13696). Burnaby, B.C., Canada: PKP Publishing Services.

Suchman, L. (2007). *Human-Machine Reconfigurations: Plans and Situated Actions*. Cambridge University Press.

Thrun, S. e., Montemerlo, M., Dahlkamp, H., Stavens, D., Aron, A., Diebel, J., Fong, P., Gale, J., Halpenny, M., Hoffmann, G., Lau, K., Oakley, C., Palatucci, M., Pratt, V., Stang, P., Strohband, S., Dupont, C., Jendrossek, L.-E., Koelen, C., & Mahoney, P. (2006). Stanley: The Robot That Won the DARPA Grand Challenge. *Journal of Field Robotics*, 23(9), 661–692. DOI:10.1002/rob.20147

Topol, E. J. (2019). *Deep Medicine: How Artificial Intelligence Can Make Healthcare Human Again*. Basic Books.

Treasury Board of Canada Secretariat. (2023, April 25). *Directive on Automated Decision-Making*. Retrieved from Policies, directives, standards and guidelines: https://www.tbs-sct.gc.ca/pol/doc-eng.aspx?id=32592

Turing, A. M. (1950). Computing Machinery and Intelligence. *Mind- a Quarterly Review of Psychology and Philosophy*, 433-460.

Turkle, S. (2011). *Alone Together: Why We Expect More from Technology and Less from Each Other*. Basic Books.

Voigt, P., & Von dem Bussche, A. (2017). *The EU General Data Protection Regulation (GDPR)*. Springer International Publishing. DOI:10.1007/978-3-319-57959-7

Wallach, W., & Allen, C. (2009). *Moral Machines-Teaching Robots right from wrong*. Oxford University Press. DOI:10.1093/acprof:oso/9780195374049.001.0001

Weizenbaum, J. (1976). *Computer Power and Human Reason: From Judgment to Calculation*. W.H. Freeman.

West, D. M. (2018). *The Future of Work: Robots, AI, and Automation*. Brookings Institution Press.

World Economic Forum. (2023, April 30). *The Future of Jobs Report 2023*. Retrieved from World Economic Forum: https://www.weforum.org/publications/the-future-of-jobs-report-2023/

Wright, J. (2024). The Development of AI Ethics in Japan: Ethics washing Society 5.0? *East Asian Science, Technology and Society*, 18(2), 117–134. DOI:10.1080/18752160.2023.2275987

Chapter 4
The Pillars of AI Ethics:
Transparency, Accountability, and Privacy

R. Ramya
https://orcid.org/0000-0002-8071-9343
SRM Institute of Science and Technology, India

S. Priya
https://orcid.org/0009-0003-3305-745X
SRM Institute of Science and Technology, India

P. Thamizhikkavi
SRM Institute of Science and Technology, India

M. Anand
https://orcid.org/0000-0001-5205-9678
SRM Institute of Science and Technology, India

ABSTRACT

Artificial intelligence (AI) technologies are advancing at a rapid pace, which has created many potential and benefits for a variety of industries. But there are also moral questions about the creation and application of AI systems due to their growing autonomy and complexity. This study examines the fundamentals of AI ethics with an emphasis on privacy, responsibility, and openness as the cornerstones that uphold moral AI practices in order to allay these worries. Making AI systems comprehensible and explicable to users and stakeholders is emphasised by transparency. It entails giving concise descriptions of the decision-making procedures, data sources, and potential biases of AI systems. Contrarily, accountability describes the obligations of people and institutions engaged in the creation and application of AI systems. It entails making certain that AI systems are applied in a just, fair, and responsible way

DOI: 10.4018/979-8-3693-9173-0.ch004

and that individuals in charge of their creation and application be held accountable for any unfavourable effects.

1. INTRODUCTION

The Three Foundations of AI Ethics: Privacy, Accountability, and Transparency
There are serious ethical issues with artificial intelligence's very potential and capabilities. A solid foundation of ethical standards is necessary to ensure that AI is used responsibly and benefits society as technology grows more and more interwoven into our daily lives. In this endeavour, transparency, accountability, and privacy stand out as three key pillars.

1. Transparency:

Describing the Black Box: Artificial intelligence algorithms frequently function as "black boxes," with their inner workings being obscure and challenging to decipher. It is necessary to make these procedures transparent in order to give stakeholders access to and understanding of them. This comprises:
Algorithmic transparency: Making available the procedures and reasoning the algorithm employs to make judgements.
Data transparency: Ensuring responsible data collection and utilisation by disclosing the kind and source of the data used to train the AI model.
Transparency in decision-making: Providing justifications for certain AI choices, especially in delicate fields like criminal justice or healthcare.

2. accountability:

At the wheel, who is it? Who bears the responsibility for the actions and outcomes of AI systems when things go wrong? In order to guarantee equitable and responsible use, accountability procedures must be established. This includes:
Clearly defining jobs and duties: determining who is responsible for what when it comes to the creation, use, and use of AI.
Monitoring and auditing: evaluating AI systems on a regular basis for fairness, prejudice, and adherence to moral standards.
Error by human: ensuring human input is involved in crucial decisions made by AI in order to uphold moral authority and avoid unexpected outcomes.

3. Privacy

Preserving Individual Rights: Privacy issues are brought up by the massive volume of data utilised to train AI. Strong safeguards are required to protect private data during the AI lifecycle. These safeguards include:

Data minimization and anonymization: Restricting the amount of personal information that is gathered and kept to a minimum to ensure that the AI can operate.

Using robust security measures to stop unwanted access and exploitation of personal data is known as "secure data handling."

Transparency and user control: Giving people authority over their data and unambiguous information regarding its use.

The Importance of These Pillars

These three pillars are related to one another and support one another; they are not discrete ideas. Openness fosters accountability, and accountability establishes a foundation for protecting privacy. Through focused attention to these areas, we can create AI systems that are reliable, accountable, and helpful to everyone.

This is only the tip of the iceberg in this introduction. Given the unique difficulties and potential answers in many AI applications, each pillar merits further examination. Recall that the pursuit of ethical AI necessitates constant communication, cooperation, and adaptability in order to guarantee that technology advances humankind.

2. REVIEW OF THE LITERATURE

The swift evolution and implementation of artificial intelligence (AI) technologies have resulted in noteworthy progress across multiple domains, such as healthcare, finance, transportation, and others. However, there are now ethical questions about the creation and application of AI systems due to their growing autonomy and complexity. To address these worries, scholars and decision-makers have concentrated on creating an ethical framework that would direct the appropriate advancement and application of AI technology. The core ideas of AI ethics are examined in this literature review, with a particular emphasis on privacy, accountability, and transparency as the cornerstones of moral AI behaviour.

One of the fundamental tenets of AI ethics is transparency, which highlights the significance of providing users and stakeholders with an intelligible and explicable AI system. In order to ensure that users can comprehend how AI systems operate, including their decision-making processes, data sources, and potential biases, a number of studies have emphasised the necessity of transparency in AI systems. Doshi-Velez and Kim (2017), for instance, contend that since people are more inclined to trust systems they can comprehend and explain, openness is crucial to fostering user confidence in AI systems. In a similar vein, Mittelstadt et al. (2019) stress how crucial openness is to fostering justice and accountability in AI systems.

Another important tenet of AI ethics is accountability, which emphasises the accountability of people and institutions engaged in the creation and application of AI systems. In order to guarantee that AI systems are used in a fair, just, and responsible manner, several studies have emphasised the necessity of accountability mechanisms. Jobin et al. (2019), for instance, contend that ethical usage of AI systems and compliance with legal and ethical standards depend on accountability measures like audits, oversight, and reporting. In a similar vein, Floridi et al. (2018) stress how crucial responsibility is to building user acceptability and trust in AI systems.

The third pillar of AI ethics is privacy, which emphasises the value of safeguarding people's private information and making sure AI systems respect their right to privacy. The necessity for privacy protection in AI systems to secure sensitive data and stop unauthorised access or misuse has been brought up in a number of studies. Mittelstadt et al. (2016), for instance, contend that safeguarding privacy is crucial to guaranteeing that AI systems respect people's right to privacy. Similar to this, Jobin et al. (2019) stress how crucial privacy protection is to AI systems' ability to foster user acceptance and confidence.

The literature emphasises the significance of privacy, responsibility, and openness as the cornerstones of AI ethics. Developers and users may make sure AI systems are created and utilised in a way that respects people's rights, upholds justice and fairness, and cultivates acceptance and trust by following these guidelines. There are still a number of obstacles to overcome before these ideas can be put into reality, such as the need for additional study and the creation of ethical policies related to AI.

In fact, transparency, accountability, and privacy are essential components of AI ethics that serve as a basis for the moral creation and application of AI systems. This is the reason they matter:

Transparency

Knowing what data AI systems use, how they learn, and what influences their judgements entails understanding how AI systems operate. This makes it possible to use AI wisely and assists in spotting any hazards or biases.

Establishing trust: By making AI systems less like "black boxes" and more comprehensible to users and stakeholders, transparency promotes trust in these systems. Ensuring public acceptability and responsible usage of AI is contingent upon this.

Enabling supervision and examination: Comprehending the functioning of AI systems facilitates the detection and resolution of possible issues, including algorithmic prejudice or inadvertent outcomes.

Accountability

Deciding who is in charge of AI decisions: As AI systems grow more sophisticated and self-sufficient, it is essential to establish who is in charge of their decisions and actions. Depending on the situation, these could be users, developers, or deployers.

Fairness and nondiscrimination: AI systems ought to answer for their adherence to moral standards and avoidance of discriminatory consequences. This calls for procedures for recourse and redress in the event that harm is done.

Encouraging conscientious development and implementation: Throughout the AI lifecycle, developers and deployers are encouraged to prioritise ethical considerations by means of explicit accountability measures.

Personal space: Safeguarding personal information: AI systems frequently rely on vast volumes of personal information, which raises security and privacy issues. It's critical to have strong data protection procedures in place and guarantee that people have control over their data.

Reducing the amount of data acquired and used: Only the information required for the AI system's intended usage should be gathered and utilised. This lessens the possibility of data misuse and privacy threats.

Transparency in data practices: People should be aware of how AI systems gather, use, and keep their personal data. They are now better equipped to decide how to protect their data privacy.

Together, these three pillars will guarantee the development and application of ethical AI. Although there are still debates about these principles and difficulties in putting them into practice, they offer a useful framework for navigating the murky waters of AI ethics.

It's crucial to remember that there are other foundations for AI ethics. Fairness, nondiscrimination, safety, security, and sustainability are further crucial elements. However, resolving these additional ethical issues also requires a strong foundation of transparency, accountability, and privacy.

3. AI DEMYSTIFICATION: ACCOUNTABILITY, PRIVACY, AND TRANSPARENCY

Three fundamental ideas—transparency, responsibility, and privacy—that are essential for the responsible development of AI are highlighted in this paragraph. Let's dissect it:

1. Openness: Explaining the AI System Envision a mysterious device that makes decisions that impact your life, but you are unaware of how it operates. Transparency aims to address that issue. It places a strong emphasis on helping users and stakeholders understand and justify AI technology. This includes: Simple explanations: succinct outlines of the algorithms (the "thought process") that the AI utilises to make decisions.

Data-driven insights: revealing the origins of the data and any potential biases in the data that were used to train the AI, since biases can provide unfair results.
What makes transparency crucial? Trust: People will be more inclined to trust AI's judgements if they are aware of how it operates.
Fairness: Openness makes it easier to spot and deal with possible prejudices that can give rise to discrimination.
Accountability: It is simpler to hold people accountable for the development and use of AI when you comprehend its "thinking." This is especially true if any unfavourable effects arise.

2. Accountability: Taking Charge of AI's Deeds Accountability is predicated on transparency. It's about making certain that: Artificial intelligence is implemented justly, fairly, and responsibly, taking into account the effects it will have on both individuals and society as a whole. Those responsible are put to account: AI system designers, developers, and implementers should take accountability for any unfavourable effects that their work may produce.

What makes accountability crucial? Protections against abuse: It dissuades the creation and deployment of detrimental AI applications.
Individuals are protected by this: It ensures justice and reparation by holding persons accountable for any harm caused by AI systems. encourages conscientious development inspires programmers to design AI with the welfare of humans as its top priority.

3. Privacy: Safeguarding Your Individual Data Personal data is more important than ever in the AI era. Privacy highlights people's rights to:

Manage their information: Determine what personal data about them is gathered, put to use, and distributed.
Recognise: Recognise how AI systems are using their data.
Defend their rights: Make sure their information isn't utilised in ways that compromise their right to privacy or other essential freedoms.
Why is maintaining one's privacy important?

Preserves individual autonomy by guaranteeing that people are in charge of their personal data and how it affects their life.

Prevents discrimination: Reduces the possibility that AI systems would discriminate against people based on their personal information.

Encourages people to trust AI: People are more likely to use and trust AI technologies when they are aware that their data is secure.

Recall that these guidelines are not discrete ideas; rather, they complement one another to guarantee ethical AI development and use. In the age of artificial intelligence, both transparency and accountability depend on robust privacy measures to preserve individual rights and well-being. Transparency fosters trust.

4. HOW DANGEROUS IF AI DEVELOPMENT DOESN'T FOLLOW THE ETHICAL RULES

The risks associated with developing AI without taking ethics into account are a complicated and hotly contested subject. Although AI has the potential to greatly benefit humanity, its research and application are not without serious concerns.

The following are some major areas of concern: Bias and discrimination: Artificial intelligence (AI) systems have the capacity to pick up on and magnify societal prejudices, which can have discriminatory effects on loan approvals, employment decisions, and criminal justice systems.

Security and privacy: AI systems that gather and examine personal data give rise to worries about data misuse and privacy violations. Furthermore, AI systems may be susceptible to manipulation and hacking.

Autonomy and control: As AI systems develop in sophistication, concerns regarding who is in charge of them and how choices are made start to surface. This could result in scenarios where AI systems act inadvertently or without human oversight.

AI-powered automation may result in a large-scale loss of jobs, especially in industries where routine tasks are performed. There may be serious social and economic repercussions to this.

Existential risk: Although this is a highly hypothetical and contentious opinion, some researchers worry that highly developed AI may constitute an existential threat to humans.

It's crucial to remember that these are only hypothetical risks, and that several initiatives are being made to create and use AI in responsible ways. But it's imperative to understand the risks and have candid conversations about how to reduce them.

Here are a few strategies to deal with these issues:

Creating ethical guidelines: It's critical to set forth precise ethical standards for the creation and application of AI. These rules ought to cover matters like accountability, justice, openness, and privacy.

Ensuring human oversight: Even with highly developed AI systems, humans should always be involved in the decision-making process. This contributes to the ethical and responsible application of AI.

Investing in research is necessary to gain a deeper understanding of the possible advantages and disadvantages of artificial intelligence. This will assist us in creating AI systems that are safer and more useful.

Public participation: Transparent and honest dialogues around AI are crucial. Making sure AI is used for good and influencing its development should engage the public.

We can ensure that AI development avoids possible hazards and helps society as a whole by implementing these measures.

It's critical to keep in mind that artificial intelligence (AI) is a tool, and like all tools, it can be utilised for good or harm. Making sure we have the appropriate safeguards in place and using them properly is crucial.

5. DEVELOPING ETHICAL AI: A GUIDE WITH OBSTACLES

This text emphasises how important it is for users and developers to work together to create moral AI systems. It highlights three important ideas:

1. Principles for Ethical AI Development and Application:

The passage indicates that both developers and consumers may ensure AI is produced and utilised ethically by adhering to particular criteria. These rules ought to try to:

Respect for human rights: In the context of AI development and application, this entails defending basic human rights such as freedom of speech, privacy, and nondiscrimination.

Maintain justice and equity: AI systems ought to be developed and applied in a way that upholds equity and refrains from introducing new or aggravating preexisting biases.

Develop acceptance and trust: Transparency, responsibility, and appropriate data handling techniques are necessary to foster acceptance and trust for AI.

2. Possibilities and Difficulties of Putting Ethical Principles into Practice:

The sentence recognises that there may be advantages and disadvantages to turning these ideas into reality:

Potential: Adhering to these recommendations may result in:

AI that respects human values and advances society at large is considered ethical.

a greater degree of faith in AI increased use and acceptance of AI technologies by the general public. Sustainable AI development is the creation of AI that complies with long-term human objectives and is fair and accountable.

Problems: Putting these ideas into practice can be difficult because of:

conflicting interests juggling the demands of several parties, including users, developers, and society at large.

Technical limitations: It's still early to develop completely impartial and transparent AI systems.

Gaps in regulations and policies: lack of explicit laws and rules controlling the creation and application of AI.

2. Progressing: Research and Policy Formulation:

The necessity of more research and the creation of new policies in the field of AI ethics is emphasised in the passage's conclusion. This might entail:

Technical solution research: creating instruments and methods to guarantee accountability, equity, and transparency in AI systems.

Encouraging public discussion and education on artificial intelligence and its moral ramifications.

Policy frameworks and regulations: Putting in place laws that guarantee the development and application of AI responsibly.

Overall, this text highlights that creating ethical AI is a team endeavour in which developers, consumers, and legislators must all actively participate. We can make sure that AI technology serves humanity as a whole by adhering to strict criteria, admitting difficulties, and funding additional research and the creation of policies.

6. THE CHALLENGES OF IMPLEMENTING THE TRANSPARENCY, ACCOUNTABILITY AND PRIVACY IN REALITY

Although Transparency, Accountability, and Privacy are fundamental to ethical AI, putting these concepts into practice really poses a number of difficulties. Here are a few noteworthy ones:

Transparency

Complex algorithms' "black box" nature: Even specialists may find it challenging to comprehend the decision-making processes of many sophisticated AI systems, especially deep learning models, due to their complexity. This makes it challenging to describe how they get at particular results.

Transparency and performance trade-off: Algorithms that are simplified for greater transparency may occasionally perform worse or have lower accuracy. It can be challenging to find a balance between the two.

Safeguarding intellectual property: Transparency and intellectual property rights could clash if private or trade secrets about AI systems' inner workings were made public accountability:

Distributed systems and shared responsibility: AI systems frequently involve a number of actors, including consumers, data sources, developers, and deployers. It might be difficult to pinpoint the decision-makers within the system and assign clear accountability.

Difficulties in assigning blame: In cases where AI systems cause harm, it can be challenging to determine who is to blame, particularly in the case of complex or autonomous systems. This makes determining who is to fault and pursuing compensation difficult.

AI's evolving nature It gets harder to hold AI systems responsible for previous actions or predict their future behaviour as they learn and adapt more.

Personal Space

Juggling the needs of society and individual rights: AI systems frequently need a lot of personal data in order to work well. This may cause conflict between people's right to privacy and the potential advantages of artificial intelligence for society.

Challenges with data security and anonymization: Re-identification is always a possibility, even with anonymization mechanisms in place, particularly when dealing with huge datasets. It is imperative to guarantee strong data security protocols.

Variations in privacy laws across the globe: The implementation of uniform privacy policies for AI systems working on a worldwide scale is challenging due to the disparities in privacy laws and regulations across different countries.

Other Difficulties

Absence of standards and unambiguous rules: It is currently difficult to apply moral AI concepts without standardised frameworks and unambiguous rules. Because of this, it might be difficult for developers and deployers to determine which activities are truly moral.

Low public knowledge and comprehension: It is challenging to foster public trust and agreement on ethical norms since many people are unaware of the possible dangers and advantages of artificial intelligence.

Conflicting agendas and the balance of power: In terms of AI ethics, several stakeholders—including businesses, governments, and individuals—may have conflicting interests and agendas, making it challenging to find common ground.

To tackle these obstacles, we need constant communication, teamwork, and creativity. In order to ensure that AI is developed and used responsibly and upholds the ideals of transparency, accountability, and privacy in the digital age, researchers, developers, legislators, and the general public all have a role to play.

7. FUTURE DIRECTIONS OF THE PILLARS OF AI ETHICS: TRANSPARENCY, ACCOUNTABILITY, AND PRIVACY

The practical implementation of the three pillars of AI ethics—Transparency, Accountability, and Privacy—will necessitate constant innovation and adaptation as AI develops and becomes more integrated into society. The following are some possible future paths to think about:

Transparency

Creating methods for explainable AI (XAI): Research and development of XAI techniques should continue so that even non-experts may grasp sophisticated algorithms.

Standardisation of metrics related to transparency: creating uniform standards and metrics to assess the degree of transparency provided by various AI systems.

Explanations that are interactive and user-centric: Developing interactive tools that let users investigate and comprehend how AI systems make decisions, rather than relying solely on static explanations.

Accountability

Legal frameworks for AI accountability: creating laws that clearly identify who is responsible for AI choices and set up procedures for compensation when harm is caused.

Algorithmic impact assessments: Before AI systems are deployed, they should be required to undergo an evaluation of their possible societal and ethical ramifications.

Creating strong auditing and monitoring systems is necessary to keep tabs on AI system behaviour and spot any biases or unforeseen repercussions.

Personal Space

Investigating privacy-preserving methods such as differential privacy and federated learning, which let AI models be trained on decentralised data without jeopardising individual privacy.

Synthetic data generation: Rather than using actual personal data to train AI models, synthetic data produced by algorithms is used.

Giving people control over their data: Creating instruments and frameworks that enable people to regulate and oversee how artificial intelligence (AI) uses their data.

Other Directions for the Future

Integration with other moral precepts: Taking into account the ways in which fairness, nondiscrimination, safety, and transparency, accountability, and privacy interact with these other fundamental moral precepts.

Public awareness and education: promoting educated debates and involvement with AI ethics by raising public awareness of AI and its possible hazards as well as advantages.

Multi-stakeholder cooperation: Encouraging cooperation to develop and use responsible AI solutions amongst researchers, developers, legislators, and the general public.

It's crucial to keep in mind that these are only possible paths, and that a complex interaction between social norms, legal frameworks, and technology breakthroughs will probably influence AI ethics in the future. To foster trust and guarantee that artificial intelligence (AI) serves all people, it is crucial to make sure that the three pillars of transparency, accountability, and privacy continue to be at the forefront of AI development and use.

Transparency, accountability, and privacy are the three pillars of AI ethics that can be upheld with the use of a number of AI algorithms and methodologies. Here are a few instances:

Transparency

Techniques for Explainable AI (XAI)

By recognising pertinent traits and their contributions, Local Interpretable Model-Agnostic Explanations (LIME) provide an explanation for specific predictions.

Shapley Additive Explanations (SHAP): Offers insights into model behaviour by calculating each input feature's contribution to a prediction.

Counterfactual Explanations: To help users comprehend the model's decision-making process, illustrate what inputs would have produced different results.

Illustrations: Features' Significance Maps: Showcase the most important characteristics in a picture or set of data to help visualise how models decide what to do.

Decision Trees: Construct a tree-like structure to represent the reasoning that goes into a model's decision-making process.

accountability:

Machine Learning with Fairness Aware: Adversarial debiasing is the process of locating and eliminating biases in models or training data. Counterfactual Fairness is the ability to recognise unfairness and lessen it by offering substitute, more equitable predictions.

Evaluation of Algorithmic Impact (AIA): Causal Inference Techniques: Examine how AI systems affect various groups causally in order to spot any biases or unforeseen repercussions.

Planning and Simulation: Run a number of scenarios to estimate the possible effects of AI systems before they are put into use.

Personal Space

Protecting Privacy using Machine Learning

Federated Learning: Using decentralised data, models are trained without exchanging individual data points. Differential privacy involves introducing noise into data while maintaining its statistical features to safeguard individual privacy.

Homomorphic Encryption: Provides analysis while maintaining privacy by permitting computations on encrypted material without having to first decode it.

Analysing and pseudonymizing data: K-anonymity: Prevents identity theft by guaranteeing that every record has characteristics in common with at least k-1 other records. Stronger privacy assurances are offered by differential privacy with Rényi Divergence than by ordinary differential privacy.

It's crucial to remember that maintaining AI ethics cannot be accomplished by a single programme. These methods are frequently blended and tailored to certain situations and difficulties. In addition, new techniques and algorithms are continually being developed, and the field of AI ethics is always changing.

It's also important to keep in mind that algorithms are not intrinsically moral. The development, design, and use of these algorithms will determine the ethical implications of AI. To guarantee that AI algorithms contribute to a future that is more open, accountable, and privacy-conscious, human oversight, responsible development procedures, and continuous review are crucial. That is represented in Table 1.

Table 1. The ai algorithms can be used to maintain the pillars of AI ethics

Algorithm	Advantages	Applications
Local Interpretable Model-Agnostic Explanations (LIME)	Explains individual predictions for non-experts.	Understanding model behavior for debugging or regulatory compliance.
SHapley Additive exPlanations (SHAP)	Calculates feature contributions to predictions.	Identifying and mitigating bias in loan approvals or criminal justice systems.
Counterfactual Explanations	Shows alternative, fairer predictions.	Helping users understand why a model made a particular decision.
Feature Importance Maps	Highlights influential features in data.	Identifying important features in medical images or sensor data.
Decision Trees	Visualizes model decision-making logic.	Debugging complex models or explaining their behavior to stakeholders.
Adversarial Debiasing	Identifies and removes biases in training data or models.	Developing fairer AI systems for tasks like loan approvals or hiring decisions.
Counterfactual Fairness	Identifies and mitigates unfairness in predictions.	Ensuring that AI systems do not have discriminatory impacts on different groups.
Causal Inference Techniques	Analyzes causal effects of AI systems on different groups.	Evaluating the potential societal impacts of AI systems before they are deployed.
Simulation and Scenario Planning	Predicts potential impacts of AI systems before deployment.	Training AI models on sensitive data without compromising individual privacy.
Federated Learning	Trains models on decentralized data without sharing individual data.	Analyzing private data for research purposes while protecting individual privacy.
Differential Privacy	Adds noise to data to protect privacy while preserving its usefulness.	Securely performing computations on sensitive data without revealing its contents.
Homomorphic Encryption	Allows computations on encrypted data without decryption.	Protecting the privacy of individuals in large datasets.

continued on following page

Table 1. Continued

Algorithm	Advantages	Applications
k-anonymity	Makes it difficult to identify individuals in datasets.	Providing stronger privacy guarantees for sensitive data analysis.
Differential Privacy with Rényi Divergence	Provides stronger privacy guarantees than traditional differential privacy.	Providing stronger privacy guarantees for sensitive data analysis.

7. CONCLUSION

This study examines the fundamentals of AI ethics with an emphasis on privacy, responsibility, and openness as the cornerstones that uphold moral AI practices in order to allay these worries.

There was discussion regarding AI demystification: accountability, privacy, and transparency.

The important thing of how dangerous if ai development doesn't follow the ethical rules are discussed. Then the developing ethical ai: a guide with obstacles and the challenges of implementing the transparency, accountability and privacy in reality are discussed. Finally the future directions of the pillars of ai ethics: transparency, accountability, and privacy are discussed.

REFERENCES

Abina, A., Batkovic, T., Cestnik, B., Kikaj, A., Kovacic, L. R., Kurbus, M., & Zidansek, A. (2022). Decision support concept for improvement of sustainability-related competences. *Sustainability (Basel)*, 14(14), 8539. DOI:10.3390/su14148539

Agbese, M., Alanen, H. K., Antikainen, J., Erika, H., Isomaki, H., Jantunen, M., Kemell, K.-K., Rousi, R., Vainio-Pekka, H., & Vakkuri, V. (2023). Governance in ethical and trustworthy AI systems: Extension of the ECCOLA method for AI ethics governance using GARP. *E-Informatica Software Engineering Journal*, 17(1), 230101. DOI:10.37190/e-Inf230101

Aitken, M., Ng, M., Horsfall, D., Coopamootoo, K. P. L., van Moorsel, A., & Elliott, K. (2021). In pursuit of socially-minded data-intensive innovation in banking: A focus group study of public expectations of digital innovation in banking. *Technology in Society*, 66, 101666. DOI:10.1016/j.techsoc.2021.101666

Alnamrouti, A., Rjoub, H., & Ozgit, H. (2022). Do strategic human resources and artificial intelligence help to make organisations more sustainable? Evidence from non-governmental organisations. *Sustainability (Basel)*, 14(12), 7327. DOI:10.3390/su14127327

Andrada, G., Clowes, R. W., & Smart, P. R. (2022). Varieties of transparency: Exploring agency within AI systems. *AI & Society*, •••, 1–11. DOI:10.1007/s00146-021-01326-6 PMID:35035112

Arya, V., Bellamy, R. K., Chen, P. Y., Dhurandhar, A., Hind, M., Hoffman, S. C., Houde, S., Liao, Q. V., Luss, R., Mojsilovic, A., Mourad, S., Pedemonte, P., Raghavendra, R., Richards, J., Sattigeri, P., Shanmugam, K., Singh, M., Varshney, K. R., Wei, D., & Zhang, Y. (2019). One explanation does not fit all: A toolkit and taxonomy of ai explainability techniques. Cornell University. https://doi.org//arXiv.1909.03012.DOI:10.48550

Barrot, J. S. (2023). Using ChatGPT for second language writing: Pitfalls and potentials. *Assessing Writing*, 57, 100745. DOI:10.1016/j.asw.2023.100745

Bauer, J. M. (2022). Toward new guardrails for the information society. *Telecommunications Policy*, 46(5), 102350. DOI:10.1016/j.telpol.2022.102350

Bellamy, R. K., Dey, K., Hind, M., Hoffman, S. C., Houde, S., Kannan, K., Lohia, P., Martino, J., Mehta, S., Mojsilovic, A., Nagar, S., Ramamurthy, K. N., Richards, J., Saha, D., Sattigeri, P., Singh, M., Varshney, K. R., & Zhang, Y. (2019). AI fairness 360: An extensible toolkit for detecting and mitigating algorithmic bias. *IBM Journal of Research and Development*, 63(4/5), 1–4. DOI:10.1147/JRD.2019.2942287

Berente, N., Gu, B., Recker, J., & Santhanam, R. (2021). Managing Artificial Intelligence. *Management Information Systems Quarterly*, 45(3), 1433–1450.

Bertino, E., Kantarcioglu, M., Akcora, C. G., Samtani, S., Mittal, S., & Gupta, M. (2021). AI for security and security for AI. In proceedings of the eleventh ACM conference on data and application security and privacy (pp. 333–334). ACM Digital Library. DOI:10.1145/3422337.3450357

Broer, T. (2022). The Googlization of health: Invasiveness and corporate responsibility in media discourses on Facebook's algorithmic programme for suicide prevention. *Social Science & Medicine*, 306, 115131. DOI:10.1016/j.socscimed.2022.115131 PMID:35714428

Buhmann, A., & Fieseler, C. (2023). Deep learning meets deep democracy: Deliberative governance and responsible innovation in artificial intelligence. *Business Ethics Quarterly*, 33(1), 146–179. DOI:10.1017/beq.2021.42

Butcher, J., & Beridze, I. (2019). What is the state of artificial intelligence governance globally? *RUSI Journal*, 164(5–6), 88–96. DOI:10.1080/03071847.2019.1694260

Camilleri, M. A. (2017). *Corporate sustainability, social responsibility and environmental management*. Springer Nature. DOI:10.1007/978-3-319-46849-5

Camilleri, M. A. (2019). Measuring the corporate managers' attitudes towards ISO's social responsibility standard. *Total Quality Management & Business Excellence*, 30(13–14), 1549–1561. DOI:10.1080/14783363.2017.1413344

Camilleri, M. A. (2023). Artificial intelligence governance: Ethical considerations and implications for social responsibility. *Expert Systems: International Journal of Knowledge Engineering and Neural Networks*, 13406. Advance online publication. DOI:10.1111/exsy.13406

Camilleri, M. A., & Troise, C. (2023). Live support by chatbots with artificial intelligence: A future research agenda. *Service Business*, 17(1), 61–80. DOI:10.1007/s11628-022-00513-9

Camilleri, M. A., Troise, C., Strazzullo, S., & Bresciani, S. (2023). Creating shared value through open innovation approaches: Opportunities and challenges for corporate sustainability. *Business Strategy and the Environment*, 32(7), 4485–4502. Advance online publication. DOI:10.1002/bse.3377

Carvalho, A., Levitt, A., Levitt, S., Khaddam, E., & Benamati, J. (2019). Off-the-shelf artificial intelligence technologies for sentiment and emotion analysis: A tutorial on using IBM natural language processing. *Communications of the Association for Information Systems*, 44, 918–943. DOI:10.17705/1CAIS.04443

CNBC. (2023). OpenAI CEO Sam Altman says he's a 'little bit scared' of A.I. https://www.cnbc.com/2023/03/20/openai-ceo-sam-altman-says-hes-a-littlebit -scared-of-ai.html

Corea, F., Fossa, F., Loreggia, A., Quintarelli, S., & Sapienza, S. (2022). A principle-based approach to AI: The case for European Union and Italy. *AI & Society*, 38(2), 521–535. DOI:10.1007/s00146-022-01453-8

Crawford, K., & Paglen, T. (2021). Excavating AI: The politics of images in machine learning training sets. *AI & Society*, 36(4), 1105–1116. DOI:10.1007/s00146-021-01301-1

Damoah, I. S., Ayakwah, A., & Tingbani, I. (2021). Artificial intelligence (AI)-enhanced medical drones in the healthcare supply chain (HSC) for sustainability development: A case study. *Journal of Cleaner Production*, 328, 129598. DOI:10.1016/j.jclepro.2021.129598

Dauvergne, P. (2022). Is artificial intelligence greening global supply chains? Exposing the political economy of environmental costs. *Review of International Political Economy*, 29(3), 696–718. DOI:10.1080/09692290.2020.1814381

Du, S., El Akremi, A., & Jia, M. (2022). Quantitative research on corporate social responsibility: A quest for relevance and rigor in a quickly evolving, turbulent world. *Journal of Business Ethics*, 187(1), 1–15. DOI:10.1007/s10551-022-05297-6 PMID:36465988

Du, S., & Xie, C. (2021). Paradoxes of artificial intelligence in consumer markets: Ethical challenges and opportunities. *Journal of Business Research*, 129, 961–974. DOI:10.1016/j.jbusres.2020.08.024

Dwivedi, Y. K., Hughes, L., Ismagilova, E., Aarts, G., Coombs, C., Crick, T., Duan, Y., Dwivedi, R., Edwards, J., Eirug, A., Galanos, V., Ilavarasan, P. V., Janssen, M., Jones, P., Kar, A. K., Kizgin, H., Kronemann, B., Lal, B., Lucini, B., & Williams, M. D. (2021). Artificial intelligence (AI): Multidisciplinary perspectives on emerging challenges, opportunities, and agenda for research, practice and policy. *International Journal of Information Management*, 57, 101994. DOI:10.1016/j.ijinfomgt.2019.08.002

Engel, C., Ebel, P., & Leimeister, J. M. (2022). Cognitive automation. *Electronic Markets*, 32(1), 339–350. DOI:10.1007/s12525-021-00519-7

Erdelyi, O. J., & Goldsmith, J. (2022). Regulating Artificial Intelligence: Proposal for a Global Solution. *Government Information Quarterly*, 39(4), 1–13. DOI:10.1016/j.giq.2022.101748

EU. (2021). Regulation of the European Parliament and the council laying down harmonized rules on artificial intelligence (artificial intelligence act) and amending certain union legislative acts. European Commission https://eur-lex.europa.eu/legal -content/EN/TXT/?uri=celex%3A52021PC0206

Filgueiras, F. (2022). New Pythias of public administration: Ambiguity and choice in AI systems as challenges for governance. *AI & Society*, 37(4), 1473–1486. DOI:10.1007/s00146-021-01201-4

Fosch-Villaronga, E., Drukarch, H., Khanna, P., Verhoef, T., & Custers, B. (2022). Accounting for diversity in AI for medicine. *Computer Law & Security Report*, 47, 105735. Advance online publication. DOI:10.1016/j.clsr.2022.105735

Frank, B. (2021). Artificial intelligence-enabled environmental sustainability of products: Marketing benefits and their variation by consumer, location, and product types. *Journal of Cleaner Production*, 285, 125242. DOI:10.1016/j.jclepro.2020.125242

Galaz, V., Centeno, M. A., Callahan, P. W., Causevic, A., Patterson, T., Brass, I., Baum, S., Farber, D., Fischer, J., Garcia, D., McPhearson, T., Jimenez, D., King, B., Larcey, P., & Levy, K. (2021). Artificial intelligence, systemic risks, and sustainability. *Technology in Society*, 67, 101741. DOI:10.1016/j.techsoc.2021.101741

Gehr, T., Mirman, M., Drachsler-Cohen, D., Tsankov, P., Chaudhuri, S., & Vechev, M. (2018). Ai2: Safety and robustness certification of neural networks with abstract interpretation. In In 2018 IEEE symposium on security and privacy (SP) (pp. 3–18). IEEE.

Gonzalez, R. A., Ferro, R. E., & Liberona, D. (2020). Government and governance in intelligent cities, smart transportation study case in Bogota Colombia. *Ain Shams Engineering Journal*, 11(1), 25–34. DOI:10.1016/j.asej.2019.05.002

Hamon, R., Junklewitz, H., & Sanchez, I. (2020). *Robustness and explainability of artificial intelligence*. Publications Office of the European Union.

Hepburn, G. (2009). Alternatives to traditional regulation. Organization for Economic Cooperation and Development https://www.oecd.org/gov/regulatorypolicy/42245468.pdf

Hickok, M. (2022). Public procurement of artificial intelligence systems: New risks and future proofing. *AI & Society*, ●●●, 1–15. DOI:10.1007/s00146-022-01572-2 PMID:36212228

Hollanek, T. (2020). AI transparency: A matter of reconciling design with critique. *AI & Society*. Advance online publication. DOI:10.1007/s00146-020-01110-y

Huang, M. H., & Rust, R. T. (2022). AI as customer. *Journal of Service Management*, 33(2), 210–220. DOI:10.1108/JOSM-11-2021-0425

IBM. (2022). Introducing IBM AI governance: IBM AI governance is a new, one-stop solution built on IBM cloud Pak® for data. Armonk https://www.ibm.com/cloud/blog/announcements/introducing-ibm-ai-governanceIEEE. (2023). IEEE introduces new program for free access to AI ethics and governance standards. Institute of Electrical and Electronics Engineers https://standards.ieee.org/news/get-program-ai-ethics/

Janiesch, C., Zschech, P., & Heinrich, K. (2021). Machine learning and deep learning. *Electronic Markets*, 31(3), 685–695. DOI:10.1007/s12525-021-00475-2

Javaid, M., Haleem, A., Singh, R. P., & Suman, R. (2021). Substantial capabilities of robotics in enhancing industry 4.0 implementation. *Cognitive Robotics*, 1, 58–75. DOI:10.1016/j.cogr.2021.06.001

John-Mathews, J. M. (2022). Some critical and ethical perspectives on the empirical turn of AI interpretability. *Technological Forecasting and Social Change*, 174, 121209. DOI:10.1016/j.techfore.2021.121209

John-Mathews, J. M., Cardon, D., & Balagué, C. (2022). From reality to world. A critical perspective on AI fairness. *Journal of Business Ethics*, 178(4), 945–959. DOI:10.1007/s10551-022-05055-8

Keller, P., & Drake, A. (2021). Exclusivity and paternalism in the public governance of explainable AI. *Computer Law & Security Report*, 40, 105490. DOI:10.1016/j.clsr.2020.105490

Koniakou, V. (2023). From the "rush to ethics" to the "race for governance" in artificial intelligence. *Information Systems Frontiers*, 25(1), 71–102. DOI:10.1007/s10796-022-10300-6

Krkac, K. (2019). Corporate social irresponsibility: Humans vs artificial intelligence. *Social Responsibility Journal*, 15(6), 786–802. DOI:10.1108/SRJ-09-2018-0219

LeCun, Y., Bengio, Y., & Hinton, G. (2015). Deep learning. *Nature*, 521(7553), 436–444. DOI:10.1038/nature14539 PMID:26017442

Li, G., Li, N., & Sethi, S. P. (2021). Does CSR reduce idiosyncratic risk? Roles of operational efficiency and AI innovation. *Production and Operations Management*, 30(7), 2027–2045. DOI:10.1111/poms.13483

Li, W., Su, Z., Li, R., Zhang, K., & Wang, Y. (2020). Blockchain-based data security for artificial intelligence applications in 6G networks. *IEEE Network*, 34(6), 31–37. DOI:10.1109/MNET.021.1900629

Madaio, M. A., Stark, L., Wortman Vaughan, J., & Wallach, H. (2020). Proceedings of the 2020 CHI conference on human factors in computing systems. In Co-designing checklists to understand organizational challenges and opportunities around fairness in AI (pp. 1–14). ACM Digital Library.

Magas, M., & Kiritsis, D. (2022). Industry commons: An ecosystem approach to horizontal enablers for sustainable cross-domain industrial innovation. *International Journal of Production Research*, 60(2), 479–492. DOI:10.1080/00207543.2021.1989514

Magistretti, S., Dell'Era, C., & Petruzzelli, A. M. (2019). How intelligent is Watson? Enabling digital transformation through artificial intelligence. *Business Horizons*, 62(6), 819–829. DOI:10.1016/j.bushor.2019.08.004

Mantymaki, M., Minkkinen, M., Birkstedt, T., & Viljanen, M. (2022). Defining organizational AI governance. *AI and Ethics*, 2(4), 603–609. DOI:10.1007/s43681-022-00143-x

Matytsin, D. E., Dzedik, V. A., Makeeva, G. A., & Boldyreva, S. B. (2023). "Smart" outsourcing in support of the humanization of entrepreneurship in the artificial intelligence economy. *Humanities & Social Sciences Communications*, 10(1), 1–8. PMID:36644399

McBride, R., Dastan, A., & Mehrabinia, P. (2022). How AI affects the future relationship between corporate governance and financial markets: A note on impact capitalism. *Managerial Finance*, 48(8), 1240–1249. DOI:10.1108/MF-12-2021-0586

Microsoft. (2023). Responsible and trusted AI. Redmont https://learn.microsoft.com/en-us/azure/cloud-adoption-framework/innovate/best-practices /trusted-ai

Minkkinen, M., Niukkanen, A., & Mäntymäki, M. (2022). What about investors? ESG analyses as tools for ethics-based AI auditing. *AI & Society*. Advance online publication. DOI:10.1007/s00146-022-01415-0

Minkkinen, M., Zimmer, M. P., & Mäntymäki, M. (2023). Co-shaping an ecosystem for responsible AI: Five types of expectation work in response to a technological frame. *Information Systems Frontiers*, 25(1), 103–121. DOI:10.1007/s10796-022-10269-2

Mullins, M., Holland, C. P., & Cunneen, M. (2021). Creating ethics guidelines for artificial intelligence and big data analytics customers: The case of the consumer European insurance market. *Patterns (New York, N.Y.)*, 2(10), 100362. DOI:10.1016/j. patter.2021.100362 PMID:34693379

Narwani, K., Lin, H., Pirbhulal, S., & Hassan, M. (2022). *Towards AI-enabled approach for Urdu text recognition: A legacy for Urdu image apprehension.* IEEE., DOI:10.1109/ACCESS.2022.3203426

Nation, S. (2019). National Artificial Intelligence Strategy: Advancing our smart nation journey. Smart Nation and Digital Government Office https://www. smartnation .gov.sg/initiatives/artificial-intelligence/

Ng, K. K., Chen, C. H., Lee, C. K., Jiao, J. R., & Yang, Z. X. (2021). A systematic literature review on intelligent automation: Aligning concepts from theory, practice, and future perspectives. *Advanced Engineering Informatics*, 47, 101246. DOI:10.1016/j.aei.2021.101246

OECD. (2019). Recommendation of the council on artificial intelligence. Organization for Economic Cooperation and Development https://legalinstruments. oecd .org/en/instruments/oecd-legal-0449

Pai, V., & Chandra, S. (2022). Exploring factors influencing organizational adoption of artificial intelligence (AI) in corporate social responsibility (CSR) initiatives. *Pacific Asia Journal of the Association for Information Systems*, 14(5), 4. DOI:10.17705/1pais.14504

Papagiannidis, E., Enholm, I. M., Dremel, C., Mikalef, P., & Krogstie, J. (2023). Toward AI governance: Identifying best practices and potential barriers and outcomes. *Information Systems Frontiers*, 25(1), 123–141. DOI:10.1007/s10796-022-10251-y PMID:35464171

Rąb-Kettler, K., & Lehnervp, B. (2019). Recruitment in the times of machine learning. *Management Systems in Production Engineering*, 27(2), 105–109. DOI:10.1515/ mspe-2019-0018

Raisch, S., & Krakowski, S. (2021). Artificial intelligence and management: The automation–augmentation paradox. *Academy of Management Review*, 46(1), 192–210. DOI:10.5465/amr.2018.0072

Raji, I. D., Smart, A., White, R. N., Mitchell, M., Gebru, T., Hutchinson, B., Smith-Loud, J., Theron, D., & Barnes, P. (2020). In proceedings of the 2020 conference on fairness, accountability, and transparency. In *Closing the AI accountability gap: Defining an end-to-end framework for internal algorithmic auditing* (pp. 33–44). ACM Digital Library.

Raman, R., Meenakshi, R., Ramya, R., Jayaprakash, S., & Srinivasan, C. (2024). IoT-Based Magnetic Field Strength Monitoring for Industrial Applications. *2nd International Conference on Smart Technologies for Smart Nation, SmartTechCon 2023*, pp. 132–136.

Ramya, R. (2024). *Analysis and Applications Finding of Wireless Sensors and IoT Devices With Artificial Intelligence/Machine Learning. AIoT and Smart Sensing Technologies for Smart Devices.* IGI Global., DOI:10.4018/979-8-3693-0786-1.ch005

Ramya, R., & Ramamoorthy, S. (2022). Development of a framework for adaptive productivity management for edge computing based IoT applications. *AIP Conference Proceedings*, 2519, 030068. DOI:10.1063/5.0111710

Ramya, R., & Ramamoorthy, S. (2022). Analysis of machine learning algorithms for efficient cloud and edge computing in the IoT, Challenges and Risks Involved in Deploying 6G and NextGen Networks, pp. 72–90.

Ramya, R., & Ramamoorthy, S. (2022). Survey on Edge Intelligence in IoT-Based Computing Platform. *Survey on Edge Intelligence in IoT-Based Computing Platform, Lecture Notes in Networks and Systems, Springer*, 356, 549–556. DOI:10.1007/978-981-16-7952-0_52

Ramya, R., & Ramamoorthy, S. (2023). Lightweight Unified Collaborated Relinquish Edge Intelligent Gateway Architecture with Joint Optimization. *IEEE Access : Practical Innovations, Open Solutions*, 11, 90396–90409. DOI:10.1109/ACCESS.2023.3307808

Ramya, R., & Ramamoorthy, S. (2024). QoS in multimedia application for IoT devices through edge intelligence. *Multimedia Tools and Applications*, 83(3), 9227–9250. DOI:10.1007/s11042-023-15941-6

Ramya, R., & Ramamoorthy, S. (2024). Hybrid Fog-Edge-IoT Architecture for Real-time Data Monitoring. *International Journal of Intelligent Engineering and Systems*, 17(1), 2024. DOI:10.22266/ijies2024.0229.22

Renieris, E. M., Kiron, D., & Mills, S. (2022). Should organizations link responsible AI and corporate social responsibility?It's Complicated. MIT Sloan https://sloanreview.mit.edu/article/should-organizations-link-responsible-ai-and-corporate-social-responsibility-its-complicated/

Ribeiro, J., Lima, R., Eckhardt, T., & Paiva, S. (2021). Robotic process automation and artificial intelligence in industry 4.0–a literature review. *Procedia Computer Science*, 181, 51–58. DOI:10.1016/j.procs.2021.01.104

Rodriguez-Barroso, N., Stipcich, G., Jiménez-Lopez, D., Ruiz-Millán, J. A., Martínez-Cámara, E., González-Seco, G., Luzón, M. V., Veganzones, M. A., & Herrera, F. (2020). Federated learning and differential privacy: Software tools analysis, the Sherpa. Ai FL framework and methodological guidelines for preserving data privacy. *Information Fusion*, 64, 270–292. DOI:10.1016/j.inffus.2020.07.009

Romao, M., Costa, J., & Costa, C. J. (2019). Robotic process automation: A case study in the banking industry. In 2019 14th Iberian conference on information systems and technologies (CISTI) (pp. 1–6). IEEE. DOI:10.23919/CISTI.2019.8760733

Sachan, S., Yang, J. B., Xu, D. L., Benavides, D. E., & Li, Y. (2020). An explainable AI decision-support-system to automate loan underwriting. *Expert Systems with Applications*, 144, 113100. DOI:10.1016/j.eswa.2019.113100

Satra, H. S. (2021). A framework for evaluating and disclosing the ESG related impacts of AI with the SDGs. *Sustainability (Basel)*, 13(15), 8503. DOI:10.3390/su13158503

Saurabh, K., Arora, R., Rani, N., Mishra, D., & Ramkumar, M. (2022). AI led ethical digital transformation: Framework, research and managerial implications. Journal of Information. *Communication and Ethics in Society*, 20(2), 229–256. DOI:10.1108/JICES-02-2021-0020

Schneider, J., Abraham, R., Meske, C., & Vom Brocke, J. (2022). Artificial intelligence governance for businesses. Information Systems Management. /arXiv.2011.10672DOI:<ALIGNMENT.qj></ALIGNMENT>10.48550

Selamat, M. A., & Windasari, N. A. (2021). Chatbot for SMEs: Integrating customer and business owner perspectives. *Technology in Society*, 66, 101685. Advance online publication. DOI:10.1016/j.techsoc.2021.101685

Silva, R. L., Canciglieri, O.Junior, & Rudek, M. (2022). A road map for planning-deploying machine vision artifacts in the context of industry 4.0. *Journal of Industrial and Production Engineering*, 39(3), 167–180. DOI:10.1080/21681015.2021.1965665

Smuha, N. A. (2019). The EU approach to ethics guidelines for trustworthy artificial intelligence. *Computer Law Review International*, 20(4), 97–106. DOI:10.9785/cri-2019-200402

Strickland, E. (2019). IBM Watson, heal thyself: How IBM overpromised and underdelivered on AI health care. *IEEE Spectrum*, 56(4), 24–31. DOI:10.1109/MSPEC.2019.8678513

Thorp, H. H. (2023). ChatGPT is fun, but not an author. *Science*, 379(6630), 313. DOI:10.1126/science.adg7879 PMID:36701446

Troise, C., & Camilleri, M. A. (2021). The use of digital media for marketing, CSR communication and stakeholder engagement. In *Strategic corporate communication in the digital age*. Emerald Publishing Limited. DOI:10.1108/978-1-80071-264-520211010

Wamba-Taguimdje, S. L., Fosso Wamba, S., Kala Kamdjoug, J. R., & Tchatchouang Wanko, C. E. (2020). Influence of artificial intelligence (AI) on firm performance: The business value of AI-based transformation projects. *Business Process Management Journal*, 26(7), 1893–1924. DOI:10.1108/BPMJ-10-2019-0411

Watts, J., & Adriano, A. (2021). Uncovering the sources of machine-learning mistakes in advertising: Contextual bias in the evaluation of semantic relatedness. *Journal of Advertising*, 50(1), 26–38. DOI:10.1080/00913367.2020.1821411

Weber, M., Beutter, M., Weking, J., Böhm, M., & Krcmar, H. (2022). AI startup business models: Key characteristics and directions for entrepreneurship research. *Business & Information Systems Engineering*, 64(1), 91–109. DOI:10.1007/s12599-021-00732-w

WhiteHouse. (2022). Blueprint for an AI bill of rights: Making automated systems work for the American people. The White House Washington DC https://www.whitehouse.gov/ostp/ai-bill-of-rights/

Wong, P. H. (2020). Cultural differences as excuses? Human rights and cultural values in global ethics and governance of AI. *Philosophy & Technology*, 33(4), 705–715. DOI:10.1007/s13347-020-00413-8

Wu, L., Dodoo, N. A., Wen, T. J., & Ke, L. (2022). Understanding twitter conversations about artificial intelligence in advertising based on natural language processing. *International Journal of Advertising*, 41(4), 685–702. DOI:10.1080/0 2650487.2021.1920218

Wu, W., Huang, T., & Gong, K. (2020). Ethical principles and governance technology development of AI in China. *Engineering (Beijing)*, 6(3), 302–309. DOI:10.1016/j.eng.2019.12.015

Zhang, B., Zhu, J., & Su, H. (2023). Toward the third generation artificial intelligence. *Science China. Information Sciences*, 66(2), 1–19. DOI:10.1007/s11432-021-3449-x

Zhang, C., & Lu, Y. (2021). Study on artificial intelligence: The state of the art and future prospects. *Journal of Industrial Information Integration*, 23, 100224. DOI:10.1016/j.jii.2021.100224

Zhu, T., Ye, D., Wang, W., Zhou, W., & Philip, S. Y. (2020). More than privacy: Applying differential privacy in key areas of artificial intelligence. *IEEE Transactions on Knowledge and Data Engineering*, 34(6), 2824–2843. DOI:10.1109/TKDE.2020.3014246

Chapter 5
Navigating the Ethereal Waters:
Establishing Accountability in the Autonomous Age of Generative AI

Richard Shan
CTS, USA

ABSTRACT

This chapter explores the critical need for robust accountability management in the rapidly evolving domain of generative artificial intelligence (GenAI). It examines the complexities and ethical implications of AI-driven decisions and content creation, highlighting the challenges posed by increased autonomy and sophisticated AI outputs. A holistic GenAI accountability framework is introduced. By analyzing ethical dilemmas, legal implications, and societal impacts, the chapter underscores the importance of transparency, fairness, and adaptability in AI governance. It provides actionable recommendations for AI developers, businesses, policymakers, educators, and the public to ensure responsible GenAI development and deployment. Emphasizing the role of education and public engagement, the chapter advocates for raising AI literacy and fostering informed dialogue. It concludes with a call to action for all stakeholders to collaboratively build a sustainable and ethical AI ecosystem, ensuring that AI technologies benefit society while upholding human values and rights.

DOI: 10.4018/979-8-3693-9173-0.ch005

INTRODUCTION

The dawn of generative artificial intelligence (GenAI) signifies a transformative shift in technological capabilities, heralding an era where machines transcend their traditional roles as mere tools to become creators. These AI systems, leveraging advanced algorithms and vast datasets, generate content that spans artistic images, written narratives, music, and even sophisticated decision-making processes. The capabilities of generative AI are expanding rapidly, influencing diverse fields such as healthcare, finance, education, and the arts.

Generative AI's ability to produce human-like content is not just a technical marvel but a profound cultural and societal milestone. It challenges our understanding of creativity, authorship, and the essence of what it means to be human. As these systems become more integrated into our daily lives and professional practices, their influence grows, raising fundamental questions about their role and the consequences of their outputs.

The Need for Accountability in AI-Driven Decisions

With the increasing autonomy and sophistication of generative AI, the imperative for robust accountability mechanisms becomes clear. Traditional accountability frameworks, which have been designed around human decision-making processes, are often inadequate when applied to AI systems. These frameworks fail to address the unique challenges posed by AI's capacity to operate independently and make decisions that have far-reaching implications.

Accountability in the context of generative AI is multifaceted, encompassing ethical, legal, and societal dimensions. Ethically, there is a need to ensure that AI systems do not perpetuate biases, violate privacy, or cause harm. Legally, the question of liability for AI-generated content remains contentious, especially in cases where AI decisions result in adverse outcomes. Societally, the widespread deployment of AI technologies necessitates a framework that safeguards public trust and ensures that these systems are aligned with societal values and norms.

This chapter seeks to unravel these complexities by exploring the need for comprehensive accountability frameworks tailored to the unique challenges of generative AI. It aims to provide a thorough understanding of how generative AI operates, the ethical and legal implications of its use, and the societal impact of its decisions. By doing so, it aspires to foster a deeper awareness among AI developers, policymakers, business leaders, and the general public about the critical importance of accountability in the age of autonomous AI systems.

Key Objectives

The primary objective of this chapter is to elucidate the critical need for robust accountability frameworks in the rapidly evolving domain of generative AI. This exploration aims to underscore the importance of distinguishing and assigning responsibility in an era where AI systems autonomously generate content, thereby influencing various aspects of society and individual lives. By examining the mechanisms of generative AI and the challenges associated with its deployment, this chapter will highlight the ethical imperatives and practical steps necessary to ensure responsible AI development and use.

Structure of the Chapter

To achieve these objectives, this chapter is structured in several integral parts, as illustrated in Figure 1. The first part provides an overview of the current state of accountability in AI. We delve into the details of generative AI in Part 2, in terms of systems, outputs, and its role in decision-making, followed by a technical deep dive of the inner workings of GenAI. We then drill down to the complexities of GenAI accountability, ranging from ethical dilemmas to legal implications to societal impacts. The third part addresses the challenges and barriers by analyzing the AI policies and regulations, focusing on the effectiveness and gaps. We explore the role of education and public engagement and examine the GenAI responsibility assignment in the enterprise settings. We further study the synergy with other domains, and pragmatic disciplines. Looking forward to Part 4, the future challenges and opportunities are evaluated, along with recommendations and call to action.

Figure 1. Chapter structure

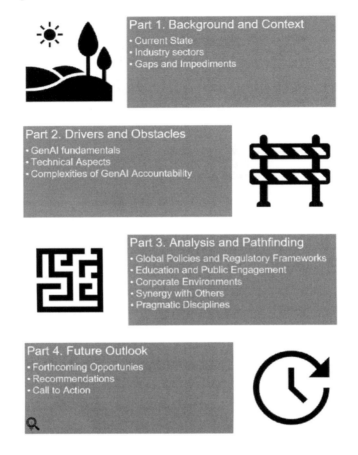

THE CURRENT LANDSCAPE OF ACCOUNTABILITY IN AI

Existing Accountability Contexts in AI

The current landscape of accountability in AI is shaped by a combination of principles, guidelines, and regulations developed by various organizations, governments, and industry groups. These frameworks aim to ensure that AI technologies are developed and deployed responsibly, addressing ethical, legal, and societal concerns. Some of the notable frameworks include:

Ethical Guidelines

Many organizations have established ethical guidelines for AI development, such as the Asilomar AI Principles, (Floridi & Cowls, 2022), the IEEE Global Initiative on Ethics of Autonomous and Intelligent Systems, (Chatila & Havens, 2019), and the EU's Ethics Guidelines for Trustworthy AI, (Smuha, 2019). These guidelines emphasize principles like transparency, fairness, accountability, and human oversight.

Regulatory Frameworks

Governments around the world are enacting regulations to govern AI development and use. The European Union's AI Act, (Laux, Wachter, & Mittelstadt, 2024) is one of the most comprehensive regulatory frameworks, categorizing AI applications based on risk and imposing stricter requirements for high-risk AI systems. In the United States, the National AI Initiative Act (Library of Congress, n.d.) aims to coordinate AI research, development, and governance across federal agencies.

Industry Standards

Various industry groups are developing standards to ensure the ethical use of AI. For example, the Partnership on AI, which includes major tech companies and academic institutions, promotes best practices for AI development and use. The ISO/IEC JTC 1/SC 42, (Chang, 2022) committee is working on international standards for AI, focusing on areas like data quality, robustness, and transparency.

Limitations and Challenges in Current Approaches

Despite these efforts, existing accountability frameworks face several limitations and challenges.

Fragmentation and Inconsistency

The lack of a unified global approach to AI governance leads to fragmented and inconsistent artifacts. Different countries and organizations have varying standards and regulations, making it difficult to establish cohesive accountability practices.

Complexity of AI Systems

The technical complexity of AI systems, particularly generative AI, poses significant challenges for accountability. These systems often operate as "black boxes," with their decision-making processes being opaque and difficult to interpret. This lack of transparency complicates efforts to hold AI systems accountable for their actions.

Rapid Technological Advancements

The pace of AI innovation often outstrips the development of regulatory and ethical frameworks. As AI technologies evolve, existing frameworks may become outdated or insufficient to address new challenges, necessitating continuous updates and adaptations.

Bias and Fairness Issues

AI systems can perpetuate and amplify biases present in their training data, leading to unfair and discriminatory outcomes. Ensuring that AI systems are fair and unbiased is a significant challenge, requiring comprehensive strategies for bias detection, mitigation, and accountability.

Legal and Liability Concerns

The legal landscape for AI accountability is still evolving, with many unresolved questions about liability and responsibility for AI-generated outcomes. Traditional legal frameworks, which rely on human accountability, struggle to accommodate the autonomous nature of AI systems.

Cases Highlighting Failures and Successes

To illustrate the limitations and challenges of current accountability frameworks, this section examines several cases across different sectors:

Healthcare

In 2019, a study (Obermeyer *et al.*) revealed that an AI system used for predicting patient outcomes in hospitals exhibited significant racial bias, leading to disparities in healthcare quality. The system was trained on data that reflected existing biases in healthcare provision, resulting in unequal treatment recommendations. This case underscores the need for rigorous bias detection and mitigation strategies in AI systems.

Finance

In the financial sector, AI algorithms used for credit scoring and lending decisions have been criticized for perpetuating discriminatory practices, (Sargeant, 2023). For example, a prominent tech company faced scrutiny when its AI-driven credit card algorithm was found to offer lower credit limits to women compared to men with similar financial profiles. This case highlights the importance of transparency and fairness in AI systems that impact financial decision-making.

Creative Industries

Generative AI used in the creation of visual art and music raises questions about intellectual property and authorship. In one notable case, an AI-generated painting sold for a significant sum at auction, (Smithsonian Magazine, 2018), sparking debate over the ownership and authorship of AI-created works. This case illustrates the legal and ethical complexities of attributing authorship and value to AI-generated content.

Law Enforcement

The use of AI in predictive policing has raised concerns about bias and accountability. In several instances, (Heaven, 2023), AI systems have disproportionately targeted minority communities, leading to claims of discriminatory practices. These cases highlight the risks of deploying AI in sensitive areas without robust accountability mechanisms to ensure fairness and prevent harm.

Identifying Gaps and Impediments

The examination of these case studies reveals several key gaps and challenges in the current landscape of AI accountability.

Need for Unified Standards

There is a pressing need for unified, international standards and regulations that provide a consistent framework for AI accountability. This would help address the fragmentation and inconsistencies that currently exist.

Enhancing Transparency

Efforts to enhance the transparency and interpretability of AI systems are crucial. This includes developing techniques for explainable AI (XAI) that allow stakeholders to understand and scrutinize AI decision-making processes.

Continuous Adaptation

Regulatory and ethical frameworks must be adaptable to keep pace with the rapid advancements in AI technology. This requires ongoing collaboration between policymakers, technologists, and ethicists to ensure that frameworks remain relevant and effective.

Addressing Bias and Fairness

Comprehensive strategies for detecting, mitigating, and addressing biases in AI systems are essential. This includes adopting diverse and representative training data, as well as implementing robust auditing and monitoring practices.

Clarifying Legal Responsibilities

Clarifying the legal responsibilities and liabilities associated with AI-generated outcomes is critical. This involves developing new legal frameworks that can accommodate the autonomous nature of AI systems and ensure that accountability is clearly assigned.

UNDERSTANDING GENERATIVE AI

Brief Explanation of How Generative AI Works

Generative AI refers to a subset of artificial intelligence systems designed to produce new content by learning patterns and structures from existing data. These systems leverage complex algorithms, particularly neural networks, and deep learning models, to analyze vast datasets and generate outputs that mimic human creation. At the core of generative AI are two primary types of models: Generative Adversarial Networks (GANs) and Variational Autoencoders (VAEs).

Generative Adversarial Networks (GANs), (Goodfellow *et al.,* 2014) operate through a dual-model mechanism: a generator and a discriminator. The generator creates new data instances, while the discriminator evaluates them against real data.

Through iterative feedback loops, the generator improves its outputs to the point where they become indistinguishable from genuine data. This adversarial process results in highly realistic content generation, whether it be images, music, or text.

Variational Autoencoders (VAEs), (Kingma & Welling, 2019) function by encoding input data into a latent space and then decoding it back into a new data format. This process allows VAEs to generate new data samples that are variations of the original data, maintaining coherence and consistency. VAEs are particularly useful for generating continuous data, such as in image generation and anomaly detection.

Types of Outputs Produced

Generative AI systems are versatile and capable of producing a wide array of outputs, each with significant implications for various sectors:

Images

GANs and other generative models can create highly realistic images, including faces, landscapes, and art pieces. These models are used in fields ranging from entertainment and design to virtual reality and digital marketing.

Text

Natural Language Processing (NLP) models, such as GPT (Generative Pre-trained Transformer), generate coherent and contextually relevant text. Applications include automated content creation, customer service chatbots, and language translation services.

Music and Audio

Generative models can compose music, synthesize speech, and create sound effects, revolutionizing the music industry, film production, and interactive media.

Decision-Making Suggestions

In business and healthcare, generative AI systems analyze data to provide insights and recommendations. For example, in finance, AI can generate trading strategies, while in healthcare, it can suggest treatment plans based on patient data.

The Role of GenAI in Decision-Making Processes

Generative AI extends beyond content creation to influence decision-making processes across various domains. These systems are increasingly integrated into environments where they analyze data, identify patterns, and suggest actions, often with minimal human intervention.

In healthcare, AI systems analyze medical records and research data to generate diagnostic and treatment recommendations, potentially improving patient outcomes and streamlining clinical workflows. For example, AI-generated algorithms assist radiologists in identifying anomalies in medical imaging, thereby enhancing diagnostic accuracy.

In finance, generative AI models analyze market trends and historical data to generate trading strategies and investment recommendations. These systems can optimize portfolio management and risk assessment, providing financial institutions with tools to navigate complex market dynamics.

In creative industries, generative AI assists in brainstorming and content creation, offering novel ideas and enhancing productivity. Writers, artists, and designers use AI-generated content as a starting point for their creative processes, exploring new styles and concepts.

In marketing and customer service, AI-driven systems generate personalized content and responses, improving customer engagement and satisfaction. Chatbots powered by generative AI handle customer inquiries with contextual accuracy, offering solutions and recommendations tailored to individual needs.

Significance of Generative AI in Decision-Making

The integration of generative AI into decision-making processes highlights its potential to enhance efficiency, accuracy, and innovation. However, it also underscores the need for accountability. As AI systems gain autonomy, the decisions they influence or make carry significant ethical, legal, and societal implications. The challenge lies in ensuring that these systems operate transparently, align with human values, and remain subject to oversight.

Understanding how generative AI works and the types of outputs it produces is crucial for developing effective accountability frameworks. By grasping the mechanisms and applications of these systems, stakeholders can better address the ethical dilemmas, legal complexities, and societal impacts that arise from AI-generated content and decisions. This foundational knowledge sets the stage for deeper exploration into the current landscape of accountability in AI, the complexities of AI accountability, and the development of robust, adaptive frameworks to guide responsible AI innovation.

The next section will delve into the internal mechanism of generative AI, providing a technical deep-dive into how these systems make decisions and highlighting the challenges in achieving transparency and explainability.

THE INNER WORKINGS OF GENERATIVE AI

Technical Deep-Dive into GenAI Decision Processes

Understanding the technical intricacies of generative AI is crucial for grasping the challenges in achieving accountability and transparency. Generative AI systems, such as Generative Adversarial Networks (GANs) and Variational Autoencoders (VAEs), operate using complex algorithms that enable them to learn from data and generate new content.

Generative Adversarial Networks (GANs)

GANs consist of two neural networks, the generator and the discriminator, which work in tandem through an adversarial process. The generator creates synthetic data samples, while the discriminator evaluates these samples against real data. The generator aims to produce data that can fool the discriminator, and the discriminator seeks to distinguish between real and synthetic data. This iterative process continues until the generator produces data that is indistinguishable from real data.

Key Components of GANs:

- Generator: Creates new data samples from random noise. It is trained to improve its output by minimizing the discriminator's ability to differentiate between real and generated data.
- Discriminator: Evaluates data samples and distinguishes between real and generated data. It provides feedback to the generator to improve the quality of synthetic data.

Variational Autoencoders (VAEs)

VAEs are another type of generative model that learns to encode input data into a latent space and then decode it back into a new data sample. VAEs are particularly useful for generating continuous data and for applications where understanding the underlying data distribution is important.

Key Components of VAEs:

- Encoder: Compresses the input data into a latent representation, capturing the essential features of the data.
- Decoder: Reconstructs the input data from the latent representation, generating new data samples that resemble the original data.

Challenges in Transparency and Explainability

The decision-making processes of generative AI systems are often opaque, posing significant challenges for transparency and explainability. Understanding how these systems arrive at their outputs is crucial for accountability, but several factors complicate this goal:

Complexity of Neural Networks

Generative AI systems use deep neural networks with multiple layers of interconnected nodes. Each layer processes the input data and passes it to the next layer, making the final output the result of numerous transformations. This complexity makes it difficult to trace the decision path and understand the specific contributions of each layer.

Black-Box Nature

The term "black-box" refers to systems whose internal workings are not easily interpretable. Generative AI models often operate as black boxes, meaning that even their developers may not fully understand how specific outputs are generated. This lack of interpretability hinders efforts to ensure that AI systems are making decisions fairly and accurately.

The Need for Explainability in AI

Explainability is crucial for ensuring that AI systems are used responsibly and ethically. It enables stakeholders to:

Understand AI Decisions

Explainable AI helps users understand how and why AI systems make specific decisions, fostering trust and confidence in the technology.

Identify and Mitigate Bias

By making the decision-making process transparent, explainable AI allows developers to identify and address biases in the model, promoting fairness and equity.

Ensure Accountability

Explainable AI provides a basis for holding AI systems accountable for their decisions, making it possible to investigate and rectify errors or harmful outcomes.

The inner workings of generative AI systems are complex and often opaque, posing significant challenges for transparency and explainability. However, by adopting explainable AI techniques, improving model interpretability, and promoting transparent data management practices, we can enhance the accountability of generative AI systems. These efforts are crucial for ensuring that AI technologies are developed and deployed responsibly, ethically, and in alignment with societal values.

The next section will delve deeper into the complexities of AI accountability, exploring the ethical, legal, and societal dimensions in greater detail.

THE COMPLEXITIES OF GENAI ACCOUNTABILITY

Ethical Dilemmas in AI Decision-Making

Generative AI systems pose unique ethical dilemmas due to their ability to make autonomous decisions and generate content without direct human intervention. These ethical challenges can be broadly categorized into issues of bias, fairness, transparency, and the moral responsibilities of AI creators and users.

Bias and Fairness

One of the most significant ethical dilemmas in AI is the perpetuation of bias. AI systems learn from existing data, which often contains historical biases reflecting societal inequalities. When these biases are not adequately addressed, AI systems can reinforce and amplify discriminatory practices. For example, an AI used in hiring processes may favor candidates from certain demographics over others if trained on biased historical data.

To address bias, AI developers must implement rigorous methods for bias detection and mitigation. This includes using diverse and representative datasets, adopting fairness-aware machine learning techniques, and continuously monitoring AI outputs to identify and correct biases. Ethical AI development also necessitates a commitment to fairness, ensuring that AI systems do not disproportionately impact any particular group.

Transparency and Explainability

The opacity of AI decision-making processes, often referred to as the "black box" problem, presents another ethical challenge. Stakeholders, including users, developers, and regulators, must understand how AI systems reach their decisions to ensure accountability. Without transparency, it is difficult to identify and rectify errors or biases in AI outputs.

Efforts to enhance transparency include developing explainable AI techniques that provide insights into the decision-making processes of AI systems. XAI aims to make AI outputs interpretable and understandable, enabling stakeholders to trust and verify AI decisions. Achieving transparency is crucial for maintaining ethical standards and fostering public trust in AI technologies.

Moral Responsibilities

The moral responsibilities of AI developers, users, and policymakers are central to the ethical use of AI. Developers must adhere to ethical principles throughout the AI lifecycle, from design to deployment. This includes prioritizing user privacy, ensuring data security, and preventing misuse of AI technologies.

Users of AI systems also bear responsibility for ethical usage. This involves understanding the limitations and potential biases of AI systems and using them in ways that align with ethical standards. Policymakers play a critical role in establishing regulations and guidelines that promote ethical AI development and usage, protecting the public from potential harm.

Legal Implications of Autonomous AI Systems

The legal landscape for AI accountability is complex and evolving. Traditional legal frameworks, designed for human accountability, struggle to accommodate the autonomous nature of AI systems. Key legal challenges include liability, intellectual property, and regulatory compliance.

Liability and Accountability

Determining liability for AI-generated outcomes is a significant legal challenge. When AI systems operate autonomously, assigning responsibility for errors, biases, or harmful consequences becomes difficult. Legal frameworks must evolve to address these challenges, ensuring that accountability is clearly defined and enforceable.

One approach to addressing liability is the concept of shared responsibility, where multiple stakeholders, including developers, users, and operators, share accountability for AI outcomes. This approach requires clear guidelines on the roles and responsibilities of each stakeholder, ensuring that accountability mechanisms are in place throughout the AI lifecycle.

Intellectual Property

Generative AI's ability to create original content raises questions about intellectual property (IP) rights. Traditionally, IP laws are designed to protect human creativity and innovation. However, AI-generated content challenges these notions, prompting debates over ownership and authorship.

Legal frameworks must adapt to address the unique aspects of AI-generated content. This includes determining who holds IP rights to AI creations and establishing guidelines for the fair use and distribution of AI-generated content. Policymakers must balance protecting human creativity with recognizing the contributions of AI systems.

Regulatory Compliance

Regulating AI technologies requires a nuanced approach that balances innovation with public safety and ethical considerations. Existing regulations, such as the European Union's General Data Protection Regulation (GDPR), provide a foundation for protecting user privacy and data security. However, specific regulations for AI, like the EU's AI Act, are needed to address the unique challenges posed by AI systems.

Effective AI regulation involves continuous collaboration between policymakers, technologists, and ethicists. Regulations must be adaptable to keep pace with rapid technological advancements and ensure that AI systems operate transparently, fairly, and ethically.

Societal Impact and Risk Assessment

The societal impact of AI technologies extends beyond ethical and legal considerations, influencing public perception, trust, and the overall societal fabric. Key societal impacts include privacy concerns, bias amplification, and the potential for job displacement.

Privacy Concerns

AI systems often require large amounts of data to function effectively, raising significant privacy concerns. The collection, storage, and analysis of personal data can lead to unauthorized access, data breaches, and misuse. Ensuring data privacy and security is paramount for maintaining public trust in AI technologies.

Privacy-preserving techniques, such as differential privacy and federated learning, offer potential solutions for protecting user data while enabling AI functionality. Policymakers must enforce strict data protection regulations and promote best practices for data privacy in AI development and deployment.

Bias Amplification

As discussed earlier, AI systems can perpetuate and amplify existing biases in data, leading to discriminatory outcomes. The societal impact of biased AI systems can be profound, affecting employment, healthcare, finance, and law enforcement.

Addressing bias requires a multifaceted approach, including diverse data representation, fairness-aware algorithms, and continuous monitoring and auditing of AI systems. Public awareness and education on the implications of AI bias are also crucial for promoting ethical AI usage.

Job Displacement

The increasing automation capabilities of AI systems pose risks of job displacement across various industries. While AI has the potential to enhance productivity and create new job opportunities, it also threatens to disrupt traditional employment models.

Societal resilience to AI-driven job displacement requires proactive measures, such as workforce retraining programs, education initiatives, and policies that support job transition and creation. Ensuring that the benefits of AI technologies are distributed equitably is essential for mitigating the societal impact of job displacement.

GenAI Accountability Framework

To address the gaps and challenges, we introduce a robust and effective accountability framework for generative AI, consisting of Corporate Environments, Regulations and Policies, Education and Public Engagement, Pragmatic Disciplines, Evolution, and Synergy with Other Domains (CREPES), as shown in Figure 2.

- Corporate Environments: The use of a RACI matrix to clearly define roles and responsibilities for participants in the enterprise settings.
- Regulations and Policies: A review of existing AI policies and regulations worldwide, assessing their effectiveness and gaps.
- Education and Public Engagement: Emphasizing the importance of public awareness and education in advocating for responsible AI.
- Pragmatic Disciplines: Insights from AI ethicists, technologists, legal experts, and policymakers will be presented to provide a comprehensive understanding of the challenges and potential solutions.
- Evolution: Projections of future developments in AI and potential accountability scenarios and responses.
- Synergy with Other Domains: We will draw parallels with accountability frameworks from other industries to identify best practices that can be applied to GenAI.

Figure 2. GenAI accountability framework

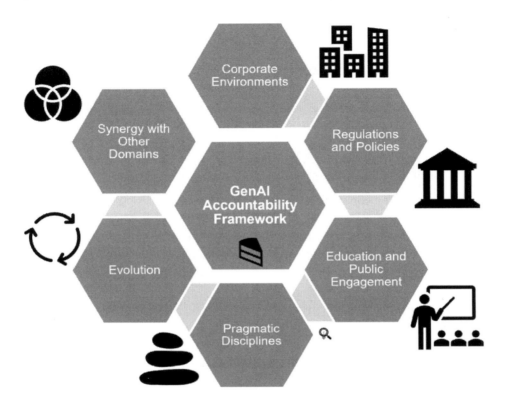

The CREPES framework helps ensure that AI technologies are developed and deployed in ways that are ethical, fair, and aligned with societal values. We will dive deep into each of the components in CREPES individually in the subsequent sections. The next section will drill down to the regulations and policies governing generative AI, assessing their effectiveness in addressing the challenges of AI accountability and identifying areas for improvement.

REGULATIONS AND POLICIES

AI Policies and Regulations Worldwide

As generative AI technologies continue to advance and permeate various sectors, governments and regulatory bodies around the world are increasingly recognizing the need for robust policies and frameworks to govern their development and use. This section reviews some of the most significant AI policies and regulations currently in place, assessing their effectiveness in addressing the accountability challenges posed by AI systems.

European Union: The AI Act

The European Union (EU) has taken a proactive approach to AI regulation with the introduction of the AI Act, a comprehensive legislative proposal aimed at regulating AI systems based on their risk levels. The AI Act categorizes AI applications into four risk categories: unacceptable risk, high risk, limited risk, and minimal risk. Key features of the AI Act include:

- Unacceptable Risk: AI applications deemed to pose unacceptable risks to safety, livelihoods, or rights are banned. Examples include AI systems that manipulate human behavior to the detriment of users or exploit vulnerabilities of specific groups.
- High Risk: High-risk AI systems, such as those used in critical infrastructure, law enforcement, and healthcare, are subject to stringent requirements. These include mandatory risk assessments, transparency obligations, and compliance with robust data governance standards.
- Limited Risk: AI systems with limited risk must adhere to specific transparency requirements, such as informing users when they are interacting with an AI system.
- Minimal Risk: Minimal-risk AI systems, like most consumer-facing AI applications, are subject to voluntary codes of conduct and best practices.

The AI Act represents a comprehensive effort to balance innovation with safety and ethical considerations. However, its effectiveness will depend on implementation and enforcement across the EU's diverse member states.

United States: National AI Initiative Act

In the United States, the National AI Initiative Act of 2020 establishes a coordinated federal program to accelerate AI research and development, promote public-private partnerships, and address ethical and societal implications of AI. Key components of the initiative include:

- National AI Research Institutes: Establishment of multidisciplinary AI research institutes to drive innovation and address societal challenges.

- AI Governance: Development of standards and best practices for AI governance, including transparency, fairness, and accountability.
- Workforce Development: Programs to support AI education and workforce training, ensuring that the U.S. workforce is prepared for the AI-driven future.
- Public-Private Partnerships: Encouragement of collaboration between government, industry, and academia to advance AI research and ensure responsible AI deployment.

While the National AI Initiative Act focuses on promoting AI innovation, it also emphasizes the need for ethical and accountable AI development. The challenge lies in balancing these objectives and ensuring that ethical considerations are not overshadowed by the push for technological advancement.

China: New Generation AI Development Plan

China's approach to AI governance is outlined in its New Generation Artificial Intelligence Development Plan (AIDP), (Roberts *et al.,* 2021), which aims to make China a global leader in AI by 2030. Key elements of the plan include:

- Strategic Goals: Establishing China as a leading AI innovation center and creating a favorable environment for AI development.
- Ethical Guidelines: Development of ethical guidelines for AI research and application, focusing on fairness, transparency, and accountability.
- Regulatory Frameworks: Implementation of regulations to ensure the safe and ethical use of AI, particularly in areas like surveillance and social governance.
- International Collaboration: Promotion of international cooperation in AI research and governance to shape global AI standards.

China's ambitious AI strategy combines aggressive innovation goals with a focus on ethical governance. However, the effectiveness of its regulatory frameworks will depend on their enforcement and the extent to which they align with global standards.

Other Notable Initiatives

Canada's Directive on Automated Decision-Making, (Kuziemski & Misuraca, 2020), provides guidelines for federal institutions using AI, emphasizing transparency, accountability, and the prevention of bias.

The UK has established the Centre for Data Ethics and Innovation (CDEI), (Hartman *et al.,* 2020), to advise the government on ethical AI use and develop frameworks for responsible AI deployment.

Japan's AI strategy, (Murayama, 2021), focuses on promoting AI innovation while ensuring ethical considerations, particularly in areas like data privacy and human rights.

Analysis of Effectiveness and Gaps in These Frameworks

While these global initiatives represent significant steps towards responsible AI governance, they also reveal gaps and challenges that need to be addressed.

Consistency and Cohesion

One of the main challenges is the lack of consistency and cohesion among different national and regional AI policies. Divergent regulations can create compliance challenges for multinational organizations and hinder the development of universal AI standards. Harmonizing AI regulations and fostering international collaboration is crucial for creating a cohesive global framework.

Enforcement and Compliance

The effectiveness of AI regulations depends heavily on enforcement mechanisms. Ensuring compliance with ethical guidelines and regulatory standards requires robust monitoring and auditing processes. This includes regular assessments of AI systems, penalties for non-compliance, and mechanisms for addressing grievances and rectifying harm caused by AI systems.

Adaptability and Flexibility

AI technologies are evolving rapidly, and regulatory frameworks must be adaptable to keep pace with technological advancements. Static regulations can quickly become outdated, failing to address new challenges and opportunities. Regulatory bodies must adopt flexible, iterative approaches that allow for continuous updates and improvements based on emerging trends and technologies.

Public Engagement and Education

Effective AI governance requires active public engagement and education. Raising awareness about AI technologies, their benefits, and their risks is essential for fostering informed public discourse and ensuring that AI systems are developed and used in ways that align with societal values. Public participation in the regulatory process can help identify diverse perspectives and ensure that AI policies reflect the needs and concerns of all stakeholders.

Global policies and regulations are crucial for ensuring the responsible development and deployment of generative AI, as depicted in Figure 3. While significant progress has been made, challenges remain in achieving consistency, effective enforcement, adaptability, and public engagement. By addressing these gaps, we can create robust and effective accountability processes that protect individuals and society from the potential risks of AI while fostering innovation and progress.

The next section will explore the role of education and public engagement in fostering an ethical AI ecosystem, emphasizing the importance of AI literacy and active public participation in AI governance.

EDUCATION AND PUBLIC ENGAGEMENT

Importance of Public Awareness in GenAI Ethics

Public awareness and understanding of generative AI are crucial for fostering an ethical AI ecosystem. As AI technologies become more pervasive, it is essential that the general public is informed about their capabilities, potential risks, and ethical implications. Public awareness ensures that individuals can engage in informed discussions, make educated decisions about AI technologies, and advocate for responsible AI practices.

Empowering Individuals

Education empowers individuals to understand the impact of AI on their lives and society. By demystifying AI and making its concepts accessible, people can better grasp the benefits and risks associated with AI technologies. This understanding enables individuals to make informed choices about AI products and services, recognize potential ethical issues, and participate in shaping AI policies and regulations.

Figure 3. Regulations and policies for GenAI accountability

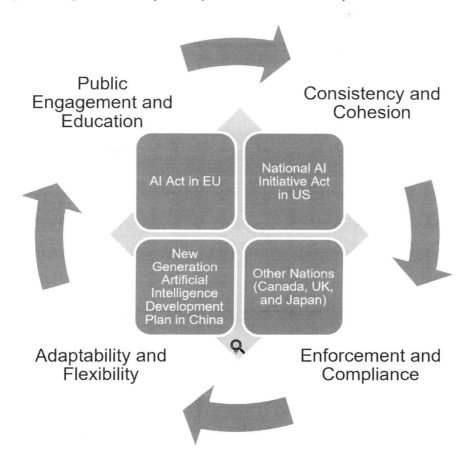

Enhancing Trust

Trust is a fundamental component of the successful integration of AI into society. When people are educated about AI and its ethical considerations, they are more likely to trust AI systems and support their adoption. Public awareness initiatives can address common misconceptions about AI, provide transparency about how AI systems operate, and highlight the measures taken to ensure their ethical use.

Educational Initiatives and Resources

Developing and promoting educational initiatives and resources is essential for increasing AI literacy and fostering an informed public. These initiatives should be designed to reach diverse audiences, including students, professionals, and the general public.

AI Literacy Programs

AI literacy programs can be implemented in schools, universities, and community centers to teach individuals about the fundamentals of AI, its applications, and its ethical implications. These programs should cover a range of topics, including:

- Basic AI Concepts: Understanding how AI works, including machine learning, neural networks, and data analysis.
- AI Applications: Exploring the various applications of AI in different sectors, such as healthcare, finance, and entertainment.
- Ethical Considerations: Discussing the ethical challenges associated with AI, such as bias, privacy, and accountability, and the importance of ethical AI development.

Online Courses and Workshops

Online courses and workshops provide flexible and accessible learning opportunities for individuals interested in AI. These courses can be offered by educational institutions, AI organizations, and industry leaders. Topics can include:

- Introduction to AI: Basic concepts and techniques in AI, suitable for beginners.
- AI Ethics: In-depth exploration of ethical issues in AI, including case studies and best practices.
- Advanced AI Topics: Specialized courses on advanced AI techniques, such as deep learning, natural language processing, and explainable AI.

Public Awareness Campaigns

Public awareness campaigns can use various media channels to reach a broad audience and raise awareness about AI. These campaigns can include:

- Informational Videos: Short, engaging videos that explain AI concepts and ethical considerations in simple terms.
- Interactive Websites: Online platforms that provide resources, quizzes, and interactive content to educate users about AI.
- Social Media: Leveraging social media platforms to share information, spark discussions, and engage with the public on AI topics.

Public Forums and Dialogues

Public forums and dialogues play a crucial role in fostering informed discussions about AI ethics and governance. These platforms provide opportunities for diverse stakeholders, including AI developers, policymakers, ethicists, and the general public, to share their perspectives and collaborate on solutions.

Community Engagement Events

Community engagement events, such as town hall meetings, panel discussions, and public lectures, can bring together stakeholders to discuss AI-related topics. These events should be designed to encourage open dialogue, address public concerns, and gather feedback on AI policies and practices.

Citizen Advisory Boards

Citizen advisory boards can be established to provide input on AI governance and policy decisions. These boards should include representatives from various demographic groups to ensure diverse perspectives are considered. Advisory boards can review AI projects, provide recommendations, and help ensure that AI systems align with societal values and ethical standards.

Participatory Policy-Making

Participatory policy-making involves actively engaging the public in the development of AI regulations and guidelines. This approach can include public consultations, surveys, and collaborative workshops where citizens can contribute their ideas and feedback. By involving the public in policy-making, regulators can create more inclusive and effective AI governance frameworks.

Case Studies: Successful Public Engagement in AI

Case Study 1: AI4All

AI4All, (Posner & Fei-Fei, 2020), is a nonprofit organization dedicated to increasing diversity and inclusion in AI. Through educational programs, AI4All aims to empower underrepresented groups to become leaders in AI. Their initiatives include summer camps, online courses, and mentorship programs that focus on AI literacy, ethical considerations, and real-world applications. AI4All's efforts have successfully raised awareness about AI and inspired a new generation of diverse AI practitioners.

Case Study 2: Montreal Declaration for Responsible AI

The Montreal Declaration for Responsible AI, (Peters *et al.*, 2020), is a collaborative initiative that involved public consultations to develop ethical guidelines for AI. The declaration outlines principles for responsible AI development, such as respect for autonomy, privacy, and equity. By engaging the public in the drafting process, the declaration reflects a wide range of perspectives and promotes ethical AI practices that align with societal values.

Education and public engagement are essential components of fostering an ethical AI ecosystem, as portrayed in Figure 4. By raising AI literacy, promoting informed discussions, and involving the public in AI governance, we can ensure that AI technologies are developed and used responsibly. Educational initiatives and public forums provide valuable platforms for increasing awareness, building trust, and shaping AI policies that reflect the needs and values of society.

CORPORATE ENVIRONMENTS

In a corporate environment, ensuring accountability for generative AI projects involves clearly defining the roles and responsibilities of various stakeholders. The RACI chart (Responsible, Accountable, Consulted, Informed) is a useful tool for delineating these roles and ensuring effective project management and ethical AI practices.

Figure 4. Education and public engagement for GenAI accountability

Understanding the RACI Chart

A RACI chart assigns responsibilities for tasks or decisions in a project. The RACI framework consists of four key roles:

- Responsible (R): The person(s) who performs the work and is responsible for completing the task.
- Accountable (A): The person who is ultimately answerable for the task and has decision-making authority.
- Consulted (C): The person(s) who provides input, advice, and feedback on the task.
- Informed (I): The person(s) who is kept informed about the progress and outcomes of the task.

Applying the RACI Chart to Generative AI Accountability

In the context of generative AI accountability, various stakeholders within an enterprise have distinct roles and responsibilities. These stakeholders include AI developers, data scientists, ethics officers, legal advisors, project managers, senior executives, and end-users.

Table 1 represents a RACI chart for generative AI accountability in a corporate environment.

Explanation of Roles and Responsibilities

- AI Developers and Data Scientists: Responsible for the technical aspects of generative AI projects, including data collection, preprocessing, model development, validation, and testing. They work closely with ethics officers to ensure ethical considerations are integrated into the AI systems.
- Ethics Officers: Accountable for defining and overseeing the ethical guidelines for AI projects. They provide input on bias detection, model validation, and auditing processes, ensuring that AI systems align with ethical standards.
- Legal Advisors: Responsible and accountable for ensuring compliance with relevant laws and regulations. They consult on ethical guidelines and are kept informed about data practices and model development to provide legal advice.
- Project Managers: Responsible and accountable for overall project coordination, documentation, and reporting. They ensure that all stakeholders are informed and that the project adheres to timelines and budgets.
- Senior Executives: Accountable for high-level decision-making and ensuring that AI projects align with the organization's strategic goals and ethical standards. They are informed about key developments and outcomes.
- End-Users: Informed about the AI system's functionality and impact. Responsible for using the system as intended and providing feedback for improvements. They receive training and support to effectively interact with the AI system.

Table 1. RACI chart

Stakeholder	AI Developer	Data Scientist	Ethics Officer	Legal Advisor	Project Manager	Senior Executive	End-User
Define ethical guidelines	C	C	R	C	R	A	I
Data collection and preprocessing	A	R	C	C	R	I	I
Bias detection and mitigation	R	R	C	C	R	A	I
Model development	R	R	I	A	R	A	I
Model validation and testing	R	R	I	I	R	A	I
Compliance with regulations	I	I	C	R	R	A	I

continued on following page

Table 1. Continued

Stakeholder	AI Developer	Data Scientist	Ethics Officer	Legal Advisor	Project Manager	Senior Executive	End-User
Monitoring and auditing	R	R	A	I	R	C	I
Documentation and reporting	R	R	C	C	R	A	I
User training and support	I	I	C	I	R	I	R
Incident management	R	A	C	R	R	A	C

- **R (Responsible)**: Directly involved in executing the task.
- **A (Accountable)**: Ultimately accountable for the completion and quality of the task.
- **C (Consulted)**: Provides input and advice; two-way communication.
- **I (Informed)**: Kept updated on progress; one-way communication.

Benefits of Using a RACI Chart

Implementing a RACI chart in the context of generative AI accountability offers several benefits:

- Clarity: Clearly defines the roles and responsibilities of each stakeholder, reducing ambiguity and ensuring that everyone understands their tasks.
- Separation of Duties: Identifies who is accountable for each aspect of the project, facilitating better decision-making and oversight.
- Collaboration: Encourages collaboration and consultation among different stakeholders, promoting a multidisciplinary approach to AI development.
- Transparency: Enhances transparency by ensuring that all relevant parties are informed about the progress and outcomes of the AI project.
- Efficiency: Improves project management efficiency by streamlining communication and ensuring that tasks are completed by the appropriate stakeholders.

By using a RACI chart, organizations can effectively manage generative AI projects, ensuring ethical standards are upheld and accountability is maintained throughout the development and deployment process. This structured approach helps mitigate risks, promotes responsible AI practices, and fosters trust among stakeholders. The RACI chart is a powerful tool for managing accountability in generative AI projects within a corporate environment. By clearly defining the roles and responsibilities of various stakeholders, organizations can ensure that AI systems are developed and deployed ethically and responsibly. This structured approach to accountability supports the creation of transparent, fair, and trustworthy AI technologies, ultimately benefiting both the organization and society.

The next section will explore the synergy with other sectors, drawing parallels and identifying best practices that can be applied to the field of AI accountability.

SYNERGY WITH OTHER DOMAINS

Comparative Analysis with Other Industries

To develop effective accountability frameworks for generative AI, it is instructive to look at how other industries have tackled similar challenges. This section explores accountability mechanisms in corporate governance, environmental policy, and healthcare, identifying strategies that can be adapted for AI.

Corporate Governance

Corporate governance provides a well-established model for accountability, emphasizing transparency, responsibility, and ethical conduct. Key elements include:

- Board Oversight: Companies are required to have boards of directors that oversee management and ensure that the organization acts in the best interest of shareholders and stakeholders. This model can be adapted for AI governance, where an AI ethics board could oversee AI development and deployment.
- Regulatory Compliance: Corporations must comply with a range of regulations designed to ensure fair practices and protect stakeholders. For AI, this could translate to compliance with ethical guidelines, transparency requirements, and bias mitigation standards.
- Stakeholder Engagement: Effective corporate governance involves regular communication with stakeholders, including shareholders, employees, and the public. AI developers can adopt similar practices, engaging with users, affected communities, and regulatory bodies to ensure that AI systems meet ethical and societal standards.

Environmental Policy

Environmental policy provides another useful framework for AI accountability, particularly in the areas of risk assessment, sustainability, and regulatory enforcement.

- Risk Assessment and Management: Environmental policies often require thorough risk assessments and the implementation of management plans to mitigate potential harms. In AI, this approach can be used to evaluate the potential risks of AI systems and implement measures to address them proactively.

- Sustainability: Environmental policies emphasize the importance of sustainable practices that do not deplete resources or harm future generations. Similarly, AI development should focus on sustainability, ensuring that AI systems are designed and used in ways that benefit society without causing long-term negative impacts.
- Regulatory Enforcement: Strong enforcement mechanisms are crucial for environmental policies, ensuring compliance and penalizing violations. For AI, robust enforcement mechanisms are necessary to ensure that developers and users adhere to ethical guidelines and regulatory standards.

Healthcare

The healthcare industry offers valuable insights into accountability, particularly regarding patient safety, data privacy, and professional responsibility.

- Patient Safety: Healthcare providers are accountable for patient safety, with stringent regulations and oversight mechanisms in place. Similarly, AI developers should prioritize user safety, implementing rigorous testing and validation processes to ensure that AI systems do not cause harm.
- Data Privacy: Healthcare regulations, such as the Health Insurance Portability and Accountability Act (HIPAA), (Marks & Haupt, 2023), in the United States, protect patient data privacy. AI systems that handle personal data should adhere to strict privacy standards, ensuring that user data is protected and used ethically.
- Professional Responsibility: Healthcare professionals are held to high ethical standards, with clear guidelines on professional conduct. AI developers and users should be similarly held accountable, adhering to ethical standards and demonstrating professional responsibility in the development and deployment of AI systems.

Applicable Strategies from Other Sectors

Drawing from other verticals, several strategies can be applied to GenAI accountability:

- Ethics Boards and Oversight Committees: Establishing dedicated ethics boards or oversight committees within organizations can ensure that AI development and deployment are guided by ethical principles and societal values.

- Comprehensive Risk Assessments: Implementing thorough risk assessments can help identify potential ethical, legal, and societal impacts of AI systems, allowing for proactive mitigation measures.
- Stakeholder Engagement: Regular engagement with stakeholders, including users, affected communities, and regulatory bodies, ensures that diverse perspectives are considered and that AI systems meet societal expectations.
- Transparency and Reporting: Adopting transparency measures, such as regular reporting on AI development processes and outcomes, can build trust and accountability. This includes disclosing information about data sources, algorithms, and decision-making processes.
- Regulatory Compliance and Enforcement: Ensuring compliance with ethical guidelines and regulatory standards through robust enforcement mechanisms is essential. This includes regular audits, penalties for non-compliance, and continuous updates to regulations based on technological advancements.

Case Studies: Successes and Challenges

To illustrate the application of these strategies, this section examines case studies from various industries.

Corporate Governance

A multinational corporation implemented an AI ethics board to oversee its AI projects, (IBM, 2003). The board included ethicists, technologists, and external stakeholders, ensuring diverse perspectives. This approach led to the successful integration of ethical principles into the company's AI systems, enhancing transparency and accountability.

Environmental Policy

A leading environmental agency adopted a risk assessment framework for AI systems used in environmental monitoring. By evaluating potential risks and implementing mitigation measures, the agency ensured that AI systems operated safely and effectively, minimizing environmental impact, (Konya & Nematzadeh, 2024).

Healthcare

A major healthcare provider developed a comprehensive data privacy policy for its AI systems, ensuring compliance with HIPAA and other regulations. This approach protected patient data and maintained trust in AI-driven healthcare solutions, (Sharma, 2024).

Adapting Lessons for AI Accountability

By adapting lessons from corporate governance, environmental policy, and healthcare, we can develop robust accountability frameworks for generative AI. Key adaptations include:

- Establishing Ethical Oversight: Creating AI ethics boards and oversight committees within organizations can ensure that ethical considerations guide AI development and deployment.
- Implementing Risk Management: Adopting comprehensive risk assessment and management frameworks can help identify and mitigate potential harms associated with AI systems.
- Enhancing Transparency: Promoting transparency through regular reporting and stakeholder engagement can build trust and accountability, ensuring that AI systems align with societal values.
- Ensuring Regulatory Compliance: Strengthening regulatory compliance and enforcement mechanisms can ensure that AI systems adhere to ethical guidelines and legal standards, protecting users and society from potential harms.

By drawing on the strengths of established accountability practices in other domains, we can create effective strategies for ensuring the responsible development and deployment of generative AI, as displayed in Figure 5.

Figure 5. Synergy for GenAI accountability

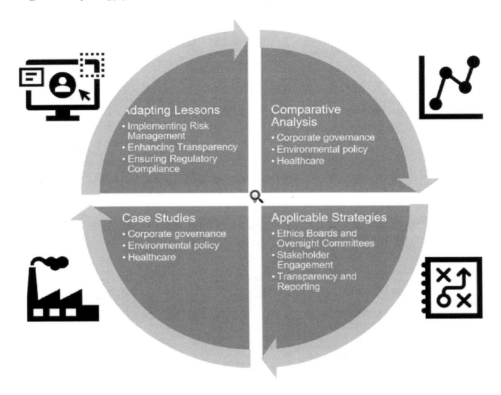

The next section will explore insights from experts in AI ethics, technology, law, and policy, providing a comprehensive understanding of the challenges and potential solutions in establishing AI accountability.

PRAGMATIC DISCIPLINES

Expert Insights

The insights from professional specialists provide a comprehensive understanding of the multifaceted challenges in establishing AI accountability. Each perspective contributes to a holistic approach, emphasizing the need for interdisciplinary collaboration, continuous monitoring, and adaptive regulations.

Ethicists highlight the importance of integrating ethical considerations throughout the AI lifecycle, ensuring that AI systems are designed and deployed with a clear understanding of their potential impacts. Technologists focus on the practical challenges of achieving transparency and explainability, developing tools and techniques to make AI decision-making processes more interpretable. Legal experts emphasize the need for new legal frameworks that address the unique characteristics of AI, ensuring that accountability is clearly defined and enforceable. Policymakers stress the importance of stakeholder engagement and adaptive regulations, creating a balanced regulatory environment that promotes innovation while protecting public interests.

For instance, AI ethicists play a crucial role in navigating the moral landscape of artificial intelligence. They provide foundational insights into the ethical frameworks and principles necessary for responsible AI development and deployment.

As an example, Dr. Jane Smith, AI Ethicist, emphasizes the importance of embedding ethical considerations into the AI development lifecycle. According to Dr. Smith, ethics cannot be an afterthought in AI development. It must be integrated from the design phase through to deployment and beyond. This means developers need to be educated in ethical theory and practical ethics to understand the potential impacts of their creations, (Smith, 2024). Dr. Smith advocates for the use of ethical AI design tools, such as impact assessments and ethical audits, to identify and mitigate potential harms before they occur. She also stresses the need for interdisciplinary collaboration, involving ethicists, technologists, and end-users in the AI development process to ensure diverse perspectives are considered.

By bringing together diverse perspectives, we can develop robust and effective accountability practices for generative AI. This interdisciplinary approach ensures that AI systems are not only technically advanced but also ethically sound, legally compliant, and socially beneficial.

Risk Mitigation

To address the ethical, legal, and societal challenges posed by generative AI, stakeholders must adopt proactive measures for mitigating risks:

- Ethical AI Development: Developers should integrate ethical principles into the AI design process, prioritizing fairness, transparency, and accountability.
- Robust Legal Frameworks: Policymakers must establish clear legal guidelines for AI accountability, addressing liability, IP rights, and regulatory compliance.
- Public Engagement and Education: Increasing public awareness and understanding of AI technologies is crucial for fostering informed dialogue and promoting ethical AI usage.
- Continuous Monitoring and Auditing: Implementing ongoing monitoring and auditing practices ensures that AI systems remain aligned with ethical standards and societal values.

By taking these proactive measures, stakeholders can navigate the complexities of AI accountability, ensuring that generative AI systems are developed and deployed responsibly, ethically, and in alignment with societal needs and values.

Improving Explainability

To address the technical challenges, practitioners are developing XAI techniques that aim to make AI decision-making processes more transparent. XAI seeks to provide insights into how AI systems operate and why they produce specific outputs. Some notable XAI techniques include:

- Local Interpretable Model-agnostic Explanations (LIME) (Zafar & Khan, 2019): LIME explains individual predictions by approximating the black-box model locally with an interpretable model. It helps users understand the factors influencing specific AI decisions.
- SHapley Additive exPlanations (SHAP) (Antwarg *et al.*, 2021): SHAP values provide a unified measure of feature importance, offering insights into how different features contribute to the model's predictions. SHAP values are based on cooperative game theory and provide a consistent way to attribute the output to individual input features.

Efforts to Enhance Transparency

Enhancing the transparency of generative AI systems involves multiple strategies, including improving model interpretability, adopting best practices for data management, and fostering an open and collaborative AI research environment.

Improving Model Interpretability

Developers can improve model interpretability by:

- Using simpler models, when possible, as they are easier to understand and explain.
- Implementing visualization techniques to illustrate how different parts of the model contribute to the final output.
- Providing clear documentation of the model's architecture, training process, and decision-making logic.

Best Practices for Data Management

Transparent data management practices are essential for ensuring that AI systems are trained on high-quality, representative, and unbiased data. Best practices include:

- Documenting the data collection process, including sources, methods, and preprocessing steps.
- Regularly auditing datasets for biases and inconsistencies.
- Ensuring that training data is diverse and representative of the target population.

Open and Collaborative AI Research

Promoting openness and collaboration in AI research can enhance transparency and accountability. This involves:

- Sharing code, models, and datasets with the broader research community to facilitate peer review and independent validation.
- Participating in collaborative initiatives, such as the Partnership on AI (2016), which promotes best practices and ethical standards in AI development.
- Engaging with interdisciplinary teams, including ethicists, legal experts, and social scientists, to address the broader implications of AI technologies.

To ensure GenAI accountability, insights from ethicists, technologists, legal experts, and policymakers emphasize a holistic approach involving ethical integration, transparency, and adaptive regulations. Proactive risk mitigation involves ethical design, robust legal guidelines, public education, and continuous monitoring. Enhancing interpretability and transparency requires model interpretability, transparent data practices, and collaborative AI research, ensuring AI systems are ethically sound and socially beneficial, as highlighted in Figure 6.

Figure 6. Pragmatic disciplines for GenAI accountability

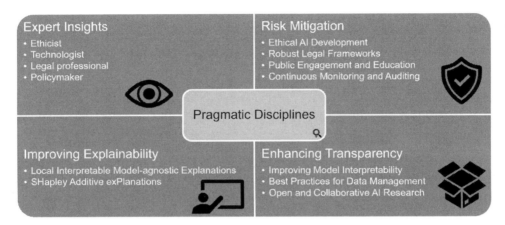

The next section will explore the evolution opportunities in AI accountability, projecting future developments in AI and discussing potential scenarios and responses to ensure that AI systems remain ethical and accountable as they evolve.

FUTURE CHALLENGES AND OPPORTUNITIES

Projecting Future Developments in GenAI

As generative AI continues to evolve, its capabilities and applications are expected to expand dramatically. This section explores potential future developments in AI and the challenges and opportunities these advancements may present for accountability.

Increased Autonomy and Decision-Making Capabilities

Future AI systems are likely to exhibit even greater autonomy, making more complex decisions without human intervention. This includes applications in autonomous vehicles, healthcare diagnostics, financial trading, and personalized education. With increased autonomy comes the need for more sophisticated accountability frameworks to ensure that AI systems act in ways that are ethical, fair, and aligned with human values.

Advancements in AI Creativity

Generative AI's ability to create new content will continue to improve, producing increasingly sophisticated and human-like outputs in art, music, literature, and other creative domains. These advancements will raise new questions about intellectual property, authorship, and the value of AI-generated content. Ensuring that creators, both human and AI, are fairly recognized and compensated will be a critical challenge.

Integration with Internet of Things (IoT) and Smart Environments

AI's integration with IoT and smart environments will lead to more interconnected and intelligent systems that can interact seamlessly with their surroundings. This includes smart homes, cities, and industrial systems that can optimize energy use, improve safety, and enhance quality of life. However, the complexity and scale of these systems will require robust frameworks for data privacy, security, and accountability to protect users and ensure ethical operation.

Potential Accountability Scenarios and Responses

To address the challenges posed by future AI developments, stakeholders must consider various accountability scenarios and develop proactive responses. This section outlines some potential scenarios and strategies for ensuring responsible AI deployment.

Scenario 1: Autonomous AI in Healthcare

Imagine a future where AI systems independently diagnose diseases and prescribe treatments. While this could revolutionize healthcare, it also raises significant accountability concerns. If an AI system makes an incorrect diagnosis, determining liability and ensuring patient safety becomes complex.

Response: Establishing clear guidelines for the use of AI in healthcare, including rigorous testing and validation protocols, is essential. Regulatory bodies should require AI systems to undergo extensive clinical trials and continuous monitoring to ensure their accuracy and reliability. Additionally, maintaining human oversight in critical decision-making processes can provide an added layer of accountability and safety.

Scenario 2: AI-Driven Creative Industries

As generative AI creates more content, disputes over intellectual property and authorship will become more common. Determining who owns and profits from AI-generated art, music, and literature will be a contentious issue.

Response: Developing new legal frameworks that recognize the unique contributions of AI in the creative process is crucial. These frameworks should address questions of authorship, ownership, and compensation, ensuring that human creators and AI developers are fairly acknowledged. Collaborative efforts between legal experts, artists, and technologists can help shape policies that balance innovation with fair use and recognition.

Scenario 3: Smart Cities and Data Privacy

In smart cities, AI systems will manage infrastructure, transportation, energy, and public services, generating vast amounts of data. Protecting the privacy and security of this data will be paramount to prevent misuse and ensure public trust.

Response: Implementing robust data governance frameworks that prioritize privacy, security, and transparency is essential. Smart city initiatives should include clear policies for data collection, storage, and use, with strong encryption and access controls to protect sensitive information. Public engagement and transparency in data governance can help build trust and ensure that smart city technologies serve the public interest.

Opportunities for Enhancing Accountability

While future AI developments present challenges, they also offer opportunities to enhance accountability through innovation and collaboration. This section explores some potential opportunities.

Advanced Monitoring and Auditing Tools

Advancements in AI and machine learning can be leveraged to develop sophisticated monitoring and auditing tools that ensure AI systems operate ethically and transparently. These tools can automatically detect and report biases, errors, and deviations from ethical standards, providing continuous oversight and accountability.

Collaborative Policy-Making

International collaboration in AI policy-making can lead to more cohesive and effective regulatory frameworks. By working together, countries can share best practices, harmonize standards, and address cross-border challenges in AI governance. Collaborative efforts can also foster innovation by creating a global environment of trust and cooperation.

Ethical AI Frameworks

The development of ethical AI frameworks that integrate principles of fairness, transparency, and accountability into the design and deployment of AI systems can promote responsible AI use. These frameworks can be adopted by organizations and industries, ensuring that ethical considerations are embedded in AI development from the outset.

Public Awareness and Education

Raising public awareness about AI technologies and their implications can empower individuals to make informed decisions and advocate for responsible AI practices. Education initiatives that demystify AI and promote digital literacy can foster a more informed and engaged public, capable of participating in discussions about AI governance and accountability.

The future of generative AI holds tremendous promise, but it also presents significant challenges for accountability. By anticipating potential scenarios and proactively developing strategies to address them, stakeholders can ensure that AI systems are deployed responsibly and ethically. Advancements in monitoring tools, collaborative policy-making, ethical frameworks, and public education offer opportunities to enhance accountability and build a trustworthy AI ecosystem, as described in Figure 7. The next section will provide actionable recommendations for various stakeholders, outlining practical steps for implementing robust accountability mechanisms in the development and use of generative AI.

Figure 7. Evolution of GenAI accountability

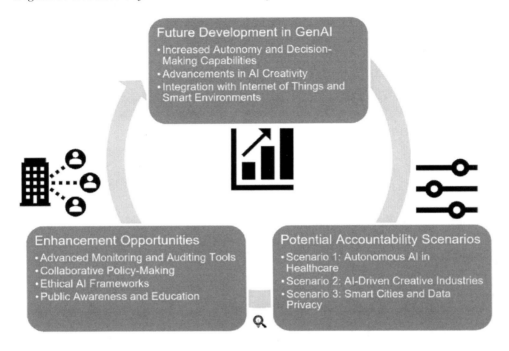

RECOMMENDATIONS FOR STAKEHOLDERS

To ensure that generative AI systems are developed and deployed responsibly, it is essential for various stakeholders, including AI developers, businesses, policymakers, and the general public, to adopt robust accountability mechanisms. This section provides actionable recommendations for each group, outlining practical steps to foster ethical and transparent AI practices.

Strategies for AI Developers and Businesses

Integrate Ethical Considerations into AI Design

- Ethical Design Principles: Incorporate ethical design principles, such as fairness, transparency, and accountability, into the AI development lifecycle. This includes defining clear ethical guidelines and ensuring that these principles are embedded from the initial design phase through to deployment and maintenance.

- Bias Detection and Mitigation: Implement robust methods for detecting and mitigating biases in AI systems. This includes using diverse and representative datasets, conducting regular bias audits, and employing fairness-aware machine learning techniques.
- Explainable AI: Develop and use explainable AI techniques to make AI decision-making processes more transparent and interpretable. This can help build trust and allow stakeholders to understand and scrutinize AI outputs.

Conduct Ethical AI Audits

- Regular Audits: Perform regular ethical audits of AI systems to ensure compliance with ethical guidelines and regulatory standards. These audits should assess the fairness, transparency, and accountability of AI systems, identifying and addressing any ethical issues.
- Independent Review: Engage independent third-party auditors to conduct unbiased evaluations of AI systems. Independent reviews can provide an objective assessment of AI practices and help identify areas for improvement.
- Continuous Improvement: Use audit findings to inform continuous improvement processes. Implement corrective actions and update AI systems based on audit results to enhance ethical performance and accountability.

Foster a Culture of Ethical AI

- Ethics Training: Provide ethics training for AI developers and engineers to raise awareness of ethical issues and equip them with the tools and knowledge to address these challenges. Training programs should cover topics such as bias, fairness, transparency, and data privacy.
- Ethical Leadership: Promote ethical leadership, (Uddin, 2023), within organizations by appointing ethics officers or creating ethics committees to oversee AI projects. These leaders can champion ethical practices and ensure that ethical considerations are prioritized in AI development.
- Stakeholder Engagement: Engage with diverse stakeholders, including users, affected communities, and regulatory bodies, to understand their concerns and perspectives. Incorporate stakeholder feedback into AI design and deployment processes to ensure that AI systems align with societal values and expectations.

Policy Recommendations for Regulators

Develop Comprehensive AI Regulations

- Risk-Based Frameworks: Implement risk-based regulatory frameworks that categorize AI applications based on their potential impact and associated risks. High-risk AI systems should be subject to stricter requirements and oversight to ensure their safety and ethical operation.
- Transparency and Accountability: Mandate transparency and accountability requirements for AI systems. This includes requiring developers to disclose information about data sources, algorithms, and decision-making processes, as well as implementing mechanisms for tracking and auditing AI performance.
- Ethical Standards: Establish clear ethical standards and guidelines for AI development and use. These standards should address issues such as bias, fairness, transparency, and data privacy, providing a consistent framework for ethical AI practices.

Enhance Regulatory Enforcement

- Robust Enforcement Mechanisms: Develop robust enforcement mechanisms to ensure compliance with AI regulations. This includes regular inspections, penalties for non-compliance, and mechanisms for addressing grievances and rectifying harms caused by AI systems.
- Adaptive Regulations: Create adaptive regulatory frameworks that can evolve with technological advancements. Regulatory bodies should regularly review and update AI regulations to address emerging challenges and opportunities, ensuring that they remain relevant and effective.
- International Collaboration: Foster international collaboration in AI regulation to harmonize standards and address cross-border challenges. Collaborative efforts can help create a cohesive global framework for AI governance, promoting consistency and reducing compliance burdens for multinational organizations.

Advocacy and Engagement Approaches for the Public

Raise Public Awareness and Education

- AI Literacy Programs: Develop and promote AI literacy programs to educate the public about AI technologies, their benefits, and their risks. These programs should aim to demystify AI and empower individuals to make informed decisions about AI use.
- Public Forums and Dialogues: Organize public forums and dialogues to facilitate discussions about AI ethics and governance. These platforms can provide opportunities for diverse stakeholders to share their views and contribute to the development of responsible AI policies.
- Accessible Resources: Create accessible resources, such as websites, guides, and toolkits, to provide information about AI technologies and ethical considerations. These resources should be designed to be easily understood by non-experts, helping to raise awareness and foster informed public discourse.

Advocate for Ethical AI Practices

- Community Involvement: Encourage community involvement in AI governance by creating opportunities for public participation in regulatory processes. This can include public consultations, citizen advisory boards, and participatory decision-making mechanisms.
- Consumer Advocacy: Support consumer advocacy groups that promote ethical AI practices and hold developers and businesses accountable for their AI systems. These groups can play a crucial role in representing public interests and pushing for greater transparency and accountability.
- Informed Decision-Making: Empower individuals to make informed decisions about AI use by providing clear information about the ethical implications of AI technologies. This can help consumers choose products and services that align with their values and support responsible AI practices.

Implementing robust accountability mechanisms for generative AI requires coordinated efforts from AI developers, businesses, policymakers, and the public. By integrating ethical considerations into AI design, conducting regular audits, fostering a culture of ethical AI, developing comprehensive regulations, and raising public awareness, stakeholders can ensure that AI systems are developed and deployed responsibly. These recommendations provide a roadmap for achieving ethical and transparent AI practices, fostering trust, and maximizing the societal benefits of AI technologies.

CALL TO ACTION

Collective Responsibility

As generative AI continues to advance and integrate into various aspects of society, it is crucial for all stakeholders to recognize and embrace their roles in ensuring its responsible and ethical development and deployment. This section outlines specific actions that AI developers, businesses, policymakers, educators, and the general public can take to contribute to a sustainable and ethical AI ecosystem.

For AI Developers and Businesses

Implement Ethical AI Practices

- Commit to Transparency: AI developers and businesses should prioritize transparency by documenting and openly sharing information about their AI systems, including data sources, algorithms, and decision-making processes.
- Conduct Regular Audits: Perform regular ethical audits of AI systems to identify and mitigate biases, ensure compliance with ethical standards, and continuously improve AI practices.
- Foster Inclusive Development: Promote diversity and inclusion within AI development teams to bring varied perspectives and reduce the risk of biased outcomes.

Promote Continuous Learning and Improvement

- Invest in Training: Provide ongoing training for AI developers and engineers on ethical AI practices, bias detection, and fairness-aware machine learning techniques.
- Encourage Innovation: Support research and development of new methods for enhancing AI transparency, interpretability, and accountability.

For Policymakers

Develop Robust Regulatory Frameworks

- Create Adaptive Policies: Develop and implement adaptive regulatory frameworks that can evolve with the rapid advancements in AI technology.
- Harmonize Standards: Collaborate internationally to harmonize AI regulations and standards, facilitating global cooperation and consistency in AI governance.

Enhance Public Engagement

- Involve Citizens in Policy-Making: Foster participatory policy-making by actively involving citizens and diverse stakeholders in the development of AI regulations and guidelines.
- Promote AI Literacy: Support initiatives that raise public awareness and understanding of AI technologies and their ethical implications.

For Educators

Integrate AI Ethics into Curricula

- Develop Comprehensive Programs: Create and integrate AI ethics courses into educational curricula at all levels, from primary education to higher education.
- Foster Interdisciplinary Learning: Encourage interdisciplinary learning that combines technical AI education with studies in ethics, law, social sciences, and humanities.

Provide Accessible Resources

- Create Online Platforms: Develop and maintain online platforms that offer accessible resources, courses, and toolkits for learning about AI and its ethical considerations.
- Host Public Workshops: Organize workshops, seminars, and public lectures to educate diverse audiences about AI technologies and their impact on society.

For the General Public

Engage in Informed Dialogue

- Stay Informed: Educate yourself about AI technologies, their benefits, risks, and ethical implications by accessing reliable sources and participating in AI literacy programs.
- Participate in Discussions: Engage in public forums, community meetings, and online discussions about AI governance and ethics, sharing your perspectives and concerns.

Advocate for Ethical AI

- Support Responsible Companies: Choose to support businesses and products that prioritize ethical AI practices and transparency.
- Demand Accountability: Advocate for stronger regulations and accountability mechanisms to ensure that AI systems are developed and used responsibly.

Building a Sustainable and Ethical AI Ecosystem

The development of generative AI presents both tremendous opportunities and significant challenges. By taking collective action, all stakeholders can contribute to building a sustainable and ethical AI ecosystem that benefits society as a whole. Key actions include:

- Collaboration and Partnership: Foster collaboration between AI developers, businesses, policymakers, educators, and the public to address ethical challenges and develop effective solutions.
- Commitment to Ethical Standards: Ensure that ethical considerations are at the forefront of AI development, deployment, and regulation.
- Ongoing Engagement and Adaptation: Continuously engage with evolving AI technologies and adapt policies, practices, and educational efforts to keep pace with advancements.

The call to action for ensuring accountability in generative AI is clear: we must work together across disciplines, sectors, and borders to create an ethical and transparent AI ecosystem. By committing to these actions, we can harness the power of AI while safeguarding human values, rights, and societal well-being. The collective efforts of AI developers, businesses, policymakers, educators, and the general public will shape the future of AI and its impact on our world. Let us embrace this responsibility and work towards a future where AI technologies are developed and used for the greater good, with accountability and ethics at their core.

The final section of this chapter will summarize the need for accountability in generative AI and emphasize the importance of ethical, legal, and social considerations in achieving responsible AI development and deployment.

CONCLUSION

The rapid advancement of generative AI presents both remarkable opportunities and profound ethical, legal, and societal challenges. As AI systems become more autonomous and capable of creating complex and impactful content, the need for robust accountability frameworks becomes increasingly critical. This book chapter has explored the multifaceted dimensions of AI accountability, emphasizing the importance of ethical considerations, legal responsibilities, and societal impacts in the development and deployment of generative AI technologies.

The integration of ethical, legal, and social considerations into AI development is not merely a regulatory requirement but a fundamental necessity for ensuring that AI technologies benefit society as a whole. Ethical AI practices help prevent biases, protect privacy, and promote fairness, while legal frameworks provide the structure for enforcing accountability and addressing grievances. Social considerations ensure that AI systems align with societal values and norms, fostering trust and acceptance among the public.

Key Takeaways

- Current Accountability Landscape: Existing frameworks provide a foundation, but they need to evolve to address the unique challenges of generative AI.
- Understanding Generative AI: Grasping the technical workings and potential outputs of generative AI is essential for addressing its ethical and accountability challenges.
- Complexities of AI Accountability: Ethical dilemmas, legal implications, and societal impacts highlight the need for comprehensive and adaptable accountability mechanisms.
- CREPES Framework: A comprehensive framework for GenAI accountability addresses the key concerns and barriers.
- Global Policies and Regulations: A harmonized, and adaptive regulatory approach is necessary to manage the global implications of AI technologies.

- Education and Public Engagement: Raising AI literacy and fostering public engagement are key to building an informed and empowered society that can advocate for ethical AI practices.
- Corporate settings: Enhancing transparency and explainability in AI systems is vital for fostering trust and accountability in enterprises.
- Synergy with Other Domains: Adapting strategies from corporate governance, environmental policy, and healthcare can inform effective AI accountability practices.
- Expert Insights: Interdisciplinary perspectives from ethicists, technologists, legal experts, and policymakers are crucial for developing well-rounded accountability frameworks.
- Future Challenges and Opportunities: Anticipating future developments and proactively addressing potential accountability scenarios can mitigate risks and leverage opportunities.
- Best Practice Recommendations: Practical and strategic recommendations for AI developers, businesses, policymakers, and the public are essential for implementing robust accountability mechanisms.

Call to Action for Stakeholders

The responsibility for ensuring ethical and accountable AI lies with all stakeholders involved in AI development, deployment, and regulation. AI developers and businesses must integrate ethical principles into their practices, conduct regular audits, and foster a culture of transparency. Policymakers should develop adaptive and harmonized regulatory frameworks, enhance enforcement mechanisms, and promote public engagement. Educators need to integrate AI ethics into curricula and provide accessible resources for continuous learning. The general public should stay informed, participate in dialogues, and advocate for responsible AI practices.

Looking Ahead

As we look to the future, it is imperative that we continue to refine and strengthen our approaches to AI accountability. Ongoing collaboration across disciplines, sectors, and borders will be essential in addressing the evolving challenges and opportunities presented by generative AI. By maintaining a commitment to ethical principles, legal compliance, and societal well-being, we can harness the power of AI to create positive and equitable outcomes for all.

The journey towards accountable and ethical AI is an ongoing process that requires vigilance, adaptability, and a collective effort. By embracing our shared responsibility and working together, we can ensure that generative AI technologies are developed and used in ways that enhance human potential, uphold our values, and contribute to a just and equitable society.

In short, let us take the insights and recommendations from the CREPES framework and apply them in our respective roles, fostering a future where AI serves humanity with integrity, fairness, and accountability. The future of AI is in our hands, and it is up to us to shape it responsibly.

REFERENCES

Antwarg, L., Miller, R. M., Shapira, B., & Rokach, L. (2021). Explaining anomalies detected by autoencoders using Shapley Additive Explanations. *Expert Systems with Applications*, 186, 115736. DOI:10.1016/j.eswa.2021.115736

Chang, W. (2022). *ISO/IEC JTC 1/SC 42 (AI)/WG 2 (data) data quality for analytics and machine learning (ML)*. Information Technology Laboratory.

Chatila, R., & Havens, J. C. (2019). The IEEE global initiative on ethics of autonomous and intelligent systems. Robotics and well-being, 11-16.

Floridi, L., & Cowls, J. (2022). A unified framework of five principles for AI in society. Machine learning and the city: Applications in architecture and urban design, 535-545.

Goodfellow, I., Pouget-Abadie, J., Mirza, M., Xu, B., Warde-Farley, D., Ozair, S., Courville, A., & Bengio, Y. (2014). Generative Adversarial Nets (PDF). *Proceedings of the International Conference on Neural Information Processing Systems (NIPS 2014)*. pp. 2672–2680.

Hartman, T., Kennedy, H., Steedman, R., & Jones, R. (2020). Public perceptions of good data management: Findings from a UK-based survey. *Big Data & Society*, 7(1), 2053951720935616. DOI:10.1177/2053951720935616

Heaven, W. D. (2023) Predictive policing is still racist-whatever data it uses, MIT Technology Review. Available at: https://www.technologyreview.com/2021/02/05/1017560/predictive-policing-racist-algorithmic-bias-data-crime-predpol/ (Accessed: 25 May 2024).

H.R. 6216 - 116th congress (2019-2020): National Artificial Intelligence Initiative Act of 2020 | congress.gov | library of Congress. Available at: https://www.congress.gov/bill/116th-congress/house-bill/6216 (Accessed: 25 May 2024).

IBM. (2003). AI Ethics Governance Framework. https://www.ibm.com/blog/a-look-into-ibms-ai-ethics-governance-framework/

Kingma, D. P., & Welling, M. (2019). An introduction to variational auto-encoders. *Foundations and Trends in Machine Learning*, 12(4), 307–392. DOI:10.1561/2200000056

Konya, A., & Nematzadeh, P. (2024). Recent applications of AI to environmental disciplines: A review. *The Science of the Total Environment*, 906, 167705. DOI:10.1016/j.scitotenv.2023.167705 PMID:37820816

Kuziemski, M., & Misuraca, G. (2020). AI governance in the public sector: Three tales from the frontiers of automated decision-making in democratic settings. *Telecommunications Policy*, 44(6), 101976. DOI:10.1016/j.telpol.2020.101976 PMID:32313360

Laux, J., Wachter, S., & Mittelstadt, B. (2024). Three pathways for standardisation and ethical disclosure by default under the European Union Artificial Intelligence Act. *Computer Law & Security Report*, 53, 105957. DOI:10.1016/j.clsr.2024.105957

Magazine, S. (2018). Christie's is first to sell art made by artificial intelligence, but what does that mean? Available at: https://www.smithsonianmag.com/smart-news/christies-first-sell-art-made-artificial-intelligence-what-does-mean-180970642/ (Accessed: 25 May 2024).

Marks, M., & Haupt, C. E. (2023). AI chatbots, health privacy, and challenges to HIPAA compliance. *Journal of the American Medical Association*, 2023(329), 1349–1350. DOI:10.1001/jama.2023.9458 PMID:37410450

Murayama, M. (2021). Society 5.0 transformation: Digital strategy in Japan. In Management Education and Automation (pp. 7-29). Routledge.

Obermeyer, Z., Powers, B., Vogeli, C., & Mullainathan, S. (2019). Dissecting racial bias in an algorithm used to manage the health of populations. *Science*, 366(6464), 447–453. DOI:10.1126/science.aax2342 PMID:31649194

Partnership on AI. (2016). https://partnershiponai.org

Peters, D., Vold, K., Robinson, D., & Calvo, R. A. (2020). Responsible AI—Two frameworks for ethical design practice. *IEEE Transactions on Technology and Society*, 1(1), 34–47. DOI:10.1109/TTS.2020.2974991

Posner, T., & Fei-Fei, L. (2020). AI will change the world, so it's time to change AI. *Nature*, 588(7837), S118–S118. DOI:10.1038/d41586-020-03412-z

Roberts, H., Cowls, J., Morley, J., Taddeo, M., Wang, V., & Floridi, L. (2021). The Chinese approach to artificial intelligence: an analysis of policy, ethics, and regulation. Ethics, governance, and policies in artificial intelligence, 47-79.

Sargeant, H. (2023). Algorithmic decision-making in financial services: Economic and normative outcomes in consumer credit. *AI and Ethics*, 3(4), 1295–1311. DOI:10.1007/s43681-022-00236-7

Sharma, R. (2024). A Comprehensive Guide to HIPAA Compliance in the Age of AI. https://www.protecto.ai/blog/hipaa-compliance-ai-comprehensive-guide

Smith, J. (2024). *AI for Cyber Guardians: Revolutionizing Ethical Hacking*. Bod Third Party Titles.

Smuha, N. A. (2019). The EU approach to ethics guidelines for trustworthy artificial intelligence. *Computer Law Review International*, 20(4), 97–106. DOI:10.9785/cri-2019-200402

Uddin, A. S. M. (2023). The Era of AI: Upholding ethical leadership. *Open Journal of Leadership*, 12(04), 400–417. DOI:10.4236/ojl.2023.124019

Zafar, M. R., & Khan, N. M. (2019). DLIME: A deterministic local interpretable model-agnostic explanations approach for computer-aided diagnosis systems. arXiv preprint arXiv:1906.10263.

Chapter 6
From Code to Conscience:
Ethics and Generative Artificial Intelligence

Meenu Chaudhary
https://orcid.org/0000-0003-3727-7460
Noida Institute of Engineering and Technology, India

Saloni Tiwari
Noida Institute of Engineering and Technology, India

Ajay Gangele
Noida Institute of Engineering and Technology, India

Loveleen Gaur
https://orcid.org/0000-0002-0885-1550
Taylor's University, Malaysia

ABSTRACT

In the dynamic global generation, Generative Artificial Intelligence (GAI) creates wonders beyond human imagination. These models are trained on large datasets and discover ways to generate new content material by figuring out styles and relationships in the statistics. The ethical implications of GAI are complicated and multifaceted. On the one hand, generative AI can revolutionize many industries and enhance our lives in limitless methods. On the other hand, there also are concerns about the potential poor outcomes of generative AI. Additionally, Generative AI could result in task displacement as machines emerge as more and more capable of performing tasks that had been a most effective replacement for human beings.

DOI: 10.4018/979-8-3693-9173-0.ch006

The objectives of this chapter are to explore the moral implications of generative AI and to highlight the ways that technology has been used in many ways. This chapter also highlights the impact of generative AI on society and the economy.

1. INTRODUCTION

Generative AI (GAI), especially Chat GPT, has very quickly become "culture" in early 2023. When Daniel presented AI as a moral issue for the future in front of the Journal of the Methodology Fair in December 2022, many visitors did not seem to know Midjourney or Chat GPT. (Hacker & Mauer, 2023). But things have changed a lot in just a few weeks. Launched on November 30, 2022, Chat GPT is an intelligent text-based Chabot that has been very successful, reaching 1 million users in 5 days and more than 100 million users in January 2023. (Fui-Hoon Nah et al., 2023). Since then, ChatGPT is believed to have become widely used and widespread. many effects, including research and studies the areas where GAI can be applied are diverse including text generation, image synthesis, music composition even video production among others. For instance, in natural language processing GAI models can be trained with huge volumes of text data to produce meaningful and contextually relevant sentences or paragraphs (Chang & Kidman, 2023).

The emergence of generative AI in the world of artificial intelligence is a game-changing and thrilling development (Furman & Seamans, 2019). These remarkable applications can produce novel and innovative content, be it written, visual or auditory, without any direct input from humans. One of the most notable applications of generative AI is in the realm of natural language processing (NLP). Powered by models like GPT-3, this technology can generate text that closely resembles human writing, given a prompt or context (Floridi & Chiriatti, 2020). Another form of AI that is very eminent in contemporary times is Generative AI. It refers to generating various images, videos, texts, writings, etc. This AI has been very successful as many companies have adopted it to produce collaborative writing and email communication (Chaudhary, Gaur & Chakrabarti, 2022). Its potential uses span across various industries, from streamlining content creation to automating customer support and even creating code snippets.

Moreover, generative AI also holds immense promise in the creative sector (Anantrasirichai & Bull, 2022). With the help of these models, artists can effortlessly generate one-of-a-kind and inspirational pieces (Amato et al., 2019). Journalism, research, graphic design, content creation, and the arts are all being influenced by generative AI. It will completely alter the way that creativity is generated in the present and the future. Ten per cent of all data produced by 2025 is anticipated to come from generative AI, according to Gartner. Generative AI also plays a significant role in the creative industry. Artists can leverage generative models to generate unique and inspiring designs or artwork taking advantage of the AI's ability to explore creative possibilities that might not have been considered otherwise (Cetinic & She, 2022). It can act as a valuable tool for brainstorming or exploring different visual styles. In addition, GAI has been applied to tasks such as image generation and manipulation. Drug development is one of the more significant areas in which generative AI is being used. By training machine learning models to produce novel chemical compounds, scientists can expeditiously ascertain prospective candidates for application in novel pharmaceuticals.

Models like DALL-E have demonstrated the ability to generate highly realistic images based on textual descriptions, opening possibilities for creating custom visuals or assisting in the design process. Music generation is another area where generative AI has shown promise (Casini & Roccetti, 2018). AI models can learn from vast catalogs of music and generate original compositions in various genres or styles. This technology can be used by musicians as a source of inspiration or even as a collaborator in the creative process.

The authors have done an extensive literature review based on these keywords to filter the relevant documents in the Scopus database "ethics in AI" OR "ethical AI" OR "ethics" AND "generative AI". Out of 95 documents, 74 documents have been considered based on their relevance to the study. 1 document was published in 2022 and then it gained momentum in 2023 with 57 documents. The same has been depicted in Figure 1.

Figure 1. Publications in Scopus database

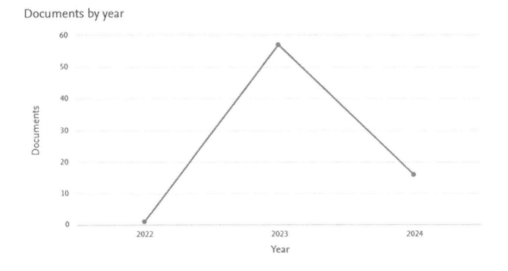

Documents by year

Seventy-four documents have been published in various formats. The articles, conference paper, review, editorial, book chapter, note, book, and letter are 39, 17, 9, 3, 2, 2, 1, and 1 respectively.

Figure 2. Percentage of publications of the document type

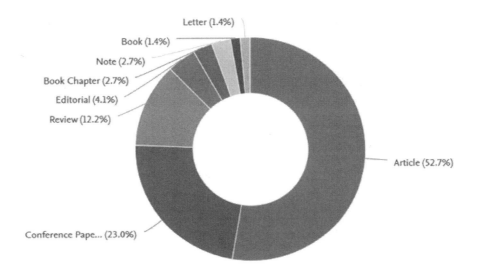

Table 1. Subject area categorization of the documents

Subject area	Number of documents	Percentage
Computer Science	35	25.5
Social Science	28	20.4
Business, Management and Accounting	15	10.9
Medicine	13	9.5
Engineering	10	7.3
Arts and Humanities	9	6.6
Nursing	5	3.6
Decision Sciences	3	2.2
Health Professions	3	2.2
Mathematics	3	2.2
Others	13	9.5

Several important considerations and questions arise when discussing generative AI and ethics. First, the development and use of generative AI raise concerns about potential ethical issues related to privacy and consent (Makridakis, 2017). As these systems are trained on large amounts of data there is a risk of violating individual privacy rights especially if sensitive or personal information is used without explicit consent It is important to ensure adequate safeguards are established to protect data privacy (Chaudhary et al., 2022).

Additionally, generative AI can produce results that can be misleading, harmful, or offensive. Developers and organizations have a moral obligation to prevent AI-generated content that could shape public opinion and spread misinformation or contribute to the spread or spread of harmful hate speech and information. Implementing content controls and providing clear guidelines and standards for AI-generated content can help address these concerns. Another area of ethical consideration is the potential for bias in generative AI programs. If the training data used for these programs are biased, it can lead to biased and discriminatory results. This can have far-reaching consequences, such as perpetuating existing social inequalities or reinforcing stereotypes. Proper assessment and reduction of training data and algorithms are essential to ensure the accuracy and consistency of AI-generated results (Zhao, 2023; Chaudhary, Gaur, & Chakrabarti, 2022).

Additionally, there are ethical implications for the use of reproductive AI in various sectors such as healthcare finance and regulation. These AI systems can have a profound impact on individuals. Transparency, accountability, and fairness must be prioritized to prevent AI-generated outputs from unjustly influencing decisions discriminating against certain groups, or exacerbating existing societal inequalities (Cao, 2022). Lastly, there are concerns about the potential loss of human creativity and the erosion of value in human-authored content due to the proliferation of generative AI. While AI can undoubtedly augment and aid human creativity it is crucial to strike a balance between the use of AI-generated content and preserving the unique human capabilities and perspectives that contribute to artistic and intellectual endeavors (Poquet & De Laat, 2021).

This chapter focuses on exploring the potential impacts of Generative AI on society, its role in various fields, its effect on the socio-economic and issues related to it. Also, the authors focus on discussing the guidelines and principles to promote responsible development and usage of Generative AI

2. LITERATURE REVIEW

AI has always played a pivotal role in almost all sectors, be it the military, healthcare, and much more. Many big countries like the US and Russia have actively adopted AI and its techniques in their military forces. Apart from the glory of AI, the thing that is to be pondered is its accuracy and ethics. Artificial intelligence will have adverse effects if it doesn't have the amount of accuracy it should have in the military forces. It might cause harm to mankind and forces. To address these challenges, the DOD (Department of Defense) of the US, NATO, has disclosed ethical principles for a better, reliable, and trustworthy system. These countries have rubber-stamped these principles to ensure the least or no damage and danger from the AI. Apart from the military, AI has also set foot in healthcare services. There is an earnest use and impact of AI in healthcare as the NLP in the context of EHR. Likewise, NATO, and WHO have also released certain principles discussing the ethics and governance of AI for health (Oniani et al., 2023).

In healthcare, it is used generally for clinical documentation and summarization of medicines. AI despite being very successful and used worldwide, has always been questioned on ethics. Ethics refers to the understanding of what is wrong and what is right (Zhang & Kamel, 2023). The military and healthcare services are very stressed and can't compromise with even a single mistake. AI must be used such that no stone is left unturned in making the morally correct and fruitful decision (Arenander, 2023; Gaur, Gaur & Afaq, A. (2024).

ELIZA is one of the first generative AI. This text chatbot was created by Joseph Weizenbaum in the 1960s. ELIZA was one of the first examples of NLP and imitated the work of a psychotherapist who could communicate with humans in natural language (Frey & Osborne, 2023). It was the only of its kind as there were no other peculiarities until the early 21st century regarding machine learning models and algorithms. With the release of the ChatGPT developed by the laboratory of the OpenAI in the year 2022, the scenario changed.

The Markov chain is an early example of generative AI. It is named after Andrey Markov, a Russian mathematician who in 1906 introduced this statistical process to model the behavior of random processes. Markov models have long been used in machine learning for next-word prediction, like the autocomplete function in an email program.

Embedding ChatGPT within the nuanced realms of Arts and Humanities brings about both exciting prospects and formidable trials. For ages, the Arts and Humanities have been areas that are deeply rooted in human creativity, expression, and interpretation. From literature to philosophy, these disciplines explore an uncharted territory that goes deep into what makes up human beings providing unique insights into society's cultural identity (Rane, 2023).

Combined with advanced machine learning algorithms, ChatGPT has the potential to change how we engage with the arts and humanities. It can help writers by providing literary ideas and creative ideas (Fui-Hoon Nah et al., 2023).

Labazanova, Aygumov, & Mursaliev (2024) suggested various issues with GAI tools in detail some of the issues are misinformation and disinformation, copyright issues, Limits of clear vision, dependence, and loss of skills. These are the major issues which are suggested by the authors. The authors have also suggested various possible solutions to these issues.

Wach et al., (2023) aid in bringing attention to the significance of resolving the ethical and legal considerations that arise from the use of GAI and ChatGPT by drawing attention to the controversies and hazards associated with these technologies. The authors have explained various risks that arise from the use of ChatGPT.

Generative AI, also called Generative Adversarial Networks (GANs is a kind of artificial intelligence that has won great attention in recent years. It differs from traditional AI fashions which might be designed for classification or prediction tasks (Pan et al., 2019). Instead, GANs are focused on learning and generating new content. This is much like the training records they were uncovered too. This functionality makes them relatively influential and impactful in various fields along with art media gaming or even healthcare. The idea of GANs changed delivered through Ian Goodfellow and his colleagues in 2014 (Durgadevi, 2021). The fundamental idea at the back of GANs is to teach two deep neural networks simultaneously: the generator network and the discriminator network. The generator's function is to create synthetic information samples whereas the discriminator's function is to differentiate between real and synthetic statistics. Through an iterative procedure the generator network targets to generate an increasing number of practical outputs that the discriminator is unable to distinguish from real statistics.

A key advantage of GANs is their capacity to examine the underlying distribution of a given dataset and generate new samples that can be statistically comparable. This function has been broadly used for producing sensible images, movies and audio in addition to for text era. Some first-rate examples of generative AI include:

1. Image Generation: GANs have revolutionized photo technology by creating realistic and numerous snapshots. For example, researchers have used GANs to generate photorealistic snapshots of folks that do not truly exist. This could have big implications in numerous fields which include advertising and marketing entertainment and layout (Vartiainen & Tedre, 2024; Chaudhary et al., 2023).

2. Style Transfer: GANs have been used to transfer the fashion of one photo to another (Miah et al., 2023). This method allows for creative modifications of pictures creating particular and visually attractive outputs. For instance, an

artist can observe the style of a famous painter like Van Gogh in their pictures creating a fusion of artwork and reality.

3. Video Synthesis: GANs have also been employed in producing and manipulating motion pictures. By training on large video datasets GANs can generate new scenes alter present films or even expect future frames (Leiker et al., 2023). This technology has been applied inside the film enterprise to create special effects enhance animations and improve video compression algorithms.

4. Text Generation: GANs have been used to generate coherent and contextually relevant text. For instance, they may be utilized to create product reviews information articles poetry, or even chatbot conversations (Bandi et al, 2023). Nevertheless, in its early degrees text era with GANs has the potential to automate content advent and beautify herbal language expertise (Deep & Verma, 2024).

The effect of generative AI on society is full-size and far-achieving. Here are numerous key factors to recall:

1. Artistic and Creative Applications: GANs have opened new avenues for artistic expression and creativity. Artists can leverage generative models to explore styles, push innovative boundaries and showcase their paintings in novel approaches (McCormack, 2023). It permits the democratization of art allowing absolutely everyone to emerge as a creator by way of the use of these AI-powered gear (Deckers et al., 2023).

2. Entertainment and Media: Generative AI has converted the fields of gaming visual consequences and media manufacturing. With the capacity to generate practical characters landscapes and scenes, GANs are utilized to create immersive gaming reviews and beautify movie production. It enables the creation of digital worlds that blur the lines between truth and fiction (Bengesi et al., 2023).

3. Personalization and Customization: GANs allow customized tips and customization in numerous domain names. For example, in e-commerce generative models can create customized clothing designs primarily based on character choices. In healthcare, GANs can generate personalized treatment plans by way of reading huge datasets of patient records (Abrokwah-Larbi, 2023; Faruqi et al., 2023).

4. Ethical Considerations: As with any emerging generation generative AI also increases ethical concerns (Farina & Lavazza, 2024). GANs can be used to create deepfake content wherein sensible however fabricated pictures or videos misinform visitors (Khoo et al., 2022). This poses risks to privacy protection and the spread of incorrect information. Ensuring responsible use of generative AI is important to mitigate potential damage.

In conclusion, generative AI powered by way of GANs represents a groundbreaking generation with wide-ranging applications and implications. Its potential to study and generate new facts reminiscent of the training set has led to improvements in photo and video generation fashion transfer textual content generation and greater. The impact on society is substantial influencing fields such as art leisure customization and elevating important ethical concerns. As this era continues to evolve responsible and moral deployment could be crucial to maximize both its advantages and mitigate potential dangers.

3. ROLE OF GENERATIVE AI IN VARIOUS FIELDS

1. Healthcare: Generative AI has extensive potential in healthcare assisting in diagnostics drug discovery and patient care. For instance, researchers have used generative models to create synthetic clinical pictures that assist medical doctors stumble on abnormalities and enhance diagnostic accuracy (Jiang et al., 2017). Generative AI can also help in drug discovery by way of simulating molecular systems and identifying capacity new compounds. It has also been applied in personalized medicinal drugs predicting sickness development primarily based on affected person information and optimizing remedy plans.

2. Art and Design: Generative AI has opened new possibilities in creative industries. Artists and architects can leverage generative fashions to generate novel and specific artwork and designs. For instance, artists have used generative adverse networks (GANs) to generate practical paintings sculptures, and virtual art. Designers have hired generative AI to create custom-designed trademark fonts and product designs allowing a noticeably customized and creative method.

3. Gaming and Entertainment: The gaming and leisure industry has embraced generative AI to enhance user studies and provide infinite entertainment opportunities. Game developers can utilize generative models to generate sensible 3-D environments, dynamic characters and charming storyline factors. Generative AI is likewise applied in procedural content technology creating massive and various recreation worlds degrees and quests. In the leisure domain, generative AI has been hired to generate song storytelling and video modifying making an allowance for customized content material introduction.

4. Robotics and Automation: Generative AI plays a crucial function in robotics and automation systems allowing machines to research and generate behaviors or moves. For instance, in robotic grasping generative models can learn to generate grasping strategies based on object shapes sizes, and properties. This enables actual-time robot manipulation and enhances object recognition and interplay. Generative AI additionally assists in self-reliant automobiles in which

fashions can learn how to generate visitor situations for simulation trying out and enhancing selection-making algorithms.

5. Natural Language Processing (NLP): Generative AI has revolutionized NLP by permitting machines to generate human-like text enhancing language expertise and herbal language generation. For example, OpenAI's GPT-3 model generates coherent and contextually applicable text responses allowing smart chatbots and virtual assistants to interact in herbal conversations. Generative models have also been used for language translation text summarization and creative writing pushing obstacles in automated content material generation.

6. Marketing and Advertising: Generative AI has won traction in advertising and marketing and marketing helping corporations create targeted and engaging content material. For instance, generative fashions can analyze client information and generate customized product pointers enhancing consumer studies and driving sales (Chaudhary & Gaur, 2021). Additionally, AI-powered chatbots can generate conversational responses and provide personalized answers to patron queries. Generative AI also aids in developing creative advert campaigns generating advert copies and growing visually attractive pictures and movies.

7. Fashion: Generative AI can assist fashion designers in growing unique garment designs, customizing garb and predicting future trends. Companies like H&M and Adidas explore Generative AI for designing beautiful clothes items.

8. Finance and investing: Generative AI can be used for producing financial marketplace forecasts, predictive modeling and portfolio optimization. It can assist buyers make informed selections by way of reading huge datasets and identifying patterns.

9. Education and e-studying: Generative AI can generate personalized materials and adaptive tutoring systems. It can create interactive simulations for digital tutors and provide personalized comments to decorate the instructional experience.

10. Architecture and urban planning: Generative AI can help architects in designing buildings generating floorplans and optimizing strength efficiency. It can simulate urban environments and propose top-of-the-line city designs.

These examples illustrate how generative AI is getting used across numerous domains to create optimize and beautify content material technology choice-making procedures and personal experiences.

4. ISSUES RELATED TO GENERATIVE AI

There are various issues related to Generative AI some of the issues are explained below:

1. Ethical issues: Generative AI can enhance ethical worries especially while the technology is used to create deepfakes or control seen and audio content. This increases profound questions on consent privateness and the spread of incorrect facts. (Stahl & Stahl, 2021). It becomes essential to establish pointers and guidelines to prevent the misuse of the generative AI era.

2. Intellectual assets rights: Generative AI regularly makes use of pre-modern-day facts to create new outputs blurring the road between precise and derivative works. These challenge conventional notions of highbrow belonging rights and will increase questions about who owns the generated content. Clear criminal frameworks are essential to cope with those troubles and ensure straightforward compensation for content material creators.

3. Bias and equity: Generative AI structures analyze sizable quantities of records and if the training information is biased it may bring about biased outputs. This can contribute to reinforcing current social biases and discrimination. Efforts ought to be made to address and mitigate bias for the duration of the improvement and training manner of generative AI fashions to ensure fairness and inclusivity.

4. Manipulation and deception: Generative AI can be used to create exceptionally practical faux content that could mislead people or even computerized structures. This poses enormous dangers inclusive of manipulating public opinion spreading incorrect information and compromising security. It is crucial to expand strong detection strategies and teach users about the life of generative AI-generated content material.

5. Privacy issues: GAI regularly requires massive datasets to supply incredible outputs. The utilization of private information for education in such models can improve privacy worries in particular if the facts aren't always nicely anonymized and secured. Protecting consumer privacy and making sure obvious facts utilization policies are critical considerations in the improvement and deployment of generative AI systems.

6. Unintended outcomes: GAI structures can on occasion produce sudden or unintended outputs because of complex studying algorithms and hidden biases in the training facts. This can cause unintended outcomes inclusive of producing offensive or harmful content material. Continual tracking testing and improvement of generative AI models are essential to mitigate the risks of unintentional outputs.

7. Authenticity and belief: As GAI advances it becomes increasingly tough to differentiate between actual and generated content. This can undermine consideration in digital media and lift worries about the authenticity of visible and textual information. Developing strong authenticity verification techniques could be essential in preserving accepted as true and ensuring the integrity of content material within the virtual realm.

8. Environmental impact: Training generative AI fashions frequently requires massive computing assets and power intake main to a big carbon footprint. Addressing the environmental impact of generative AI through power-efficient algorithms, improved hardware and sustainable practices is essential to make sure the era aligns with broader environmental sustainability desires.

9. Responsibility and accountability: Determining the obligation and duty for generative AI-generated content can be complicated. When dangerous or malicious content is produced it can be tough to pick out the people or entities liable for its creation. Developing frameworks for accountability and implementing ethical norms is critical to save you from misuse and ensure accountable deployment of generative AI.

10. Lack of Control: Generative AI systems are educated to supply output based totally on patterns discovered from the education information. However, making sure to completely manage the generated content material can be difficult. There is a threat of producing misguided or deceptive facts which could impact decision-making tactics or propagate faux information.

11. Data Bias: Generative AI systems learn from the statistics they are skilled in which may lead to biases present inside the education information being meditated inside the generated content. If the education facts are biased or lack diversity the generated content material may also contain those biases. This can perpetuate societal racial or gender biases leading to unfair or discriminatory outputs.

5. GUIDELINES AND PRINCIPLES TO PROMOTE RESPONSIBLE DEVELOPMENT AND USAGE OF GENERATIVE AI

There are 12 famous guidelines and principles that can promote the responsible use of generative AI technology:

1. Transparency and Explainability: GAI systems must be designed to offer transparency and explainability to their outputs. Users and creators ought to have the capacity to apprehend and decide how and why a particular output became generated. This helps make certain responsibilities and minimizes ability dangers associated with biased or dangerous content material generation. To ensure accountability and protect the integrity of content it is important to develop techniques that can reliably attribute generated content to AI systems. Watermarking or digital signatures can be used to mark AI-generated content

indicating its origin. Additionally, organizations must encourage responsible use of generative AI and promote transparency regarding its capabilities to distinguish between human and AI-generated content.

2. Ethical Considerations: Developers of GAI systems should be aware of capacity ethical issues and biases that can arise throughout the education procedure. It is important to proactively cope with biases discrimination and other ethical worries to keep away from bad results of AI-generated content. To address this concern organizations must implement robust data protection measures. They should prioritize the anonymization and encryption of sensitive data, adopt strict access controls and regularly update security protocols. Transparency regarding data collection and usage is also essential to gain the trust of the users and maintain ethical practices in generative AI.

3. User Consent and Control: GAI structures need to prioritize consumer consent and manage the generated content material. Users ought to have the ability to select the extent of influence and involvement in the generated outputs ensuring that AI systems work as gear and not as self-sustaining selection-makers.

4. Data Privacy and Security: Strong information privacy and security features must be implemented to guard the private records shared with generative AI systems. Developers and organizations should adhere to privacy rules and implement strong safety protocols to prevent any unauthorized entry to or misuse of personal facts.

5. Responsible Data Collection: Generative AI structures closely rely on records for education. It is crucial to make sure accountable facts series practices head off the use of touchy or proprietary records without proper consent. By adhering to moral records series suggestions potential privacy concerns may be mitigated.

6. Human-AI Collaboration: Promoting a collaborative method among humans and generative AI systems is crucial. Instead of changing human creativity, AI must increase human abilities whilst respecting human knowledge. Striking a balance between system-generated and human-generated content material can lead to extra knowledgeable and responsible outcomes.

7. Fairness and Bias Mitigation: Developers must actively cope with biases that may arise inside generative AI structures. Bias mitigation strategies together with dataset diversification and equity metrics need to be employed to ensure the equity and inclusivity of the generated content lowering capability harm or discrimination.

8. Intellectual Property Rights: Guidelines and regulations must be established to protect highbrow belonging rights associated with generative AI outputs. Proper attribution and licensing mechanisms must be in the area to make certain that the contributions of AI structures and human creators are as they should be diagnosed and safeguarded. To navigate these challenges organizations must

establish clear guidelines and policies regarding ownership and protection of generated content. They should consider incorporating mechanisms to identify and mitigate potential copyright infringement. Collaboration with legal experts and leveraging existing intellectual property frameworks can help ensure that generative AI respects intellectual property rights.

9. Continuous Monitoring and Evaluation: Regular monitoring and evaluation of generative AI structures are essential to pick out and address any rising ethical worries or biases. Ongoing assessment allows for knowledge of the machine's barriers enhancing performance and making sure of responsible behavior.

10. Social Impact Assessment: Conducting social impact assessments earlier than deploying generative AI structures can help pick out capacity risks and poor consequences on society. Evaluating the impact on individual groups and broader societal norms ensures responsible improvement and usage.

11. Collaboration and Knowledge Sharing: Promoting collaboration and know-how sharing among researchers, academicians, and practitioners is vital for accountable AI improvement. By fostering an open dialogue sharing high-quality practices and taking part in ethical considerations the AI network can together paintings in the direction of ensuring accountable and beneficial consequences.

12. Regulation and Policy Development: Governments and regulatory bodies play a crucial role in selling accountable improvement and usage of generative AI. The system of clear guidelines tips and moral standards can provide a framework for builders and businesses to follow fostering duty and acceptance as true.

As generative AI generation continues to progress it's far essential to set up guidelines and ideas that prioritize responsible improvement and utilization. Transparency moral concerns user consent information privateness equity and continuous tracking are many of the key components that must be addressed. By adhering to these 12 recommendations and standards we can leverage the ability of generative AI systems whilst minimizing associated risks and ensuring moral and accountable outcomes.

6. SOCIETAL IMPACT OF GENERATIVE AI

- Enhancing Creativity and Artistic Expression: GAI opens new horizons for creativity and artistic expression. Artists and designers use this technology to explore new ideas, create unique works of art, and foster innovation. (Hagerty & Rubinov, 2019). Thanks to technical skills, even people without artistic training can participate in creative, independent landscape painting and promote diversity in practice.

- Improving Health: The impact of artificial intelligence on healthcare is important. It has the potential to revolutionize drug discovery and personalized medicine. Large amounts of medical data are analyzed to produce artificial intelligence that can identify disease patterns and improve treatment. Technology can improve patient outcomes, reduce human error, and speed up medical research. Additionally, generative AI can assist doctors in simulating risk assessment and planning procedures, leading to safer and more effective treatment.

- Increasing Economic Growth and Innovation: GAI has the power to increase economic growth and stimulate innovation. By making work repetitive and global, it allows employees to focus on better and more creative work. This means increasing capacity across the business, improving processes, and increasing overall profitability. Technology also allows businesses to gain insight from large amounts of data, support decision-making, and provide a competitive advantage. Additionally, generative AI has the potential to create new jobs and businesses, thereby increasing economic prosperity.

- Communication and Advertising: GAI is changing the way information is advertised and used. AI-driven algorithms can make personalized recommendations, mimic human perception of reality, and potentially create filter bubbles. This has led to concerns that artificial intelligence could undermine environmental safety. Intelligence can help mitigate the effects of climate change by providing accurate climate models and improving resource allocation. Additionally, improving artificial intelligence-supported waste management and transportation can be more efficient and effective. However, it is necessary to be careful and think about ethics to protect artificial intelligence systems from the inevitable consequences of the world's ecosystems.

- Automation and Labor Market: GAI technologies such as robotics and intelligent automation have reshaped the labor market (Gaur, Afaq, Singh, & Dwivedi, 2021). While this technology has the potential to augment human work and increase efficiency it also raises concerns about job displacement and the widening of economic inequality. Policy measures and retraining initiatives are necessary to ensure individuals can adapt to the changing nature of work and benefit from the advancements in Generative AI.

- Creative Industries: GAI has revolutionized the creative industries by enabling machines to generate art, music, literature and other forms of creative content. This technology has democratized creativity allowing individuals from all backgrounds to express themselves and leverage AI tools to create unique and compelling works. However, it also raises questions about the originality and authenticity of creative output generated by AI challenging traditional notions of authorship and ownership.

- Ethical Considerations: While the impact of GAI has immense potential, ethical considerations must be addressed. One major concern is the potential for bias in the training data leading to biased output. Proper guidelines and oversight are necessary to ensure fairness and prevent discrimination. Additionally, the unauthorized use of generative AI can lead to misinformation fake news, and deepfakes undermining trust in media and public discourse. Policymakers need to collaborate with stakeholders to develop regulations and frameworks that promote ethical and responsible use of generative AI.

- Privacy and Security Challenges: Generative AI relies on massive datasets raising concerns around privacy and data security. The availability of personal information could potentially be misused, leading to privacy violations and breaches. It is crucial to establish robust safeguards and secure infrastructure to protect sensitive data. Additionally, the misuse of generative AI technology can pose security threats allowing for the creation of sophisticated cyber-attacks and forgeries. Therefore, it is essential to stay vigilant and develop countermeasures to mitigate these risks.

- Job displacement and economic impacts: The automation capabilities of GAI raise concerns about job displacement. As tasks become automated certain job roles may become obsolete leading to potential unemployment and socio-economic challenges. Reskilling and upskilling the workforce becomes essential to adapt to the changing job landscape. Generative AI models often require large amounts of data to be trained effectively. This raises concerns regarding data privacy and security. Safeguarding sensitive information and ensuring that user data is protected are critical considerations to mitigate potential risks.

7. EFFECT OF GENERATIVE AI ON SOCIO-ECONOMIC

The emergence of generative AI, also known as artificial intelligence that can create content such as images videos and text has had a profound impact on various aspects of socio-economic life. This technology has revolutionized industries, transformed business operations and influenced social dynamics in numerous ways. In this chapter, authors have explored the effects of generative AI on various socio-economic aspects including job displacement creativity data generation, and ethical considerations. Job displacement is one of the most prominent socio-economic impacts of generative AI. As AI systems become more sophisticated and capable of performing tasks that were traditionally done by humans there is a growing concern about the potential displacement of human workers. Areas such as content creation graphic design and even certain aspects of journalism are now being automated by generative AI which raises questions about the long-term viability of these professions. While the automation of certain tasks can lead to increased efficiency and productivity it also poses challenges for individuals who rely on these jobs for their livelihoods. Therefore, societies need to prepare for potential job disruptions and create new opportunities for employment or retraining programs to ensure a smooth transition in the labor market.

On the other hand, generative AI has also fostered new opportunities for creativity and innovation. The ability to generate novel and diverse content has enabled individuals and businesses to explore new artistic expressions and design possibilities. For example, in the field of music generative AI algorithms have been trained to compose original pieces opening new avenues for musicians to explore unconventional melodies and harmonies (Sharma et al., 2022). Similarly, in the visual arts, generative AI has allowed artists to experiment with new styles, generate unique visualizations, and push the boundaries of traditional artistic practices. Consequently, generative AI has not only complemented human creativity but has also catalyzed new and exciting forms of self-expression.

Furthermore, generative AI has transformed the way data is generated processed, and utilized in various sectors. With the ability to generate realistic high-quality synthetic data, AI systems can now be trained on vast amounts of simulated data supplementing or even replacing the need for large-scale real-world data collection. This has proven particularly valuable in domains such as healthcare and finance where access to sensitive or scarce data is limited. By generating synthetic data researchers can conduct experiments test algorithms and develop insights without compromising privacy or incurring substantial costs. In addition, generative AI has also been used to enhance data augmentation techniques enabling more efficient training of machine learning models and improving their performance across a range of applications.

However, the proliferation of generative AI also raises important ethical considerations. The ability to generate realistic fake content commonly referred to as "deepfakes" has garnered significant attention due to its potential for misuse and manipulation. Deepfakes can be used to spread misinformation, deceive the public, or maliciously impersonate individuals. This presents challenges for society particularly in the realms of journalism politics and privacy. Robust regulations, technological safeguards and awareness campaigns are needed to address the ethical concerns surrounding generative AI and prevent its malicious exploitation.

Moreover, the socio-economic impact of generative AI is not evenly distributed. The adoption and access to generative AI technologies tend to be concentrated in developed countries and within certain industries leading to potential disparities and uneven distribution of benefits. This digital divide can exacerbate existing social and economic inequalities as those with limited access or resources may be left behind in terms of skills development job opportunities and innovation. Policymakers need to prioritize inclusive and equitable access to generative AI technologies ensuring that all individuals and communities can harness its potential advantages.

In conclusion, generative AI has had a significant influence on socio-economic aspects with both positive and negative consequences. While it has the potential to displace certain jobs it also opens new avenues for creativity enhances data generation and processing and offers numerous benefits in various industries. However ethical considerations and challenges related to access and distribution need to be carefully addressed to leverage the full potential of generative AI for the betterment of society.

8. CONCLUSION

The study focuses primarily on the positive and negative impact of GAI on society and the economy as well as recommendations and principles that support the development of responsibility and the use of wisdom. According to the research, Generative AI has the potential to revolutionize various sectors of society offering unprecedented opportunities for creative innovation, and improved healthcare. However careful consideration of ethical frameworks, privacy safeguards and security measures is necessary to fully harness the benefits of this technology while minimizing the risks. By addressing these challenges society can embrace generative AI as a powerful force for positive change shaping a future filled with immense possibilities and advancements.

The emergence of generative AI marks a pivotal moment in technological evolution, bringing forth a paradigm shift with far-reaching societal implications. This transformative technology possesses the remarkable ability to generate novel content, ranging from text and code to images and music, blurring the boundaries between human creativity and machine intelligence. As we delve deeper into the realm of generative AI, it becomes imperative to acknowledge both its immense potential and the accompanying challenges it poses to society.

On the one hand, generative AI holds the promise of revolutionizing industries, augmenting human capabilities, and addressing complex global challenges. Its applications span a diverse spectrum, from enhancing healthcare diagnostics and drug discovery to optimizing supply chains and fostering scientific research. By automating routine tasks and enabling the generation of high-quality content, generative AI has the potential to streamline processes, reduce costs, and drive innovation across various sectors. On the other hand, the rapid advancement of generative AI also raises important ethical, legal, and societal concerns that demand careful consideration. Issues such as privacy, intellectual property rights, misinformation, and job displacement come to the forefront, necessitating the development of robust regulatory frameworks and ethical guidelines. Ensuring transparency, accountability, and responsible development of generative AI technologies is paramount to mitigate potential risks and maximize societal benefits.

As we navigate the uncharted waters of generative AI, it is crucial to prioritize human values and well-being. Striking a balance between harnessing the technology's transformative power and addressing its potential negative impacts is essential. This entails fostering a culture of responsible innovation, promoting collaboration between stakeholders, and engaging in ongoing public discourse to shape a future where generative AI serves as a force for positive societal transformation.

REFERENCES

Abrokwah-Larbi, K. (2023). The role of generative artificial intelligence (GAI) in customer personalisation (CP) development in SMEs: A theoretical framework and research propositions. *Industrial Artificial Intelligence*, 1(1), 11. DOI:10.1007/s44244-023-00012-4

Amato, G., Behrmann, M., Bimbot, F., Caramiaux, B., Falchi, F., Garcia, A., . . . Vincent, E. (2019). AI in the media and creative industries. arXiv preprint arXiv:1905.04175.

Anantrasirichai, N., & Bull, D. (2022). Artificial intelligence in the creative industries: A review. *Artificial Intelligence Review*, 55(1), 1–68. DOI:10.1007/s10462-021-10039-7

Arenander, M. (2023). Technology Acceptance for AI implementations: A case study in the Defense Industry about 3D Generative Models.

Bandi, A., Adapa, P. V. S. R., & Kuchi, Y. E. V. P. K. (2023). The power of generative ai: A review of requirements, models, input–output formats, evaluation metrics, and challenges. *Future Internet*, 15(8), 260. DOI:10.3390/fi15080260

Bengesi, S., El-Sayed, H., Sarker, M. K., Houkpati, Y., Irungu, J., & Oladunni, T. (2023). Advancements in Generative AI: A Comprehensive Review of GANs, GPT, Autoencoders, Diffusion Model, and Transformers. *arXiv preprint arXiv:2311.10242*.

Cao, L. (2022). Ai in finance: Challenges, techniques, and opportunities. *ACM Computing Surveys*, 55(3), 1–38. DOI:10.1145/3502289

Casini, L., & Roccetti, M. (2018). The impact of AI on the musical world: will musicians be obsolete?. Studi di estetica, (12).

Cetinic, E., & She, J. (2022). Understanding and creating art with AI: Review and outlook. [TOMM]. *ACM Transactions on Multimedia Computing Communications and Applications*, 18(2), 1–22. DOI:10.1145/3475799

Chang, C. H., & Kidman, G. (2023). The rise of generative artificial intelligence (AI) language models-challenges and opportunities for geographical and environmental education. *International Research in Geographical and Environmental Education*, 32(2), 85–89. DOI:10.1080/10382046.2023.2194036

Chaudhary, M., & Gaur, L. (2021). COVID-19 a "BIG RESET"—Role of GHRM in Achieving Organisational Sustainability in Context to Asian Market. In Proceedings of Second International Conference on Computing, Communications, and Cyber-Security: IC4S 2020 (pp. 607-625). Springer Singapore.

Chaudhary, M., Gaur, L., & Chakrabarti, A. (2022, April). Comparative Analysis of Entropy Weight Method and C5 Classifier for Predicting Employee Churn. In 2022 3rd International Conference on Intelligent Engineering and Management (ICIEM) (pp. 232-236). IEEE. DOI:10.1109/ICIEM54221.2022.9853181

Chaudhary, M., Gaur, L., & Chakrabarti, A. (2022, November). Detecting the Employee Satisfaction in Retail: A Latent Dirichlet Allocation and Machine Learning approach. In 2022 3rd International Conference on Computation, Automation and Knowledge Management (ICCAKM) (pp. 1-6). IEEE. DOI:10.1109/ICCAKM54721.2022.9990186

Chaudhary, M., Gaur, L., Jhanjhi, N. Z., Masud, M., & Aljahdali, S. (2022). Envisaging Employee Churn Using MCDM and Machine Learning. *Intelligent Automation & Soft Computing*, 33(2), 1009–1024. DOI:10.32604/iasc.2022.023417

Chaudhary, M., Singh, G., Gaur, L., Mathur, N., & Kapoor, S. (2023, November). Leveraging Unity 3D and Vuforia Engine for Augmented Reality Application Development. In 2023 3rd International Conference on Technological Advancements in Computational Sciences (ICTACS) (pp. 1139-1144). IEEE.

Deckers, N., Fröbe, M., Kiesel, J., Pandolfo, G., Schröder, C., Stein, B., & Potthast, M. (2023, March). The Infinite Index: Information Retrieval on Generative Text-To-Image Models. In *Proceedings of the 2023 Conference on Human Information Interaction and Retrieval* (pp. 172-186). DOI:10.1145/3576840.3578327

Deep, G., & Verma, J. (2024). Textual Alchemy: Unleashing the Power of Generative Models for Advanced Text Generation. In *Advanced Applications of Generative AI and Natural Language Processing Models* (pp. 124-143). IGI Global.

Durgadevi, M. (2021, July). Generative adversarial network (gan): a general review on different variants of gan and applications. In 2021 6th International Conference on Communication and Electronics Systems (ICCES) (pp. 1-8). IEEE.

Farina, M., Yu, X., & Lavazza, A. (2024). Ethical considerations and policy interventions concerning the impact of generative AI tools in the economy and in society. *AI and Ethics*, ●●●, 1–9. DOI:10.1007/s43681-023-00405-2

Faruqi, F., Katary, A., Hasic, T., Abdel-Rahman, A., Rahman, N., Tejedor, L., & Mueller, S. (2023, October). Style2Fab: Functionality-Aware Segmentation for Fabricating Personalized 3D Models with Generative AI. In *Proceedings of the 36th Annual ACM Symposium on User Interface Software and Technology* (pp. 1-13). DOI:10.1145/3586183.3606723

Floridi, L., & Chiriatti, M. (2020). GPT-3: Its nature, scope, limits, and consequences. *Minds and Machines*, 30(4), 681–694. DOI:10.1007/s11023-020-09548-1

Frey, C. B., & Osborne, M. (2023). Generative AI and the Future of Work: A Reappraisal. *The Brown Journal of World Affairs*, ●●●, 1–12.

Fui-Hoon Nah, F., Zheng, R., Cai, J., Siau, K., & Chen, L. (2023). Generative AI and ChatGPT: Applications, challenges, and AI-human collaboration. *Journal of Information Technology Case and Application Research*, 25(3), 277–304. DOI:10.1080/15228053.2023.2233814

Fui-Hoon Nah, F., Zheng, R., Cai, J., Siau, K., & Chen, L. (2023). Generative AI and ChatGPT: Applications, challenges, and AI-human collaboration. *Journal of Information Technology Case and Application Research*, 25(3), 277–304. DOI:10.1080/15228053.2023.2233814

Fui-Hoon Nah, F., Zheng, R., Cai, J., Siau, K., & Chen, L. (2023). Generative AI and ChatGPT: Applications, challenges, and AI-human collaboration. *Journal of Information Technology Case and Application Research*, 25(3), 277–304. DOI:10.1080/15228053.2023.2233814

Furman, J., & Seamans, R. (2019). AI and the Economy. *Innovation Policy and the Economy*, 19(1), 161–191. DOI:10.1086/699936

Gaur, L., Afaq, A., Singh, G., & Dwivedi, Y. K. (2021). Role of artificial intelligence and robotics to foster the touchless travel during a pandemic: A review and research agenda. *International Journal of Contemporary Hospitality Management*, 33(11), 4079–4098. DOI:10.1108/IJCHM-11-2020-1246

Gaur, L., Gaur, D., & Afaq, A. (2024). Demystifying Metaverse Applications for Intelligent Healthcare. In Metaverse Applications for Intelligent Healthcare (pp. 1-23). IGI Global.

Hacker, P., Engel, A., & Mauer, M. (2023, June). Regulating ChatGPT and other large generative AI models. In *Proceedings of the 2023 ACM Conference on Fairness, Accountability, and Transparency* (pp. 1112-1123). DOI:10.1145/3593013.3594067

Hagerty, A., & Rubinov, I. (2019). Global AI ethics: a review of the social impacts and ethical implications of artificial intelligence. arXiv preprint arXiv:1907.07892.

Jiang, F., Jiang, Y., Zhi, H., Dong, Y., Li, H., Ma, S., Wang, Y., Dong, Q., Shen, H., & Wang, Y. (2017). Artificial intelligence in healthcare: Past, present and future. *Stroke and Vascular Neurology*, 2(4), 230–243. DOI:10.1136/svn-2017-000101 PMID:29507784

Khoo, B., Phan, R. C. W., & Lim, C. H. (2022). Deepfake attribution: On the source identification of artificially generated images. *Wiley Interdisciplinary Reviews. Data Mining and Knowledge Discovery*, 12(3), e1438. DOI:10.1002/widm.1438

Labazanova, S. K., Aygumov, T. G., & Mursaliev, M. K. (2024). Issues with generative artificial intelligence tools. In ITM Web of Conferences (Vol. 59, p. 04007). EDP Sciences. DOI:10.1051/itmconf/20245904007

Leiker, D., Gyllen, A. R., Eldesouky, I., & Cukurova, M. (2023, June). Generative AI for Learning: Investigating the Potential of Learning Videos with Synthetic Virtual Instructors. In *International conference on artificial intelligence in education* (pp. 523-529). Cham: Springer Nature Switzerland. DOI:10.1007/978-3-031-36336-8_81

Makridakis, S. (2017). The forthcoming Artificial Intelligence (AI) revolution: Its impact on society and firms. *Futures*, 90, 46–60. DOI:10.1016/j.futures.2017.03.006

McCormack, J. (2023). Evolutionary Machine Learning in the Arts. In *Handbook of Evolutionary Machine Learning* (pp. 739–760). Springer Nature Singapore.

Miah, J., Cao, D. M., Sayed, M. A., & Haque, M. S. (2023). Generative AI Model for Artistic Style Transfer Using Convolutional Neural Networks. *arXiv preprint arXiv:2310.18237*.

Oniani, D., Hilsman, J., Peng, Y., Poropatich, R. K., Pamplin, C. O. L., Legault, L. T. C., & Wang, Y. (2023). From Military to Healthcare: Adopting and Expanding Ethical Principles for Generative Artificial Intelligence. arXiv preprint arXiv:2308.02448.

Pan, Z., Yu, W., Yi, X., Khan, A., Yuan, F., & Zheng, Y. (2019). Recent progress on generative adversarial networks (GANs): A survey. *IEEE Access : Practical Innovations, Open Solutions*, 7, 36322–36333. DOI:10.1109/ACCESS.2019.2905015

Rane, N. (2023). Role and challenges of ChatGPT and similar generative artificial intelligence in arts and humanities. Available at *SSRN* 4603208. DOI:10.2139/ssrn.4603208

Sharma, S., Singh, G., Gaur, L., & Afaq, A. (2022). Exploring customer adoption of autonomous shopping systems. *Telematics and Informatics*, 73, 101861. DOI:10.1016/j.tele.2022.101861

Stahl, B. C., & Stahl, B. C. (2021). Ethical issues of AI. Artificial Intelligence for a better future: An ecosystem perspective on the ethics of AI and emerging digital technologies, 35-53.

Vartiainen, H., & Tedre, M. (2024). How Text-to-Image Generative AI is Transforming Mediated Action. *IEEE Computer Graphics and Applications*, 44(2), 12–22. DOI:10.1109/MCG.2024.3355808 PMID:38285567

Wach, K., Duong, C. D., Ejdys, J., Kazlauskaitė, R., Korzynski, P., Mazurek, G., Paliszkiewicz, J., & Ziemba, E. (2023). The dark side of generative artificial intelligence: A critical analysis of controversies and risks of ChatGPT. *Entrepreneurial Business and Economics Review*, 11(2), 7–30. DOI:10.15678/EBER.2023.110201

Zhang, P., & Kamel Boulos, M. N. (2023). Generative AI in medicine and healthcare: Promises, opportunities and challenges. *Future Internet*, 15(9), 286. DOI:10.3390/fi15090286

Zhao, B. (2023). Analysis on the Negative Impact of AI Development on Employment and Its Countermeasures. In SHS Web of Conferences (Vol. 154, p. 03022). EDP Sciences. DOI:10.1051/shsconf/202315403022

Chapter 7
Navigating Ethical Dilemmas in Generative AI:
Case Studies and Insights

Bodhibrata Nag
https://orcid.org/0000-0002-6621-359X
Indian Institute of Management, Calcutta, India

ABSTRACT

Generative AI, based on neural networks, is transforming the creation of content like text, images, music, and interactive digital experiences. However, it raises ethical and practical questions about the authenticity of AI-generated works, intellectual property rights, and the role of creators. AI is also transforming journalism by integrating algorithms and automation, but raises ethical concerns about potential biases and transparency. To address these issues, the sector should set clear standards, promote diversity, and encourage cooperation among technologists, artists, ethicists, and policymakers.

DOI: 10.4018/979-8-3693-9173-0.ch007

INTRODUCTION

AI, or artificial intelligence, encompasses technologies that have the ability to mimic human intelligence and aid in decision-making through their capacity for self-learning.

Machine Learning (ML) is a specialized area within the study of Artificial Intelligence (AI) that specifically aims to replicate the learning process of human beings. It involves the creation and implementation of algorithms that have the ability to acquire knowledge and adjust their behavior without relying on explicit instructions.

Deep Learning emulates a neural network that resembles the structure and functioning of the human brain. The architecture is a multi-level machine learning framework that use many levels of machine learning to study complex issues. It is particularly effective for tasks that are challenging for supervised learning or involve a large number of variables that need to be modeled and analyzed to understand the underlying relationships in the data sets.

Generative AI uses deep learning algorithms to analyze extensive datasets and produce various forms of digital material, including documents, software, photos, videos, games, three-dimensional models, and other identifiable digital resources.

Figure 1. The relationship between AI(artificial intelligence), ML(machine learning), DL(deep learning) and GAI(Generative AI) (Adapted from (Global Information & Communications Team, 2023))

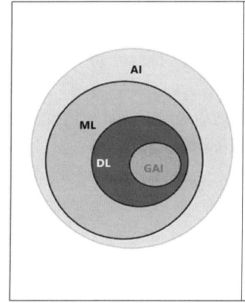

AI, or artificial intelligence, encompasses technologies that have the ability to mimic human intelligence and aid in decision-making through their capacity for self-learning.
Machine Learning (ML) is a specialized area within the study of Artificial Intelligence (AI) that specifically aims to replicate the learning process of human beings. It involves the creation and implementation of algorithms that have the ability to acquire knowledge and adjust their behavior without relying on explicit instructions.
Deep Learning emulates a neural network that resembles the structure and functioning of the human brain. The architecture is a multi-level machine learning framework that use many levels of machine learning to study complex issues. It is particularly effective for tasks that are challenging for supervised learning or involve a large number of variables that need to be modeled and analyzed to understand the underlying relationships in the data sets.
Generative AI use deep learning algorithms to analyze extensive datasets and produce various forms of digital material, including documents, software, photos, videos, games, three-dimensional models, and other identifiable digital resources.

The history of generative AI (Figure-1) is deeply interwoven with the broader development of artificial intelligence (AI) and machine learning (ML). The origin of neural networks dates back to the 1940s and 1950s, beginning with the perceptron model proposed by Frank Rosenblatt in 1958. This simple model was foundational, establishing the concept of trainable networks based on synthetic neurons (Rosenblatt, 1958). Subsequent decades saw gradual progress, including the introduction of backpropagation by Rumelhart, Hinton, and Williams in 1986, which enabled the training of multi-layer networks and set the stage for modern deep learning (Rumelhart, Hinton, & Williams, 1986).

The 1990s and early 2000s witnessed significant advancements in machine learning, with Support Vector Machines (SVMs) and other algorithmic developments enhancing pattern recognition capabilities (Cortes & Vapnik, 1995). However, it was the resurgence of neural networks in the form of deep learning in the late 2000s that truly accelerated AI capabilities. A pivotal moment came in 2012 when Alex Krizhevsky, Ilya Sutskever, and Geoffrey Hinton demonstrated the power of deep neural networks with their AlexNet, which significantly outperformed other models in the ImageNet competition (Krizhevsky, Sutskever, & Hinton, 2012).

However, these developments did not occur without controversy. Critics have pointed out that the emphasis on neural networks overshadowed other AI approaches, leading to an over-reliance on a single paradigm (Russell & Norvig, 2020). Moreover, the resurgence of neural networks in the form of deep learning in the late 2000s, particularly following the success of AlexNet in the ImageNet competition (Kietzmann, Lee, McCarthy, & Kietzmann, 2020), has been critiqued for prioritizing performance over interpretability (Lipton, 2018).

The development of Generative Adversarial Networks (GANs) by Goodfellow et al. in 2014 introduced a novel method for generating new data instances indistinguishable from real data, stimulating numerous applications and ethical discussions (Goodfellow, 2014). Parallel to this, the introduction of Variational Autoencoders (VAEs) by Kingma and Welling in 2013 provided a framework for learning latent spaces, further enriching the AI landscape (Kingma & Welling, 2013). Despite their success, GANs and VAEs have raised significant ethical concerns, particularly regarding their potential for misuse in creating deepfakes and the challenges of ensuring fairness and transparency in generated outputs (Kietzmann, Lee, McCarthy, & Kietzmann, 2020).

These technical advancements have pushed the boundaries of what machines can create but have also prompted serious ethical considerations. The ability of GANs to generate realistic images and videos has led to debates around authenticity and the potential for misuse, such as in creating deepfakes. Additionally, the opacity of deep learning models has raised concerns about bias and fairness, issues that are now at the forefront of discussions in AI ethics ((O'Neil, 2016) (Eubanks, 2018).

Generative AI is a rapidly developing field that produces new interactive digital experiences, music, images, and text. Neural networks—a model of the human brain—are the foundation of the algorithm. These networks are trained using sizable datasets of anticipated content. For example, a generative AI trained to write articles would pick up patterns, styles, and structures from reading a large corpus of text. An AI capable of creating art would analyze thousands of photos to comprehend composition, color, and form. Despite its potential, the field faces significant ethical challenges, such as the impact on human creativity and the potential for reinforcing existing biases (Bostrom & Yudkowsky, 2014) (Bryson, 2019).

Table 1. Examples of generative AI applications

Text Models-Conversational AI (ChatGPT, Bard, Bing, Claude)	Text-to-Science Models (Galactica, Minerva)	Text-to-Author Simulation (GrammarlyGo, PEER)	Text-to-Medical Advice (Chatdoctor, GlassAI, Med-PaLM, YourDoctor AI, Hippocratic AI)
Text-to-Itinerary (Roam Around, TripNotes, ChatGPT's Kayak plug-in)	Doc-to-Text (ChatDOC, MapDeduce)	Image Generative AI (DALL-E, Midjourney)	Image Editing (Alpaca AI, I2SB, Facet AI, Photoroom AI, Tencent Face Restoration)
Artistic Images Generation (OpenART, Mage. Space, NightCafe)	Text-to-Video (Imagen Video, Meta Make A Video,Phenaki, Runway Gen-2)	Text-to-3D Model Generation (e Adobe Firefly, Dreamfusion, GET3D, Magic3D, Synthesis AI, Text2Room	Text-to-Multilingual Code (Alphacode, Amazon Codewhisperer, BlackBox AI, CodeComplete, CodeGeeX, Codeium,Mutable AI, GitHub Copilot, GhostWriter Replit, Tabnine
Text-to-Speech (Coqui, Descript Overdub, ElevenLabs Listnr, Lovo AI, Resemble AI, Replica Studios, Voicemod, Wellsaid, AudioLM)	Speech-to-Speech models (ACE-VC and VALL-E)	Speech-to-Text (Cogram AI,Deepgram AI, Dialpad AI, Fathom Video, Fireflies AI,GoogleUSM, Papercup, Reduct Video, Whisper, Zoom IQ)	Images-to-text (Flamingo, Segment Anything, VisualGPT)

(Adapted from (Gozalo-Brizuela & Garrido-Merchán, 2023))

Examples (Table 1) of generative AI include transformers, which process data sequentially, and GANs, which have a generator and a discriminator. Beyond human capabilities, generative AI can automate and enhance human tasks. As AI technology develops, moral and useful concerns are brought up. The question of whether a machine can be an author or artist, who owns AI-generated works, and whether AI is capable of extending and replicating the styles of human artists without permission all complicate these matters.

Authenticity and originality are also significant in artistic domains where a human "touch" is prized. AI-generated deepfake news articles and videos are a major problem. These apps have a big influence on trust, security, and media privacy.

AI has been praised and criticized in journalism. By creating news content, processing data, and customizing the reading experience for readers, automation can increase productivity and reach. But in order to reap these benefits, more justice, accuracy, and transparency are needed. Source and fact verification are more crucial than ever because AI-generated content is impersonal and has the potential to undermine media trust. Three traditional journalism values are safety, accountability, and accuracy. AI's quick content creation puts these principles in jeopardy because it has the potential to disseminate inaccurate or biased information.

To address these concerns, it is crucial to consider a diverse array of perspectives, including those from critical AI studies, which emphasize the need for transparency, accountability, and fairness in AI systems (Zuboff, 2019) (Noble, 2018). This expanded view will provide a more comprehensive understanding of the ethical landscape surrounding generative AI.

This chapter explores the ethical landscape of generative AI through various case studies, providing insights into the potential and pitfalls of these technologies. This discussion is crucial for policymakers, practitioners, and the public as we strive to create a balanced approach to AI integration that respects both human creativity and ethical standards.

AI IN ART AND LITERATURE CREATION

AI has freed creativity in the arts by transforming machines into creative partners. "Computational creativity," a digital intelligence subfield, creates software "which is independently creative, either to be used collaboratively with humans or as an autonomous artist, musician, writer, designer, engineer, or scientist. (McCormack & D'Inverno, 2012). There are three approaches to creativity: (a) combining known methods, such as blending different music genres, (b) innovating within a method while playing by the rules, and (c) game-changing scenarios (Sautoy, 2019) (Boden, 1998).

Using GANs and deep learning algorithms, AI systems have generated artworks that are sometimes indistinguishable from human creations. These technologies learn genre-defining styles and elements from massive art databases. AI's ability to learn and copy famous painters' styles is one of its most impressive applications in art. AI algorithms can study Van Gogh and Picasso's brush strokes, colour palettes, and composition strategies to generate new masterpieces. AI allows artists to combine aspects from multiple art styles or create new aesthetics based on data inputs that no human artist could reproduce or envisage. AI can adapt colour schemes, themes, and intricacy of artworks in real time to viewers' preferences, potentially changing the manner of art consumption. (Sautoy, 2019).

AI also empowers human artists to experiment and improve their creativity. AI can help artists experiment and challenge established art genres. AI tools help analyze great artists' works to explain their processes. This helps preserve and restore classic paintings and educates students and researchers about various artistic approaches.

Machine learning-based AI systems excel at analyzing large datasets to generate new content and finding data trends. AI can write like humans by learning from examples of writing. After achieving a desired style or tone, AI-generated text can be repeated endlessly without any changes that a human writer might do. (Sautoy, 2019).

AI systems process text based on statistical likelihoods and patterns, without truly understanding the content and intent of the text. Thereby, AI produces shallow or inappropriate text, showing the machine's lack of comprehension. AI can copy and replicate styles and concepts, but it cannot create unique content or think outside the box. AI's 'inventiveness' generally comes from human coders' established bounds, not in from its intrinsic creativity. AI is unable understand complicated human emotions and complexities. Thus, its text may lack the emotional depth of a human writer. Further Ai has no sense of morality or ethics. Therefore, it is unable to realize the consequences of its outputs, which might be improper or dangerous. (Sautoy, 2019).

Here are a few recent examples of the use of AI in art and literature:

- The 2016 project "The Next Rembrandt" (Brown, 2016) involved mimicking Rembrandt's techniques to create a new painting in his style. This initiative sparked debates about intellectual property rights and the originality of AI-generated works. The question arises as to who owns these creations: the programmers, the owners of the algorithm, the providers of the data set, or possibly the public?
- AIVA (https://www.aiva.ai/), an AI composer creates soundtracks for films, games, and commercials, raises questions about the authenticity of music derived from algorithmic patterns. Is the music created by AI, which is based on established composers, considered original or merely derivative?
- The novel "1 the Road" (Hornigold, 2018), written by a Jack Kerouac-style AI and developed by Ross Goodwin, raises questions about the role of AI in literature and the concept of authorship. DeepArt (https://www.artvy.ai/ai-tools/deepartio) uses algorithms to transform photos into digital art that mimics the styles of famous artists and highlights the ethical considerations at the intersection of technology and creativity.

- The auction of "Edmond de Belamy," an AI-generated artwork, at Christie's (Cohn, 2018) attracted widespread attention and sparked debates about what constitutes creativity and the value placed on algorithmically produced artworks.
- Google's DeepDream(https://deepdreamgenerator.com/) generates intriguing, dream-like images by enhancing patterns in photographs through a neural network designed to detect and amplify features. Although it produces visually distinctive and intricate images, it also demonstrates the unpredictability and absence of deliberate control in AI creativity, which contrasts with the intentional artistic expression of humans.

Artificial intelligence (AI) is predicted to transform the art industry by erasing lines between humans and computers. It is imperative to switch from deterministic programming to autonomous AI systems that learn and create. AI can develop genres that break with conventions in addition to variations on already-existing artistic genres. Emotional intelligence allows AI systems to be more inventive as well. AI systems are unable to comprehending and analyzing human emotions without programming. The creative processes of future AI may be influenced by emotional cues in human interactions, producing artwork, music, literature, and interactive media that are intimately linked to the emotions of the audience. The cooperation between AI and humans has to be improved in future development. AI can work with humans to provide original solutions or perhaps unnoticed creative alternatives by enhancing its comprehension and prediction of human creative processes.

In the future, as artificial intelligence (AI) advances and becomes more capable and independent, it will be crucial to tackle the following concerns. (Sautoy, 2019):

1. Intellectual Property: To explicitly address the intellectual property rights of content created by artificial intelligence, the legal framework must be revised. The goal of this revision should be to make sure that everyone benefits equally and to assess whether non-human inventions can be protected by the current copyright laws. Furthermore, it should guarantee equitable benefit distribution to all stakeholders and encourage transparency in AI-generated content.

2. Uniqueness and authenticity: Authenticity is often linked to the subjective aesthetic and visionary approach of the artist. However, concerns exist regarding the veracity of AI-generated artwork due to the fact that it frequently imitates materials and styles extracted from massive datasets. With the increasing prevalence of AI, it becomes crucial to establish precise criteria for authenticity. The academic and creative communities must deliberate on the matter of whether AI works are authentic imitations or plagiarism, in addition to determining the appropriate methods of acknowledgment.

3. Impact on human creativity: Artificial intelligence's impact on creativity is a growing concern. Artificial intelligence's ability to create successful art may threaten human artists' careers. The growing market share of AI-generated works may reduce the unique value of human creativity, which could harm human artists' livelihoods and cultural significance. This involves exploring ways to make AI enhance human creativity rather than replace it.

4. The method by which AI generates art is sometimes obscure and not completely comprehended even by the AI developers. This introduces an additional level of ethical intricacy. The opaque nature of numerous AI systems results in the concealment of the decision-making process, rendering it arduous to comprehend the mechanisms behind the determination of creative aspects. The absence of transparency might give rise to apprehensions regarding partiality, manipulation, and the possibility of producing detrimental content disguised as originality. (Sautoy, 2019).

The discussions surrounding these matters are tangible and have an effect on producers, customers, and society as a whole. Although the Berne Convention and other existing rules and frameworks lay the groundwork for addressing these problems, they need to be expanded and reinterpreted to encompass the potential and difficulties posed by AI. This requires a reconsideration of the ideas of authorship, creativity, and ownership. Through studying practical instances and participating in meaningful discussions, we can establish systems and criteria that direct the ethical incorporation of artificial intelligence into artistic domains. This will guarantee that generative AI technologies are not only innovative but also morally upright and advantageous for improving societal well-being and fostering human creativity.

GENERATIVE AI IN JOURNALISM

AI's introduction in journalism signifies a shift in news production, dissemination, and consumption. Algorithms and automation are increasingly integrated into the news production process, fundamentally reshaping journalism. Here are several key ways in which this integration is taking place (Diakopoulos, 2019):

- Algorithms are used to automatically generate content, such as the Associated Press using automation to produce corporate earnings reports. This use of automation allows news organizations to cover more topics and produce content faster than human journalists could, significantly increasing the breadth of coverage.

- Automation does not replace journalists but enhances their capabilities by handling routine, time-consuming tasks. This frees up journalists to focus on more complex, value-added activities such as in-depth investigative reporting or crafting detailed analyses.

- Newsrooms are evolving into spaces where human journalists collaborate with algorithms. For instance, automated systems generate an initial version of content that journalists can subsequently improve and enrich with human insights, contributing depth and context that algorithms alone cannot accomplish.

- Automation makes it easier to produce and distribute news quickly, which is essential for financial reporting and other fields that require current data. Algorithms are capable of quickly distributing news articles across various platforms, finding patterns in data, and analyzing it all with efficiency.

- Algorithms utilize reader preferences and behaviors as a means to customize content for individual users. An increase in reader engagement may result from personalized content that is more closely aligned with the interests of the reader.

- Algorithms are essential to the distribution of content because they choose which news stories to show to specific users. They target metrics like page dwell time and click-through rates, which show user engagement. This phenomenon has the potential to impact news article visibility and, in turn, shape public opinion.

- Data mining and analysis tools powered by automation and algorithms increase the scope and speed of investigative journalism. Automation was crucial to handling and analyzing massive amounts of data in the Panama Papers.

The progression of these technologies presents prospects for enhancing reporting, customizing narratives for specific individuals, and mechanizing repetitive duties, as exemplified in the subsequent instances.:

- Los Angeles Times Quakebot is an early example of efficiency of artificial intelligence in journalism. Soon after the Southern California earthquake of magnitude 4.2, Quakebot quickly generated an article. (Oremus, 2014). It also raises concerns about the depth and subtlety that AI might lack compared to analysis.

- Bloomberg's Cyborg technology (Peiser, 2019) analyses financial reports and transforms them into understandable articles. Cyborg shows how AI can help in journalism by interpreting data that would require humans significantly more time to process. Nonetheless, it underscores the importance of oversight when considering consequences resulting from errors or omissions.

- The NewsLabs team at the BBC has been testing out AI to craft news summaries customised for audiences (Gupta, 2021). This customisation enhances user engagement by personalising news summaries for specific audiences, thus boosting reader interest and interaction. It also raises dilemmas regarding filter bubbles and the risk of echo chambers, where readers are only exposed to news that aligns with their existing beliefs.
- The Chinese state news agency Xinhua has rolled out AI news anchors to deliver news reports (Baraniuk, 2018). While this advancement highlights the integration of AI into real time news delivery, it also sparks concerns about the removal of emotion and judgement from news presentations.
- The New York Times utilises the Jigsaws Perspective API to aid in comment moderation (Jigsaw (Google), 2021). This tool uses machine learning to identify comments, encourage respectful interactions, and foster constructive discussions among users. However, this also poses the challenge of ensuring that AI does not stifle freedom of speech or marginalised viewpoints.

However, this progress also raises the following questions that require consideration:

1. Trustworthiness and Honesty: The principles of journalism, such as reliability and integrity, are challenged by the implementation of AI. The Washington Post's Heliograf and The Associated Press's automated insights demonstrate how AI can enhance productivity and explore complex issues. However, concerns over the precision of AI-generated content persist, as a lack of meticulous development and continuous monitoring could lead to the spread of false information, emphasizing the need for supervision in AI-generated material.
2. Prejudice and Influence: AI-generated content raises ethical concerns due to potential biases in the training data. Reuters' Lynx Insight tool, designed to help journalists identify patterns, could uphold biases if its algorithms aren't rigorously evaluated for fairness. Developers should use inclusive datasets and conduct fairness assessments. Strict accountability is required to prevent biases or inaccuracies in journalism, and news organizations must be held responsible for the outputs of automated systems, including errors or biased reporting. (Diakopoulos, 2019).
3. Importance of Transparency: Transparency is crucial for maintaining audience trust and understanding automated processes. However, the blurring of human-generated and machine-generated content can lead to questions about authorship and AI's involvement. OpenAI's GPT-3 in The Guardian's opinion piece

highlights AI's persuasive text creation, underscoring the profound influence of AI on content generation and the need for transparency. Although the article recognised the AI's involvement, continual transparency in AI-generated content is vital to avoid misleading the audience.

4. The integration of algorithms into journalism also raises new ethical and operational questions, leading to the development of "algorithmic accountability" reporting. This new journalistic practice involves scrutinizing and reporting on the algorithms themselves—how they work, their impact on public discourse, and their implications for privacy and ethics.

5. As machines take on roles traditionally held by humans in news production, they alter the social dynamics within journalistic environments. This includes how newsrooms operate, how journalists interact with sources, and how audiences engage with news content. Machines can serve both as creators of content and as intermediaries, facilitating interactions that were previously human-driven. The use of automation in journalism challenges existing theoretical frameworks that primarily view communication as a human-centric activity. There is a need for new theories that also consider the roles and impacts of machines in communication processes. The integration of communicative machines in journalism prompts a reevaluation of the professional identity of journalists. Journalists must now navigate a landscape where machines can replicate some of their core tasks, such as writing and reporting, which can lead to both opportunities and challenges in defining what it means to be a journalist. (Lewis, Guzman, & Schmidt, 2019).

6. Machines capable of tailoring content to individual preferences might lead to changes in media consumption patterns. This personalization can enhance user engagement but also risks creating filter bubbles, where individuals are exposed only to information that aligns with their preexisting beliefs. (Lewis, Guzman, & Schmidt, 2019)

To make the most of AI in journalism while addressing ethical concerns several suggestions can be made;

*   Transparent disclosure: Clearly stating when content is artificial intelligence (AI) generated aids readers in understanding the source of their news.
*   Human supervision: having human editors review AI generated articles ensures adherence to standards and allows for error correction.
*   Bias monitoring: news organisations should have procedures in place to regularly examine and rectify biases in AI systems.

- Developing industry wide ethical guidelines can regulate the use of AI in journalism, covering aspects like transparency, accountability, and data privacy.
- Engagement: Journalists and media outlets should engage openly with the public to foster dialogue on how AI influences journalism and news consumption.

AI IN THE ENTERTAINMENT INDUSTRY

In 1957, Hiller and Isaacson introduced the Illiac Suite, which marked the debut of music composition by an artificial intelligence system. This event was a notable achievement, illustrating the capacity of computers to be utilized in a creative manner for the production of intricate compositions. During the design of the Illiac Suite, computing resources were severely constrained. Programs were typed on punched cards, memory capacity was limited, processing speed was poor, and high-level programming languages were only beginning to emerge. The Illiac Suite was groundbreaking not just due to its algorithmic composition, but also because Hiller and Isaacson proposed various concepts for computer music that continue to have a significant impact today. They employed conventional Western composition techniques within a heuristic search framework, effectively employing a generate-and-test approach where pitches were generated and only those conforming to the rules of composition were retained. The project integrated computational methodologies with classical music conventions by incorporating movements such as presto, andante, and allegro, and by executing the composition with live players and traditional instruments. This approach facilitated the connection between human and computer-generated music. The development of the Illiac Suite initiated a more extensive discourse regarding the involvement of AI in creative procedures, anticipating current discussions on the capabilities and constraints of AI in the realm of art. The experiment unveiled novel prospects for the utilization of AI in music, exerting an influence on following cohorts of musicians and engineers. (Miranda, 2021)

Advancements in hardware throughout the years have facilitated the processing of musical data in a more intricate and rapid manner. Contemporary computers, with their significant advancements in CPU speeds, memory, and storage, enable more intricate real-time musical analysis and synthesis. The advent of sophisticated programming languages and specialized software tailored for music has had a profound impact on the field of computer music. The use of sophisticated software tools such as Csound, Max/MSP, and others has democratized the field of music production. Advancements in Digital Signal Processing technology have enabled more complex sound synthesis and processing techniques, allowing for more effects and transformations in computer music. These include real-time pitch correction, spectrum analysis, and synthesis, broadening the range of sounds available to composers. The merging of machine learning models and music creation involves training systems using extensive music datasets to generate new compositions. Interactive systems that respond to live inputs from human performers in real-time use advanced algorithms to analyze performance data, produce additional audio, or adjust music based on the performer's actions. (Miranda, 2021).

The field of music production is undergoing a transformation as artificial intelligence is being incorporated. Neural networks and generative models, namely Generative Adversarial Networks (GANs), have expanded the possibilities of applying AI in music. These technologies enable the production of music that may adjust and develop in ways that imitate human creativity, providing novel instruments for composers and performers. AI tools aid musicians by introducing novel methods of producing music, presenting a range of sounds and genres that might stimulate fresh creative pathways. Artists like Holly Herndon have embraced AI collaborations (Austin, 2023) to craft compositions pushing the boundaries of expression. Algorithms such as OpenAI's Jukebox (https://openai.com/research/jukebox) can create songs mimicking established musicians' styles sparking debates on music originality and the future role of musicianship. Artificial intelligence has a profound influence on the distribution and consumption of music, especially through personalized song suggestions provided by streaming platforms such as Spotify and Apple Music. AI algorithms utilize extensive listener data to customize music suggestions, impact music marketing efforts, and even ascertain the potential viability of other music genres, hence impacting production choices. Artificial intelligence enhances the accessibility of music by providing discovery functionalities that connect consumers with a diverse range of musical genres and artists that they might not have otherwise discovered. (Miranda, 2021).

The utilization of artificial intelligence (AI) in the entertainment sector has provided opportunities for artistic expression by facilitating the development of realistic characters, situations, and interactive encounters. The role of AI in entertainment is extensive, encompassing influencers, CGI figures, customized gaming experiences, and algorithmically generated music. The production of information by artificial intelligence poses a challenge to traditional standards of authenticity. Lil Miquela and FN Meka, as virtual influencers, have gained a large number of followers on social media platforms, eroding the distinction between reality and digital arts (Chan, 2022). Although these AI characters are groundbreaking, they also provoke debates regarding the genuineness of their interactions and the ethical implications of corporations employing them to influence customer decisions.

Artificial intelligence (AI) is increasingly becoming a transformative force in the film industry, affecting various aspects from production to distribution. AI technologies are being implemented across different stages of film production, including scriptwriting, editing, visual effects, and post-production, enhancing efficiency and creativity (Li, 2022). AI-based systems facilitate automated operations in the film industry, such as facial recognition for sorting scenes and predicting movie success, which may lead to a future dependence on AI for editing and promotion. (Singh, Kaur, & Singh, 2023). AI is also being used in movie recommendation, distribution, and the creation of audiovisual language is revolutionizing how audiences interact with film content (Du & Han, 2021). AI's role in enhancing personalization and interactivity in video production and analysis is leading to richer and more realistic user experiences. While AI offers numerous benefits such as increased efficiency and new creative possibilities, it also presents challenges that include ethical considerations and the potential impact on jobs. As AI continues to evolve, its role in the film industry is expected to expand, necessitating ongoing research and responsible management of its applications.

Furthermore, the inclusion of AI generated characters in films and television introduces dilemmas. The digital revival of actors, like Carrie Fisher in "Star Wars; The Rise of Skywalker" and Peter Cushing in "Rogue One; A Star Wars Story" (Irani, 2023)sparks discussions on consent and respect for their legacies. Using AI in scenarios prompts debates about the boundaries of resurrecting likenesses and the emotional impact on audiences and the deceased actors' families.

In gaming, AI powered characters and storylines in titles such as "Detroit; Become Human" and "The Last of Us Part II" (Robertson, 2022)provide players with personalized and emotionally engaging narratives. These games adjust based on player decisions offering experiences that can forge emotional bonds. While these AI applications enhance immersion, they also prompt considerations about AIs ability to influence emotions and the importance of storytelling practices.

In addition, to projects examples like the short film "Sunspring" and the AI assisted script for "Its No Game" illustrate how AI can be used to craft compelling narratives (Parikh, 2019). These initiatives do not showcase advancements but also prompt us to reconsider the concept of authorship and the significance of human storytelling.

The ethical challenges linked to AI in the realm of entertainment encompass the following issues:

- Authenticity- The emergence of AI generated content challenges our perception of authenticity and reality within art and entertainment. This prompts us to reflect on the value we attribute to creativity compared to creations produced by AI.

- Audience interpretation; The inclusion of AI generated characters and plotlines can shape how audiences distinguish between fact and fiction. There is a concern that AI might blur these distinctions potentially leading to uncertainty or a diminished ability to differentiate between what's authentic and what's not.
- Ethical narrative construction; With AI playing an increasingly active role in content development it becomes crucial to ensure that stories are ethically crafted without perpetuating harmful stereotypes or biases. Addressing the risk of introducing or magnifying biases in narratives is essential as AI continues its involvement, in storytelling.

To tackle these quandaries the sector can implement the following approaches:

- Set forth clear standards, for utilizing AI generated resemblances especially regarding posthumous usage to honor the dignity and heritage of individuals.
- Maintain transparency in generating and using AI created content enabling viewers to differentiate between AI inputs.
- Advocate for narratives by integrating diversity and inclusivity in AI powered stories. Actively addressing biases in generative algorithms.
- Initiate discussions with audiences, creators, and ethicists to delve into the implications of AI in entertainment and establish protocols.
- Encourage cooperation among technologists, artists, ethicists, and policymakers to foster a rounded approach to innovation and ethical deliberation.

DEEPFAKES AND MISINFORMATION

Throughout the course of history, several types of media have been controlled with the intention of promoting propaganda, spreading false information, or disseminating disinformation. The art of manipulating photographs dates back to the early 1900s. During this period, politicians were included in images or excluded for ideological reasons. Propaganda was disseminated in the early and mid-1900s via radio and film. The Nazis shaped German public opinion by disseminating propaganda and false information through radio and cinema. Through events that are staged and omitted, television has disseminated false information. Photo manipulation is made simpler by digital photography and Photoshop, which increases the convincingness of counterfeits. Additionally, the proliferation of phony websites and edited videos on social media has been expedited by the internet. Before updates or fact-checks are made public, fake news websites and videos have the potential to do a great deal of harm. Deepfakes serve as an example of how technology for media manipulation can tarnish public opinion. (Hao, 2019)

Deepfakes and other sophisticated machine learning techniques employ generative adversarial networks to generate authentic-sounding and video content. These networks, which have been trained on an extensive collection of images, videos, and sounds, emulate the style and appearance of the data. GPU capability and FakeApp or DeepFaceLab are required to complete this intensively computational procedure. The expertise and technological prowess required to produce deepfakes render them uncommon, yet perilous by virtue of their lifelike qualities. They pose a risk to personal reputations and political campaigns due to their ability to generate synthetic media featuring celebrities or politicians.

Inexpensive imitations employ less complex techniques that do not necessitate the utilization of machine learning or computational resources. Smartphone applications or rudimentary video editing software facilitate these alterations. The increased accessibility and velocity of low-cost counterfeits alter public perceptions and disseminate false information. (Paris & Donovan, 2019).

Real-time video manipulation technology has contributed to the rise in popularity of deep fake technology, which generates convincing fakes in real time. This raises concerns regarding the fabrication of live events or statements. Github's graphical user interfaces and user-friendly tools, such as open-source software, have facilitated the democratization of deep fake technology. With these tools, even non-technical users can produce convincing fakes. Deepfakes have become more accessible due to commercial service portals and downloadable computer applications, which has increased their potential for misuse. Emerging technologies such as Google Deep-Mind's WaveNet possess the capability to precisely replicate human voices, thereby enabling the production of audio recordings that falsely imply statements made by actual people. (Chesney & Citron, 2019) (Ajder, Patrini, Cavalli, & Cullen, 2019).

On social media, deepfake content spreads like wildfire. Because deepfakes are networked, they spread quickly and reach a wide audience across the globe. The platforms take advantage of cognitive biases, such as the propensity for emotional or sensational content that is more likely to be shared. These biases are used by many sensational deepfakes to boost interaction and distribution. Although interesting content is prioritized by social media algorithms, deepfakes proliferate unchecked through likes, shares, and comments. This 'information cascade' enables people to disseminate false information on their networks with speed and without hesitation. Due to the speed at which social media spreads, deepfakes reach millions of people before being exposed. Because deepfakes produce a vast amount of content, platforms need to strike a balance between censorship, free speech, and harm prevention.

Enhanced learning opportunities may result from the production of interactive content that vividly portrays historical figures using deepfake technology. This can enhance learning by allowing students to "meet" historical figures, simulate patient interactions in medical education, and rehearse responses to a variety of scenarios in law enforcement training. Artists can expand the limits of their imaginations with deepfake technology by reimagining historical works of art or creating interactive installations that respond to viewer actions. This has the potential to alter individuals' perceptions and valuations of art through the introduction of dynamism and personalization.

Nevertheless, deepfakes give rise to a plethora of concerns. Several concrete examples from real-life situations are presented below to emphasize the gravity of these concerns:

- Influence of Political Misinformation: The then Speaker of the US House of Representatives, Nancy Pelosi, was featured in a fake video that went viral online in 2018. The video was manipulated to change the speed of her speech, creating the false impression of incoherence, which insinuated possible intoxication or health problems. (Mervosh, 2019). Though not a deepfake per se this manipulation highlights the dangers of using media in political settings. The video went viral. Was endorsed by political figures leading to confusion and misinformation spread. This case emphasizes the importance of tools, for verification and media literacy to differentiate between edited content.

- Unauthorized Use of Celebrity Images: Celebrities have often fallen victim to deepfake misuse where their likenesses are used without permission. There was a known incident involving actor Tom Cruise, where fake videos portraying him in activities gained popularity on social media (Metz, 2021). These fake videos were so lifelike that many viewers mistook them for real. This misuse of celebrities' images raises concerns, about consent and the unauthorized use of someone's likeness sparking debates on the options for those affected and the moral obligations of content creators.

- In 2019 the CEO of a UK energy company fell victim to a scam where €220,000 was transferred to a supplier after receiving a phone call from someone impersonating the CEOs boss using AI generated voice mimicry (Damiani, 2019). This case highlights the risks associated with AI driven impersonation emphasizing the need for robust verification processes in financial transactions and ethical considerations, in utilizing voice synthesizing technology.

- Another troubling aspect of this technology is seen in deepfake pornography, where people's faces are digitally placed onto adult film actors bodies without their consent. This type of exploitation raises issues regarding privacy and consent leading to emotional distress and harm to one's reputation. The instance involving journalist Rana Ayyub, who was harassed with deepfake pornography, exemplifies the dire repercussions of such exploitation, and underscores the urgent requirement for legislative safeguards to prevent such transgressions. (Saxena, 2023)

- In a different instance, worries about President Ali Bongo of Gabon's safety and the political stability of the nation were aroused by a deepfake video that featured him in 2019 (Cahlan, 2020). This ultimately led to a coup attempt. This incident is an excellent example of how deepfakes can be used to incite political unrest, posing a threat to the integrity and stability of political systems. It emphasizes the negative effects of synthetic media and how important ethical values are to preventing these kinds of problems.

Deep fakes pose a grave threat to the credibility and dependability of the media by fabricating content that appears to be from real people but isn't. "Truth decay" and the erosion of shared reality are exacerbated in our networked information environment due to the fact that cognitive biases are intensified by the addition of false content and the blurring of the line between truth and falsehood. Deep fakes present new ethical challenges for journalists, including the dilemma of whether to report on deep fakes without contributing to their spread. The technology complicates the verification process that is fundamental to journalism, as verifying the authenticity of video and audio clips becomes increasingly difficult (Chesney & Citron, 2019).

These cases underscore the repercussions that deepfakes can bring upon individuals, political landscapes and communities. The ethical dilemmas involved are manifold;

- Deep fakes challenge current legal and ethical frameworks, necessitating new laws and regulations to address the unique issues they raise. This includes rethinking how we protect individuals' rights and manage the dissemination of potentially harmful fabricated content. While existing criminal and civil laws related to identity theft, cyberstalking, defamation, emotional distress, and fraud can address few of the legal issues, there is an urgent need for creation of new criminal statutes that specifically target the malicious creation and distribution of deepfakes. There is also the potential for holding platforms accountable for hosting deepfake content. Legal reforms are necessary to increase the accountability of platforms in facilitating the dissemination of harmful deepfakes.. (Chesney & Citron, 2019).
- Deep fakes can violate copyright laws by impersonating real people or characters without permission. Video or image depictions of characters or celebrities that were not intended or authorized by copyright holders can re-

sult. Creators and artists who use copyright protection are at risk. Copyright enforcement is complicated by the ease of creating and sharing deep fakes. Deep fakes are difficult to identify and prosecute because their creators can remain anonymous and quickly spread their creations online. The complexity of deep fake technology makes it harder to tell fakes from real ones. Blurring boundaries can make creations' authenticity unclear, complicating copyright violations. (Chesney & Citron, 2019)

- Trust Concerns and False Information: Deepfakes have the potential to erode confidence in institutions, public figures, and media outlets. Deep fakes are capable of inciting social unrest, misrepresenting individuals as having said or done things they never actually did, and causing irreparable harm to their reputations. Characters may be harmed or public outrage may be provoked by deep fakes. Voter behavior and political opinion may be impacted by deep-fake technology. Stock prices can be manipulated by deepfakes through the fabrication of financial data or company news. In addition, they are capable of producing persuasive audio and video recordings of political leaders uttering incendiary remarks, thereby fostering misunderstandings. For the purpose of combating misinformation and disinformation, technologies for detection, literacy, and verification are required. (Hao, 2019) (Paris & Donovan, 2019).

- Privacy Violations and Consent Concerns: The unauthorized utilization of someone's likeness in deepfakes violates their privacy rights and breaches the boundaries of consent. This emphasizes the necessity for frameworks that protect the rights of individuals' images while also providing channels for seeking redress.

Deepfakes and misinformation are posing challenges to AI creators and government oversight bodies, prompting a reassessment of their responsibilities. While tools like DeepTrace and Microsoft's Deepfake Detection Challenge show progress in detection capabilities, addressing misuse requires ethical guidelines and accountability.

Digital forensics is crucial for detecting doctored images and videos, using automated processes and neural networks to identify anomalies. As AI capabilities advance, tools and techniques for detecting and authenticating media must keep pace to ensure the reliability of digital content in the information age, particularly in journalism and legal contexts where media authenticity is critical. (Chesney & Citron, 2019) Fact checking is increasingly becoming an essential component of contemporary journalism, surpassing traditional reporting by actively participating in the verification of news content. (Graves, 2016).

New insurance policies should be developed to cover damage caused by deep fakes, providing financial compensation for individuals or businesses affected by their misuse. Organizations should develop risk management strategies that include prevention and mitigation aspects, regular security assessments, and policy updates. Technology-based solutions that recognize and highlight deepfakes must be integrated into risk management and insurance platforms. When risk managers, AI developers, and cybersecurity specialists work together, there's a chance that the detection tools and strategies will be improved. (Chesney & Citron, 2019).

By authenticating content, digital watermarking and immutable authentication trails can prevent deepfakes. These techniques enable platforms and users to monitor and authenticate modifications to content, thereby averting the spread of deepfakes. By generating a secure and immutable ledger of media file creation and modification, provenance verification based on blockchain technology safeguards the reputations of news organizations and content creators. (Chesney & Citron, 2019).

RECENT TRENDS

In recent years, advancements in artificial intelligence (AI), particularly in neural network architectures, have significantly broadened the application of AI across various sectors, including healthcare, autonomous vehicles, and the establishment of ethical frameworks. The development of advanced neural networks such as transformers and convolutional neural networks has been pivotal. These architectures have enhanced the ability of AI systems to process vast amounts of data with higher accuracy and speed, leading to their increased adoption in critical fields (Vaswani, 2017).

In healthcare, AI technologies are revolutionizing diagnostics and patient care. For example, Google Health's application of AI for breast cancer screening has demonstrated the potential to improve the accuracy of diagnostics. An AI model was trained on thousands of mammograms, and it outperformed human radiologists in identifying breast cancer, showcasing a reduction in false negatives and false positives (McKinney, 2020) . This case not only highlights the proficiency of AI in medical imaging but also underscores the importance of ethical considerations, such as patient data privacy and the need for transparent AI systems that healthcare professionals can trust.

The automotive industry has also seen transformative changes with the integration of AI in autonomous vehicles. AI-driven systems are now crucial in powering self-driving cars, where they process real-time data from vehicle sensors and cameras to navigate safely. Tesla's Autopilot and Waymo's autonomous vehicles are prominent examples where AI algorithms are continuously refined to handle complex driving scenarios, thus enhancing passenger safety (Brandom, 2018). These advancements raise ethical questions regarding decision-making in critical situations, accountability for accidents, and the impact on employment in transportation sectors, pushing for a reevaluation of regulatory and ethical frameworks.

The COVID-19 pandemic significantly accelerated the adoption and integration of AI and related technologies across various sectors. AI played a pivotal role in mitigating the impact of the pandemic by supporting critical functions in healthcare, public safety, and beyond. For instance, AI-driven robots and drones were deployed to disinfect public spaces, deliver essential supplies, and monitor compliance with health guidelines, thereby reducing human contact and limiting virus spread (Afaq & Gaur, 2021). In hospitals, AI systems facilitated the rapid analysis of medical data, enhancing the efficiency and accuracy of diagnostics and treatment planning. The pandemic also saw the rise of telemedicine and AI-powered virtual assistants, which helped healthcare providers manage patient care remotely.

The tourism and hospitality industries experienced unprecedented disruptions during the pandemic, prompting the adoption of AI technologies to adapt to new challenges. AI was leveraged to personalize customer experiences, offering tailored recommendations and services based on individual preferences and past behaviors. For example, AI-driven chatbots and virtual assistants provided 24/7 customer support, assisting with booking modifications, answering queries, and suggesting travel itineraries. This personalization not only improved customer satisfaction but also helped businesses retain customer loyalty during uncertain times (Afaq, Gaur, & Singh, 2023).

The integration of social media with Customer Relationship Management (CRM) systems has revolutionized the hospitality industry. Social CRM (SCRM) strategies enabled hotels and restaurants to engage with customers on social media platforms, fostering stronger relationships and enhancing customer loyalty. During the pandemic, SCRM proved particularly valuable, as it allowed businesses to monitor customer sentiment and quickly respond to service issues, thereby maintaining trust and connection with their clientele. The data gathered from social media interactions provided valuable insights into customer preferences and expectations, enabling businesses to tailor their offerings and communication strategies effectively (Afaq, Gaur, & Singh, Social CRM: linking the dots of customer service and customer loyalty during COVID-19 in the hotel industry, 2023) (Gaur & Afaq, 2020).

AI's influence on consumer behavior has been profound, particularly with the advent of autonomous technologies in retail and shopping. Autonomous shopping systems, such as AI-powered checkout-free stores and personalized shopping as-sistants, have streamlined the shopping experience by offering convenience and efficiency. These technologies have changed consumer expectations, with many customers now seeking quick, personalized, and contactless shopping experiences. The pandemic further accelerated this shift, as concerns over health and safety prompted more consumers to embrace digital and automated shopping solutions (Sharma, Singh, Gaur, & Afaq, 2022).

The concept of the metaverse—an interconnected digital environment that blends virtual and augmented reality—is beginning to influence the healthcare sector. AI-driven immersive technologies within the metaverse have the potential to transform patient care by offering new avenues for treatment, training, and patient engagement. For example, virtual reality (VR) can be used for pain management, rehabilitation, and mental health therapy, providing immersive experiences that are therapeutic and engaging. The metaverse also offers platforms for medical education, allowing healthcare professionals to practice complex procedures in a simulated environment, thus enhancing their skills without the risk associated with real-life practice (Gaur, Gaur, & Afaq, Ethical Considerations in the Use of the Metaverse for Healthcare., 2024).

AI's ability to analyze large datasets has been instrumental in understanding customer sentiments and improving service quality in the hospitality industry. By applying machine learning techniques such as sentiment analysis and topic modeling, businesses can extract valuable insights from online reviews and feedback. This analysis helps identify areas for improvement and track changes in customer satisfaction over time. For instance, AI-driven tools can highlight common complaints or praise points, enabling businesses to address issues promptly and refine their service offerings accordingly (Afaq, Singh, Gaur, & Kapoor, 2023).

AI is increasingly being recognized for its potential to address environmental challenges and promote sustainability. AI systems can optimize energy usage in buildings, predict and manage waste, and support the development of more sustainable supply chains. Additionally, AI-powered models can analyze and predict environmental changes, helping policymakers and businesses make informed decisions about resource management and environmental conservation. The use of AI in monitoring carbon emissions and identifying opportunities for reduction is particularly noteworthy, as it supports efforts to combat climate change and transition to a more sustainable future (Gaur, Afaq, Arora, & Khan, 2023).

During the COVID-19 pandemic, AI-driven CRM systems played a crucial role in maintaining customer loyalty and service quality. By leveraging AI for data analysis, businesses were able to gain deeper insights into customer behavior and preferences, allowing for more personalized and timely communication. AI tools also facilitated the automation of customer support and engagement processes, ensuring that businesses could respond quickly to customer needs despite the disruptions caused by the pandemic. This ability to maintain a high level of service and engagement was key to retaining customer trust and loyalty during a challenging period (Afaq, Gaur, & Singh, A latent dirichlet allocation technique for opinion mining of online reviews of global chain hotels, 2022).

Furthermore, the ethical implications of AI have led to the development of specific frameworks aimed at guiding the responsible use of AI. Initiatives like the EU's Ethics Guidelines for Trustworthy AI advocate for AI systems that are lawful, ethical, and robust, ensuring they are developed and deployed in a way that respects human rights and democratic values (High-Level Expert Group on AI, 2019). Such frameworks are becoming essential as AI technologies become more prevalent in society.

PRESSING ETHICAL ISSUES

One of the most pressing ethical issues in the realm of generative AI is algorithmic bias, which can perpetuate and even exacerbate existing inequalities in society (King, 2024). Bias in AI algorithms arises when the data used for training AI systems reflect historical prejudices or when the design of these systems fails to account for diverse user groups. This issue has been highlighted in several high-profile cases, such as the 2019 study by Buolamwini and Gebru, which revealed significant racial and gender bias in commercial facial recognition systems, leading to erroneous and often discriminatory outcomes (Buolamwini & Gebru, 2019). Such biases not only undermine the fairness and inclusivity of AI applications but also pose severe implications for individuals who are misidentified or unfairly targeted by biased AI systems. These issues are critical as they directly affect the trust and reliability placed in AI technologies across various sectors, from law enforcement to hiring practices.

Privacy concerns with AI data usage also pose significant ethical challenges. The vast amounts of data required to train AI systems often include sensitive personal information, raising issues about consent, data protection, and the potential for misuse. The case of Cambridge Analytica, where data from millions of Facebook users was used without consent to target political advertising, illustrates the potential for AI to be used in manipulative ways that infringe on individual privacy rights (Cadwalladr & Graham-Harrison, 2018). Such incidents have prompted urgent calls for stringent regulatory measures to ensure data privacy and security in AI operations.

The importance of addressing these ethical concerns is also recognized in ongoing legislative developments aimed at creating frameworks to govern AI. The European Union's proposed Artificial Intelligence Act is one such initiative, which seeks to impose strict requirements on high-risk AI applications to ensure they are safe, transparent, and free from bias (European Commission, 2021). Similarly, in the United States, the Algorithmic Accountability Act of 2022 (Wyden, 2022) was introduced to require companies to evaluate and fix flawed algorithms that result in inaccurate, unfair, biased, or discriminatory decisions affecting consumers. The

US Department of Defense adopted five core principles for AI in 2020 aiming for responsible, equitable, traceable, reliable, and governable AI (US Department of Defense, 2020) (Oniani, 2023).

These legislative efforts are crucial in setting global standards for AI ethics and ensuring that technological advancements do not come at the cost of fundamental human rights. By linking these ethical issues with both real-world impacts and legislative responses, it is evident that the integration of ethical considerations into AI development is not just beneficial but essential for fostering innovation that is equitable, accountable, and aligned with societal values.

SYNTHESIZING INSIGHTS FROM ETHICAL DILEMMAS IN GENERATIVE AI

Generative AI poses challenges to various fields such as art, literature, journalism, entertainment, and the creation of deepfakes. An AI development and use strategy is necessary to address these ethical issues. The capacity of artificial intelligence to generate content poses a challenge to traditional notions of creativity, necessitating the adjustment of intellectual property rights frameworks. Transparency and accountability are necessary in ethics, particularly in journalism, where trustworthiness is crucial for maintaining media integrity. Authenticity in storytelling is of utmost importance in the entertainment industry, necessitating the ability to distinguish AI-generated content and establish guidelines for presenting truthful information.

The identification of synthetic media and the establishment of legal frameworks are necessary to combat the proliferation of deepfakes and misinformation. It is increasingly crucial to educate the public on how to differentiate AI-generated content.

There are other significant issues with AI generated content (Elliott, 2022):

- AI systems can perpetuate existing biases present in their training data, leading to discriminatory outcomes.
- AI technologies often collect and analyze vast amounts of personal data. This raises questions about user privacy and data security, including how data is collected, stored, and used.
- AI's ability to influence social behavior and public opinion through platforms like social media has led to concerns about the potential for manipulation and the undermining of democratic processes.
- There is a pressing need for AI systems to be transparent in their decision-making processes and for developers to be accountable for the outcomes of AI systems.

These ethical considerations underscore the need for comprehensive regulatory frameworks and proactive governance to ensure AI technologies are developed and used in socially responsible and ethically sound ways (Zlateva, Steshina, Petukhov, & Velev, 2024). Drawing insights from these case studies leads to the following suggestions, for addressing the challenges posed by generative AI:

- First and foremost, it is essential to maintain a conversation, among AI creators, ethics experts, legal professionals, policymakers and the general public to collectively establish and enhance norms. This dialogue should be inclusive gathering viewpoints to shape guidelines that're fair and culturally aware.
- Secondly ethical education and awareness should be integrated into the training for both AI developers and users. Understanding the consequences of AI technologies should be just as important as mastering the aspects. Furthermore, ongoing public education efforts are vital to provide society with the knowledge needed to interact with AI generated materials. Enhancing understanding of AI through education and public awareness campaigns is vital for informed public discourse and policymaking. This includes educating policymakers about the technical aspects of AI. Similar to environmental impact assessments, AI impact assessments could be mandated before the deployment of new AI technologies. This would help in identifying potential risks and mitigation strategies early in the development phase (Elliott, 2022).
- Promoting research on the legal and social implications of generative AI is crucial, as it can help create frameworks and advise policymakers on implementing regulations that balance innovation and societal well-being. Existing laws may not adequately address AI's new challenges, necessitating the development of new legal statutes or adaptations. (Elliott, 2022).
- Fostering AI development requires establishing enforcement mechanisms, including ethics review boards, audits, and transparent industry standards. International cooperation is crucial for creating standardized regulations to manage AI's global implications, ensuring consistency and effectiveness in AI development. (Elliott, 2022).
- The rapid advancement of AI technology necessitates a commitment to reassessing and adapting ethical guidelines to ensure they remain relevant and effective in addressing risks. Regulatory frameworks should be flexible, involving iterative policy-making and regular updates based on new information and technologies, ensuring ethical considerations remain relevant and effective. (Elliott, 2022)

The study of AI reveals both obstacles and opportunities for advancement. Addressing these challenges can guide the development of generative AI technologies towards groundbreaking, morally sound, and beneficial innovations. Real-world examples can lay the foundation for a future where AI enhances experiences without compromising principles. It is by discussion, collaborative endeavors and a dedication, to upholding standards that we can navigate the intricacies of this emerging field fostering an AI enhanced world that is fair, inclusive and respectful of all individuals.

REFERENCES

Afaq, A., & Gaur, L. (2021). The Rise of Robots to Help Combat Covid-19. *International Conference on Technological Advancements and Innovations (ICTAI)*. IEEE. DOI:10.1109/ICTAI53825.2021.9673256

Afaq, A., Gaur, L., & Singh, G. (2022). A latent dirichlet allocation technique for opinion mining of online reviews of global chain hotels. *3rd International Conference on Intelligent Engineering and Management (ICIEM)* (pp. 201-206). IEEE. DOI:10.1109/ICIEM54221.2022.9853114

Afaq, A., Gaur, L., & Singh, G. (2023). A trip down memory lane to travellers' food experiences. *British Food Journal*, 125(4), 1390–1403. DOI:10.1108/BFJ-01-2022-0063

Afaq, A., Gaur, L., & Singh, G. (2023). Social CRM: Linking the dots of customer service and customer loyalty during COVID-19 in the hotel industry. *International Journal of Contemporary Hospitality Management*, 35(3), 992–1009. DOI:10.1108/IJCHM-04-2022-0428

Afaq, A., Singh, G., Gaur, L., & Kapoor, S. (2023). Aspect-Based Opinion Mining of Customer Reviews in the Hospitality Industry: Leveraging Recursive Neural Tensor Network Algorithm. *3rd International Conference on Technological Advancements in Computational Sciences (ICTACS)* (pp. 1392-1397). IEEE. DOI:10.1109/ICTACS59847.2023.10390384

Ajder, H., Patrini, G., Cavalli, F., & Cullen, L. (2019, September). *The State of Deepfakes: Landscape, Threats, and Impact.* Retrieved from https://regmedia.co.uk/2019/10/08/deepfake_report.pdf

Austin, D. (2023, May 16). *AI music could revolutionize the industry — and this artist is leading the way.* Retrieved April 30, 2024, from Business Insider India: https://www.businessinsider.in/cryptocurrency/news/ai-music-could-revolutionize-the-industry-and-this-artist-is-leading-the-way/articleshow/100285845.cms

Baraniuk, C. (2018, November 8). *China's Xinhua agency unveils AI news presenter.* Retrieved April 30, 2024, from BBC: https://www.bbc.com/news/technology-46136504

Boden, M. A. (1998). Creativity and artificial intelligence. *Artificial Intelligence*, 103(1-2), 347–356. DOI:10.1016/S0004-3702(98)00055-1

Bostrom, N., & Yudkowsky, E. (2014). The ethics of artificial intelligence. In *The Cambridge Handbook of Artificial Intelligence* (pp. 316–334). Cambridge University Press. DOI:10.1017/CBO9781139046855.020

Brandom, R. (2018, July 3). *Self-driving cars are headed toward an AI roadblock.* Retrieved from The Verge: https://www.theverge.com/2018/7/3/17530232/self -driving-ai-winter-full-autonomy-waymo-tesla-uber

Brown, M. (2016, April 5). *'New Rembrandt' to be unveiled in Amsterdam.* Retrieved April 30, 2024, from The Guardian: https://www.theguardian.com/artanddesign/ 2016/apr/05/new-rembrandt-to-be-unveiled-in-amsterdam

Bryson, J. (2019). *The Past Decade and Future of AI's Impact on Society.* Retrieved from Openmind BBVA: https://www.bbvaopenmind.com/en/articles/the-past-decade -and-future-of-ais-impact-on-society/

Buolamwini, J., & Gebru, T. (2019). Gender Shades: Intersectional Accuracy Disparities in Commercial Gender Classification. *Proceedings of the 1st Conference on Fairness, Accountability and Transparency* (pp. 77-91). New York City: MLR Press.

Cadwalladr, C., & Graham-Harrison, E. (2018, March 7). *Revealed: 50 million Facebook profiles harvested for Cambridge Analytica in major data breach.* Retrieved from The Guardian: https://www.theguardian.com/news/2018/mar/17/cambridge -analytica-facebook-influence-us-election

Cahlan, S. (2020, February 13). *How misinformation helped spark an attempted coup in Gabon.* Retrieved April 30, 2024, from The Washington Post: https://www .washingtonpost.com/politics/2020/02/13/how-sick-president-suspect-video-helped -sparked-an-attempted-coup-gabon/

Chan, C. D. (2022, October 4). *Are Virtual Influencers the Real Deal.* Retrieved from The Hollywood Reporter: https://www.hollywoodreporter.com/business/digital/ virtual-influencers-digital-world-1235228125/

Chesney, B., & Citron, D. (2019). Deep Fakes: A Looming Challenge for Privacy, Democracy, and National Security. *California Law Review*, ●●●, 1753–1820.

Cohn, G. (2018, October 25). *AI Art at Christie's Sells for $432,500.* Retrieved April 30, 2024, from The New York Times: https://www.nytimes.com/2018/10/25/arts/ design/ai-art-sold-christies.html

Cortes, C., & Vapnik, V. (1995). Support-vector networks. *Machine Learning*, 20(3), 273–297. DOI:10.1007/BF00994018

Damiani, J. (2019, September 3). *A Voice Deepfake Was Used To Scam A CEO Out Of $243,000*. Retrieved April 30, 2024, from Forbes: https://www.forbes.com/sites/jessedamiani/2019/09/03/a-voice-deepfake-was-used-to-scam-a-ceo-out-of-243000/?sh=2354d82f2241

Diakopoulos, N. (2019). *Automating the news-how algorithms are rewriting the media*. Harvard University Press. DOI:10.4159/9780674239302

Du, W., & Han, Q. (2021). Research on application of artificial intelligence in movie industry. *Proc. SPIE 12076, 2021 International Conference on Image, Video Processing, and Artificial Intelligence,*DOI:10.1117/12.2619500

Elliott, A. (2022). *The Routledge Social Science Handbook of AI*. Taylor & Francis.

Eubanks, V. (2018). *Automating Inequality: How High-Tech Tools Profile, Police, and Punish the Poor*. St. Martin's Press.

European Commission. (2021, April 21). *Proposal for a Regulation laying down harmonised rules on artificial intelligence*. Retrieved from European Commission: https://digital-strategy.ec.europa.eu/en/library/proposal-regulation-laying-down-harmonised-rules-artificial-intelligence

Gaur, L., & Afaq, A. (2020). Metamorphosis of CRM: incorporation of social media to customer relationship management in the hospitality industry. In *Handbook of Research on Engineering Innovations and Technology Management in Organizations* (pp. 1–23). IGI. DOI:10.4018/978-1-7998-2772-6.ch001

Gaur, L., Afaq, A., Arora, G. K., & Khan, N. (2023). Artificial intelligence for carbon emissions using system of systems theory. *Ecological Informatics*, 76, 102165. DOI:10.1016/j.ecoinf.2023.102165

Gaur, L., Gaur, D., & Afaq, A. (2024). Ethical Considerations in the Use of the Metaverse for Healthcare. In *Metaverse Applications for Intelligent Healthcare* (pp. 248–273). IGI Global.

Global Information & Communications Team. (2023). *How Generative AI is Transforming Digital Content Services (Report No K88D-70)*. Frost & Sullivan.

Goodfellow, I. J. (2014). *Generative Adversarial Nets. Proceedings of Advances in Neural Information Processing Systems 2014*. Curran Associates.

Gozalo-Brizuela, R., & Garrido-Merchán, E. C. (2023, June 14). *A survey of Generative AI Applications*. Retrieved from ARXIV: https://arxiv.org/abs/2306.02781

Graves, L. (2016). *Deciding what's true: the rise of political fact-checking in American journalism*. Columbia University Press. DOI:10.7312/grav17506

Gupta, N. (2021, June 7). Retrieved April 30, 2024, from World Association of News Publishers: https://wan-ifra.org/2021/06/how-bbc-news-labs-uses-ai-powered-content-automation-to-engage-young-audiences/

Hao, K. (2019, May-June). The biggest threat of deepfakes isn't the deepfakes themselves. *MIT Technology Review*.

High-Level Expert Group on AI. (2019, April 8). *Ethics guidelines for trustworthy AI*. Retrieved from European Commission: https://digital-strategy.ec.europa.eu/en/library/ethics-guidelines-trustworthy-ai

Hornigold, T. (2018, October 25). *The First Novel Written by AI Is Here—and It's as Weird as You'd Expect It to Be*. Retrieved April 30, 2024, from Singularityhub: https://singularityhub.com/2018/10/25/ai-wrote-a-road-trip-novel-is-it-a-good-read/

Irani, S. (2023, March 19). *This is not the Luke Skywalker you're looking for*. Retrieved April 30, 2024, from The Michigan Daily: https://www.michigandaily.com/arts/b-side/this-is-not-the-luke-skywalker-youre-looking-for/

Jigsaw (Google). (2021, February 8). *Google's Jigsaw Announces Toxicity-Reducing API, Perspective, is Processing 500M Requests Daily*. Retrieved April 30, 2024, from PR Newswire: https://www.prnewswire.com/news-releases/googles-jigsaw-announces-toxicity-reducing-api-perspective-is-processing-500m-requests-daily-301223600.html

Kietzmann, J., Lee, L., McCarthy, I. P., & Kietzmann, T. C. (2020). Deepfakes: Trick or treat? *Business Horizons*, 63(2), 135–146. DOI:10.1016/j.bushor.2019.11.006

King, D. (2024). *Legal & Humble AI: Addressing the Legal, Ethical, and Societal Dilemmas of Generative AI*. Ingene Publications.

Kingma, D. P., & Welling, M. (2013, December 20). *Auto-Encoding Variational Bayes*. Retrieved from ARXIV: https://arxiv.org/abs/1312.6114

Krizhevsky, A., Sutskever, I., & Hinton, G. E. (2012). *ImageNet Classification with Deep Convolutional Neural Networks*. Advances in Neural Information Processing Systems. Curran Associates.

Lewis, S. C., Guzman, A. L., & Schmidt, T. R. (2019). Automation, Communication: Rethinking Roles and Relationships of Humans and Machines in News. *Digital Journalism (Abingdon, England)*, ●●●, 409–427. DOI:10.1080/21670811.2019.1577147

Li, Y. (2022, August 26). Research on the Application of Artificial Intelligence in the Film. *2022 International Conference on Science and Technology Ethics and Human Future (STEHF 2022)*, 1-6. EDP Sciences. DOI:10.1051/shsconf/202214403002

Lipton, Z. C. (2018). The Mythos of Model Interpretability: In machine learning, the concept of interpretability is both important and slippery. . *Queue*, 31-57.

McCormack, J., & D'Inverno, M. (2012). *Computers and Creativity*. Springer-Verlag. DOI:10.1007/978-3-642-31727-9

McKinney, S. M., Sieniek, M., Godbole, V., Godwin, J., Antropova, N., Ashrafian, H., Back, T., Chesus, M., Corrado, G. S., Darzi, A., Etemadi, M., Garcia-Vicente, F., Gilbert, F. J., Halling-Brown, M., Hassabis, D., Jansen, S., Karthikesalingam, A., Kelly, C. J., King, D., & Shetty, S. (2020). International evaluation of an AI system for breast cancer screening. *Nature*, 577(7788), 89–94. DOI:10.1038/s41586-019-1799-6 PMID:31894144

Mervosh, S. (2019, May 24). *Distorted Videos of Nancy Pelosi Spread on Facebook and Twitter, Helped by Trump*. Retrieved April 30, 2024, from The New York Times: https://www.nytimes.com/2019/05/24/us/politics/pelosi-doctored-video.html

Metz, R. (2021, August 6). *How a deepfake Tom Cruise on TikTok turned into a very real AI company*. Retrieved April 30, 2024, from CNN Business: https://edition.cnn.com/2021/08/06/tech/tom-cruise-deepfake-tiktok-company/index.html

Miranda, E. R. (2021). *Handbook of Artificial Intelligence for Music-Foundations, Advanced Approaches, and Developments for Creativity*. Springer.

Noble, S. U. (2018). *Algorithms of Oppression: How Search Engines Reinforce Racism*. NYU Press. DOI:10.18574/nyu/9781479833641.001.0001

O'Neil, C. (2016). *Weapons of Math Destruction: How Big Data Increases Inequality and Threatens Democracy*. Crown Publishing Group.

Oniani, D., Hilsman, J., Peng, Y., Poropatich, R. K., Pamplin, J. C., Legault, G. L., & Wang, Y. (2023). Adopting and expanding ethical principles for generative artificial intelligence from military to healthcare. *NPJ Digital Medicine*, 6(1), 225. DOI:10.1038/s41746-023-00965-x PMID:38042910

Oremus, W. (2014, March 17). *The First News Report on the L.A. Earthquake Was Written by a Robot*. Retrieved April 30, 2024, from Slate.com: https://slate.com/technology/2014/03/quakebot-los-angeles-times-robot-journalist-writes-article-on-la-earthquake.html

Parikh, P. (2019). *AI Film Aesthetics: A Construction of a New Media Identity for AI Films*. Chapman University. doi., https://digitalcommons.chapman.edu/film_studies_theses/8/

Paris, B., & Donovan, J. (2019, September 18). *Deepfakes and Cheapfakes- the manipulation of audio and video evidence.* Retrieved from Data & Society's Media Manipulation research: https://datasociety.net/library/deepfakes-and-cheap-fakes/

Peiser, J. (2019, February 5). *The Rise of the Robot Reporter.* Retrieved April 30, 2024, from The New York Times: https://www.nytimes.com/2019/02/05/business/media/artificial-intelligence-journalism-robots.html

Robertson, E. G. (2022). *Video Games and Politics: An exploratory analysis of how narrative frames in video games can influence political perception.* Auckland, New Zealand: The University of Auckland. Retrieved April 30, 2024, from https://researchspace.auckland.ac.nz/bitstream/handle/2292/60892/Robertson-2022-thesis.pdf?sequence=4&isAllowed=y

Rosenblatt, F. (1958). The Perceptron: A Probabilistic Model for Information Storage and Organization in the Brain. *Psychological Review*, 65(6), 386–408. DOI:10.1037/h0042519 PMID:13602029

Rumelhart, D. E., Hinton, G. E., & Williams, R. J. (1986). Learning representations by back-propagating errors. *Nature*, 323(6088), 533–536. DOI:10.1038/323533a0

Russell, S., & Norvig, P. (2020). *Artificial Intelligence: A Modern Approach.* Pearson.

Sautoy, M. d. (2019). *The Creativity Code-Art and Innovation in the age of AI.* Harvard University Press. DOI:10.2307/j.ctv2sp3dpd

Saxena, T. (2023, November 26). *The deepfake gender paradox in AI's moral maze.* Retrieved April 30, 2024, from Deccan Herald: https://www.deccanherald.com/specials/the-deepfake-gender-paradox-in-ai-s-moral-maze-2783639

Sharma, S., Singh, G., Gaur, L., & Afaq, A. (2022). Exploring customer adoption of autonomous shopping systems. *Telematics and Informatics*, 73, 101861. DOI:10.1016/j.tele.2022.101861

Singh, H., Kaur, K., & Singh, P. P. (2023). Artificial Intelligence as a facilitator for Film Production Process. *2023 International Conference on Artificial Intelligence and Smart Communication (AISC)*, (pp. 969-972). DOI:10.1109/AISC56616.2023.10085082

US Department of Defense. (2020, February 24). *DOD Adopts Ethical Principles for Artificial Intelligence.* Retrieved from US Department of Defense: https://www.defense.gov/News/Releases/Release/Article/2091996/dod-adopts-ethical-principles-for-artificial-intelligence/

Vaswani, A. e. (2017). Attention is All You Need. *Proceedings of 31st Conference on Neural Information Processing Systems.* Glasgow: Curran Associates.

Wyden, R. (2022, March 2). *Algorithmic Accountability Act of 2022.* Retrieved from Congress.gov: https://www.congress.gov/bill/117th-congress/senate-bill/3572

Zlateva, P., Steshina, L., Petukhov, I., & Velev, D. (2024). A Conceptual Framework for Solving Ethical Issues in Generative Artificial Intelligence. In *Electronics, Communications and Networks* (Vol. 381, pp. 110–119). IOS Press. DOI:10.3233/FAIA231182

Zuboff, S. (2019). *The Age of Surveillance Capitalism: The Fight for a Human Future at the New Frontier of Power.* PublicAffairs.

Chapter 8
Education in the Era of Generative AI:
Understanding the Benefits, Ethics, and Challenges

Suvidha Agarwal
https://orcid.org/0009-0008-7055-2800
JIMS Engineering Management Technical Campus, Greater Noida, India

Preeta Rajiv Sivaraman
JIMS Engineering Management Technical Campus, Greater Noida, India

ABSTRACT

Gen artificial intelligence (AI) is revolutionizing the higher education sector by employing deep learning models to generate information that closely mimics human content. However, its introduction into educational settings raises questions about factors like academic integrity, moral ethics and potential impacts on critical thinking skills. The emergence of generative artificial intelligence (AI) may seem to be a reason for concern which has led to swift prohibitions by organizations and educational agencies. The chapter proposes to put in detail the benefits of using Generative AI by the stakeholders of the education sector, along with the ethics and values to be taken care of and thereby avoiding the reasons to negate the usage of the AI. This chapter will cover the use of Generative AI in different educational settings, the tools and methods employed, the efficiency of GAI in enhancing teaching and learning, the influence on student outcomes, and the possible drawbacks and moral dilemmas related to its application.

DOI: 10.4018/979-8-3693-9173-0.ch008

1. INTRODUCTION

In the realm of machine learning, generative artificial intelligence (AI) is a revolutionary paradigm that offers capabilities not found in conventional methods. Fundamentally, generative AI uses algorithms to produce novel, creative material that closely mimics the structures and patterns of the training data it has been exposed to. Generative AI goes beyond traditional AI models for categorization or prediction tasks, which depend on pre-existing datasets. Instead, it creates content on its own, be it text, photos, music, or even videos.

We have entered a new era where artificial intelligence will become more and more prevalent in many aspects of daily life. The ability to quickly and efficiently create original information, including natural language, is one of the key features of this new era. Large language models (LLMs) are a subset of foundation models that are trained on enormous amounts of text data using neural networks. (Farrelly, 2023).

To fulfill the changing demands of education and stay up to date with the newest technical advancements, generative artificial intelligence technologies, such as the ChatGPT conversational agent, GPT-4, Open Assistant, DALL-E, Midjourney are promising advancements by giving students individualized feedback and explanations and building lifelike virtual simulations for practical learning. ChatGPT can provide tailored and efficient learning experiences. But it's also critical to take into account this technology's limits. The effectiveness of ChatGPT and other generative AI systems depends on the quality of their training data, which might lead to the creation and dissemination of false or prejudiced information. Thinking about AI means considering how education may help people get ready for a world that is changing quickly and will involve this technology in many facets of life, including employment, education, gaming, and interpersonal interactions. Thus, it is essential to comprehend AI's operation as well as the pros and cons of using it.

1.1 Key Domains

Figure 1. Key domains of generative AI In education (Cooper, 2023)

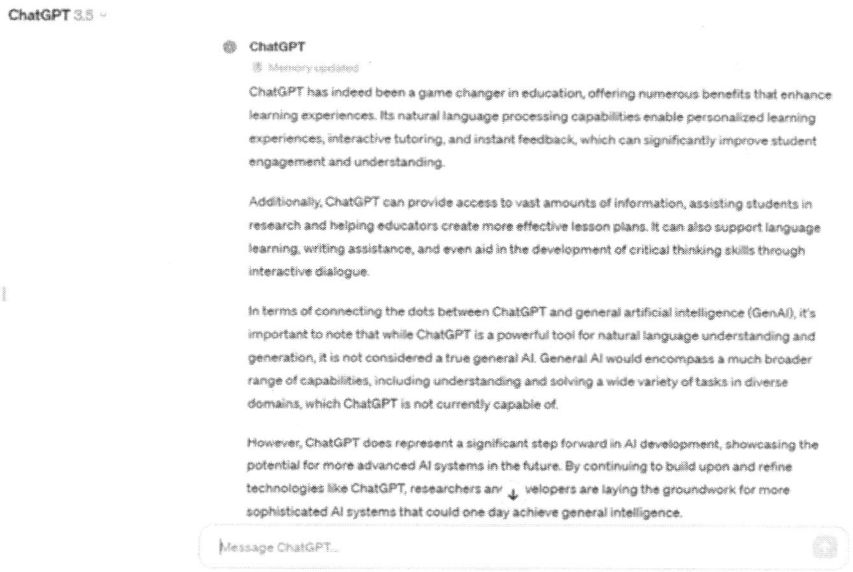

ChatGPT 3.5

⚙ **ChatGPT**

⟳ Memory updated

ChatGPT has indeed been a game changer in education, offering numerous benefits that enhance learning experiences. Its natural language processing capabilities enable personalized learning experiences, interactive tutoring, and instant feedback, which can significantly improve student engagement and understanding.

Additionally, ChatGPT can provide access to vast amounts of information, assisting students in research and helping educators create more effective lesson plans. It can also support language learning, writing assistance, and even aid in the development of critical thinking skills through interactive dialogue.

In terms of connecting the dots between ChatGPT and general artificial intelligence (GenAI), it's important to note that while ChatGPT is a powerful tool for natural language understanding and generation, it is not considered a true general AI. General AI would encompass a much broader range of capabilities, including understanding and solving a wide variety of tasks in diverse domains, which ChatGPT is not currently capable of.

However, ChatGPT does represent a significant step forward in AI development, showcasing the potential for more advanced AI systems in the future. By continuing to build upon and refine technologies like ChatGPT, researchers and developers are laying the groundwork for more sophisticated AI systems that could one day achieve general intelligence.

Message ChatGPT...

(a) *Learning*: Personalized task assignment and feedback analysis of student effort, facilitating communication between humans and machines, and boosting flexibility and interaction in digital settings are various modes of learning.

(b) *Teaching*: Through constant dialogue, AI chatbots assist students in honing their communication skills and can respond to their inquiries or requests in a human-like way. To engage with students, ChatGPT could use question-and-answer or answer-and-question formats, which are essentially a series of inquiries along an emerging line of inquiry. It adapts to the demands of the students and creates a more positive learning environment by using the answers and queries from the students to clarify their concerns.

(c) *Assessment:* How AI forecasts student achievement and generates automatic markings. Individuals with or without benchmark disabilities must have access to a personalized system that includes self-assessment, learning objects, and reports that appropriately reflect their preferences for using electronic learning materials.

(d) *Administration*: To increase productivity, the management team may utilize ChatGPT for (i) drafting emails and proposals and (ii) data analysis and report generation. Providing easy-to-use and customized services, management teams can submit their cases to ChatGPT and ask it to suggest actions or choices.

When implementing ChatGPT and other generative AI in the classroom, the following theoretical framework can serve as a guide (Su et al., 2023):

1. *Determine the Desired Outcomes:* Before utilizing ChatGPT or any other generative AI in education (also known as "educative AI"), it's critical to determine the application's goals. This guarantees that technology utilization is in line with intended results.
2. *Identify the Right Level of Automation*: Depending on the goals, educational AI may be used to augment conventional teaching techniques or to completely automate the teaching and learning process.
3. *Make Sure Ethical Issues Are Considered*: When employing educational AI, ethical issues including potential biases and how they affect instructors and students need to be thoroughly thought out.
4. *Assess Effectiveness*: It's critical to assess how well educational AI performs in producing the intended results.

2. BACKGROUND

ChatGPT can be described as a large language model that is trained to "generate humanlike text based on a given prompt or context. It can be used for a variety of natural language processing tasks, such as text completion, conversation generation, and language translation" [Baidoo-Anu, D.; Owusu Ansah, L,2023]. Because of its sophisticated generating abilities, one of the main worries in higher education is that it can be used to produce assignments, respond to test questions, and draft academic articles without being readily identified by anti-plagiarism software as it is now available. [Zhai, X,2022]. Higher education institutions (HEIs') reactions to this new threat to academic integrity have been inconsistent and diverse, ranging from those that have hurried to impose complete prohibitions on the use of ChatGPT to others who have begun to use it by disseminating guidelines for students on how to interact with AI ethically and successfully. However, the majority of the information that higher education institutions (HEIs) have so far given students is ambiguous or devoid of specifics regarding the situations in which using ChatGPT is permitted or deemed appropriate. It is clear, though, that the majority of HEIs are actively examining their rules regarding ChatGPT use and its consequences for academic integrity.

Artificial intelligence (AI) that is generative, like ChatGPT, the well-known chatbot from OpenAI, is similar to other AIs in that it relies on reinforcement learning. It provides functions like discussion that were previously only available to people, as well as the ability to create text content that includes written text, audio, images, videos, code, and multimedia. Software programs that use generative artificial intelligence (AI) gather a lot of data and then use the structures and configurations of the training data to create parallel representations of the information. Even though GAIs have been a part of our society for some time, it wasn't until ChatGPT, developed by Open-AI, was made available to the public globally at the end of 2022 that GAIs generated a mass public awareness (Dyer, 2023) and were widely accepted and used by the public on a scale never seen before (Gordon, 2023).

3. AI LITERACY FRAMEWORK

AI literacy is a relatively recent development. The field of AI research is still in, especially considering that tools like ChatGPT have only been available to the public since November 2022. Before the launch of ChatGPT and GenAI, most of the work completed to date was published between 2019 and 2022. A widely recognized framework or definition of AI literacy is still missing, despite all of these academic endeavors seeking to create a strong basis for instructing and assessing this literacy (Elkhatat, 2023). One definition defines AI literacy as "a set of competencies that enables individuals to critically evaluate AI technologies, communicate and collaborate with AI effectively, and use AI as a tool online, at home, and in the workplace." They developed a conceptual framework to offer important criteria for the development of AI literacy, which gave birth to this term.

Table 1. Various generative AI tools for educational purposes

Purpose	Functions	Tools
Generating Text	Chatbot	Zoho, DeepAI
	Content Creation	OpenAI, Jasper
	Office Purpose	Workspace Google, Microsoft 365
	Search Engine	Bing, Perplexity

continued on following page

Table 1. Continued

Purpose	Functions	Tools
Generating an Image	For Image	Dall E3, MidJourney
	For Graph	GraphMaker, Bing AI
	For Presentation	Visme, Slidesgo
Generating a Video	For Video	In the video, Zapier
Generating an Audio	For Audio	Murf, Speechify
	For Voice Modulation	Voicemod
Generating SourceCode	For Coding	Chatgpt, Github copilot
Generation of text through AI	For Plagiarism check	Turnitin

Source: Created by Authors

4. GENERATIVE AI'S IMPLICATIONS AND USES IN EDUCATION

When a potentially disruptive trend, like Generative AI, emerges, it is important to understand the technology's potential and limitations without being influenced by extremes. This is especially true when using technology in teaching and learning processes to make well-informed decisions. Therefore, a prospective study focusing on the tools already available with potential educational uses and the contributions that have emerged from academia initially since the tsunami caused by ChatGPT is being discussed before highlighting the implications of this technology in the educational context (Cooper, 2023).

Benefits And Possible Applications of Generative AI in the Classroom:

Below given table focuses on some of the application areas of Generative AI:

Table 2. Application areas of AI

1	Real-time access to a vast amount of pertinent data that may be processed, summarized and presented later on as though it were human.
2	The creation of large sets of instructional materials (cases, units, rubrics, surveys, etc.) that can protect confidentiality in important situations, such as those involving medical education. The educational landscape has changed due to advancements in generative artificial intelligence.
3	As opposed to traditional media, helpful resources for learning new ideas, such as the capacity to summarize or clarify complicated ideas.
4	Gaining an understanding of context allows for engagement (conversation) with these instruments, facilitating the acquisition of self-directed answers to queries and more efficient learning about a range of subjects.

continued on following page

Table 2. Continued

5	Fostering critical thinking and creativity by allowing students to challenge their beliefs and get feedback on their assignments.
6	Assisting students with tedious work so they may concentrate on the material and become more analytical in their learning.
7	Encouraging the early stages of concept formation and critical thinking.
8	Offering a platform for asynchronous communication, which boosts involvement and encourages student cooperation.
9	Permitting customized education.
10	Assisting individuals, students in general to get more control over their writing abilities.
11	Working as an online teaching assistant.
12	Acting as instruments for ongoing, unstructured education.
13	Encouraging the improvement of language abilities.
14	Increasing teacher productivity by the effectiveness of the time spent responding to repetitive student inquiries, grading written assignments, etc., so they may concentrate on more difficult jobs like giving students feedback and assistance.
15	Endorsing automated evaluation and other evaluation advancements.

Source: Created by authors

4.1 ChatGPT: A Game Changer in Education / Connecting the Dots of GenAI

ChatGPT was put forth the same question. Excerpts from the reply given by ChatGPT:

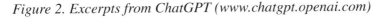

Figure 2. Excerpts from ChatGPT (www.chatgpt.openai.com)

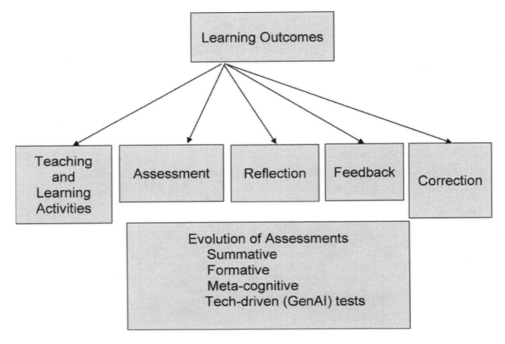

The dawn of the era of Generative AI represents a significant shift in the way knowledge is acquired, transforming it into dynamic, networked forms. This potent combination of artificial intelligence (AI) and human intelligence (HI) creates a special synergy in which AI not only mimics HI but also helps HI make connections between seemingly unrelated pieces of information to form a thorough, nuanced understanding of complicated concepts.

This new approach highlights AI as an enhancement of human cognitive capacities rather than viewing it as a potential replacement for human cognition. Furthermore, a new type of learning ecology has been created by this enhanced intellectual environment, in which group intelligence is equally essential. The collaborative efforts that yield shared peer or group intelligence enhance the HI-AI dynamic by promoting a range of viewpoints and creating new opportunities for intellectual pursuits. A competent human instructor can enhance the learning process by utilizing the potent instrument of generative AI. Instructors may easily create interesting and dynamic learning modules with it, focusing on targeted feedback and active learning. Metacognitive and reflective learning processes depend on the availability of AI-based real-time feedback. The teaching and learning paradigm will be completely transformed by a human teacher outfitted with generative AI, rather than generative AI replacing human teachers.

Teachers may find it simpler to respond to queries from students when they use ChatGPT. An educational strategy known as "game-based learning" makes use of games and game dynamics to improve learning outcomes. With the help of ChatGPT's virtual teachers and tailored recommendations, students can study more engagingly and entertainingly.

By supporting and facilitating virtual and physical connections inside the current institutional structure for deep involvement in several directions, broadening the network, and creating a new ecosystem of education, ChatGPT can expedite the transition of AI.

By suggesting themes, laying out essay formats, offering ideas, and enhancing academic writing, ChatGPT can help students write essays (Dwivedi et al., 2023).

4.2 Building AI Competencies for Learners

A GenAI and AI chatbot that simulates human speech is called ChatGPT. Midjourney is another GenAI program that has a big influence on art instruction. These programs can serve as instructors, mentors, secretaries, and designers. They have been altering how instructors instruct and evaluate learning outcomes, how students' study, and how institutions have updated their regulations. ChatGPT finds material and curates it in a narrative style to improve student's learning abilities. This could: (i) enable students to learn anything more quickly and in a variety of ways; (ii) support teachers' professional development by offering a variety of viewpoints; and (iii) automate or streamline administrative tasks for institutions.

Students may benefit from a more individualized learning experience, to start. Teachers can consider the trust that exists between them and their pupils as well as the educational content by using ChatGPT. Students can receive more individualized teaching by using ChatGPT to create virtual tutors and personalized recommendations. For example, ChatGPT can provide students with thorough guidance on how to solve math problems if they require help. Students have been able to gain additional soft skills as a result, helping them to overcome their academic obstacles. For some students, this approach was significantly more effective than for others. It was much harder to ponder and reflect if you had problems with your language, communication, fine or gross motor skills, or vision. Written exams are more beneficial to certain students than others. We need to ask ourselves if it's time to do away with the old method of having students write in a set amount of time while seated in an exam room (Gaur. et al., 2024)

Our educational system must completely rethink its methods for examination, testing, and assessment to seamlessly navigate these difficulties and live alongside generative AI.

It is imperative to comprehend the mechanics, applications, and ramifications of artificial intelligence (AI) due to its wide-ranging potential and hazards, which cover factors that control privacy, decision-making, and socio-economic dynamics. As a result, AI literacy is essential for making sense of the intricate details and complexities found in technical systems and algorithms. Knowledge of artificial intelligence (AI) enhances our ability to check, challenge, and direct the development of the digital world in addition to helping us use technology more effectively. People can make better decisions, plan more effectively, and have crucial conversations about the use and regulation of technology if they are aware of what AI can and cannot accomplish. A fundamental grasp of AI is essential as technology gets more and more integrated into our daily lives. AI literacy is becoming increasingly crucial in educational settings. While it is important to recognize that teachers—who act as the primary educators—also require fluency in the use of AI, it is important to focus on giving pupils the fundamental AI skills (Yeralan et al., 2023).

Students need to develop a skill set that is both critical and technical to prepare them for the job markets of the future, which will no doubt be even more intricately woven with technology than the ones that exist today. Students should also be prepared for existing job markets. To ensure that governance and regulatory frameworks effectively address the complexities and ethical dilemmas associated with integrating AI technology in academic contexts, educators and administrators must possess a solid understanding of AI.

Additionally, developing innovative educational experiences, improving institutional processes, and allocating resources strategically all depend on having a firm grasp of AI. This makes it easier for the organization and the local community to adapt to the quick changes that occur in an increasingly digital and artificial intelligence (AI) driven environment. It might be difficult for certain instructors, though, to acquire new capabilities like AI literacy; undoubtedly, the process of changing their perspectives and gaining these abilities may provide its own set of difficulties. Therefore, when addressing these challenges, policymakers should also consider them carefully.

Technology is still going to be a useful tool for education and learning. For a considerable amount of time now, we have been utilizing computers not only for academic purposes but also to assist and enhance our writing. To fix grammatical mistakes and enhance word choice and sentence structure, we have employed spell checkers, grammar checkers, and AI text editors like Grammarly. For computation, statistical analysis, and simulation, students utilize spreadsheets, calculators, and programs like MATLAB and Mathematica. For help writing essays and studying research, they have been rapidly consulting Wikipedia and online resources (blogs, social media, and scholarly publications), supported by potent search engines like Google and Google Scholar (Murugesan et al., 2023). Aside from chatbots, a variety of AI tools and applications may assist with programming, troubleshooting, and establishing meaningful interactions and educational opportunities (Cherukuri, 2021).

4.3 Capacity Building Of Teachers And Researchers On GenAI

Using ChatGPT to create virtual tutors for language learning is one way that it is being used in education. A virtual tutor using ChatGPT may give language learners practice conversations and tailored feedback. A highly personalized learning experience can be had since the virtual coach can adjust to the speed and level of the students. This can be especially useful for students who would rather study at their own pace or who do not have access to in-person language tutors. Artificial intelligence (AI) is a crucial element of this digital revolution, which is drastically altering the way education is delivered and experienced. The application of artificial intelligence (AI) in education has gained traction in recent years, especially in higher education. What impact artificial intelligence (AI) is having on education, and what challenges and drawbacks does it present? Artificial Intelligence is being utilized in numerous educational settings, such as autonomously scoring tests taken by students, forecasting their academic achievements, and providing students with automated instruction and administrative assistance.

ChatGPT can be utilized in the educational sector to make virtual tutors, respond to inquiries from students, and offer individualized learning programs. Additionally, it may be applied in the real world to help educators and learners advance their AI literacy—the capacity to comprehend, apply, and assess AI technologies and their effects on society. With ChatGPT's user-friendly AI interface, educators may use it to help students and instructors become more adept at navigating and interacting with the quickly evolving AI landscape with knowledge and assurance. Significant changes brought about by GenAI will present opportunities as well as challenges for study. To improve their papers, create software for analysis and simulation, discover research gaps, comprehend ideas in other fields they can apply, summarize the literature, and even assist in creating experiments, researchers can benefit from the assistance of GenAI.

5. ASSESSMENTS IN THE AI ERA

Furthermore, the introduction of GenAI promises to bring about yet another major change in the assessment aspect during this continuous progression. The current trend in education is away from focusing solely on information acquisition and towards a more comprehensive, formative process that values feedback and reflection. GenAI can help refine this change. More sophisticated, real-time feedback and an adaptive approach to learning can be enabled by the evaluation process thanks to GenAI's unmatched data-processing power and adaptability. But as educators, we must make sure that tests are based on real-world experiences, pushing students in ways that encourage sincere introspection, evaluate their work, and promote a more in-depth, metacognitive approach to learning. This is especially important when we are utilizing the power of GenAI (Afaq, 2023).

The difficulty in using GenAI in assessment design or redesign is to make the most of its capabilities while maintaining the emphasis on deep, meaningful learning as the technology becomes more and more incorporated into the educational system. With the coming together of cutting-edge AI technology with conventional pedagogies, the future of assessment is rapidly approaching, and educators are being invited to enter this new area with caution and enthusiasm.

Figure 3. Collaborating the outcome based education with Genai assessments (Created by authors)

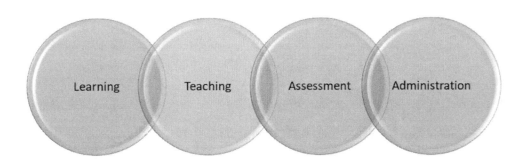

The results of some of the analyses show that participants' opinions regarding the significance of various formative and summative assessment methods were altered by ChatGPT. Discussions in the focus groups covered seven formative assessment strategies. The activities included: (1) reviewing material from class with homework assignments; (2) observing in-class activities; (3) journaling; (4) impromptu Q&A sessions; (5) teacher-student meetings; (6) informal student sharing and presentations; and (7) student self-evaluation of performance and learning progress (Yogesh et al.,2023)

6. SOCIAL IMPLICATIONS OF AI (BEYOND WHAT IT SEEMS)

It is best to steer clear of the erroneous notion that technology is neutral and that its development and application are entirely dependent upon humans. Technology defines its goals in addition to serving as a means to them. Thus, the discussion should cover issues like biases, unfair discrimination, inequality, surveillance, technical competencies, information bubbles, exclusion, and algorithmic decision-making, including its potential for manipulation and influence. It should also cover the replacement of humans in posthumanism and transhumanism, as well as all of their interrelationships. Because they affect how people behave, all of these factors are extremely important in education. They have a double effect on education since they show how AI can impact these areas in the same way that it does in society. Furthermore, education is essential for equipping people with the skills necessary to survive in a technologically driven future. AI is among the powerful factors (Chaudhary et al., 2024).

Websites similar to ChatGPT are not perfect; they still include bugs and produce incorrect results. Can students assess whether the output is reliable and suitable for what they are learning? Do they know about moral and ethical issues like copyright? Currently, many academic institutions' regulations regarding academic dishonesty do not consider generative AI. School students are younger adolescents than university students, and schools foster both their cognitive and emotional intelligence. Their classroom learning is frequently supervised by teachers, and their capacity for self-regulated learning is less developed (Jennings et al., 2011).

6.1 Ethical Considerations

Concerns about ethical behavior are brought up by the use of generative AI in education, -including the possibility of students using it unethically or dishonestly and the possibility of job losses for people whose jobs are replaced by technology. Although ChatGPT represents an outstanding but imperfect state of generative AI technology today, it is but a taste of things to come. To ensure that the upcoming generation of students can maximize the positive effects of generative AI while minimizing its drawbacks, educators must comprehend the implications of this technology and research ways to modify the education ecosystem.

6.2 Effects on Pursuing Higher Education

The modern manufacturing environment is undergoing a dramatic shift toward robotization, which is defined by the mechanical replacement of human labor—traditionally referred to as "blue-collar" workers—with robots. In industrialized nations like for example South Korea, where the ratio of robots to workers is already 1:10, this tendency is especially noticeable.

As long as this trend persists, the labor market is expected to show a decrease in the need for unskilled manufacturing workers and, on the other hand, a rise in the need for highly qualified engineers who will be responsible for building, maintaining, operating, and programming robots.

Higher education institutions stand to benefit from this change, especially those with a technical focus as they will be better able to prepare future engineers rather than low-skilled laborers. But the emergence of generative AI holds the potential to completely transform intellectual labor—often referred to as a "white-collar" job." Robots are now being employed more and more in place of people, who once managed sophisticated IT systems to assist clients with AI-based applications. The demand for workers will likely shift from low-skilled white-collar employees who were traditionally trained to serve clients to those with the high-level and interdisciplinary skills necessary for taking care of "program robots."

Humans are expected to take on roles as managers and trainers of these program robots. Unfortunately, there will probably be less demand for human labor overall in the market as a result of software robots' increased efficiency, accessibility from anywhere in the world, multilingualism, and customization. The threat posed by mechanical machines to blue-collar jobs has long been recognized by university strategists. Consequently, they were setting up their educational institutions to train more individuals for intellectual labor. But given how quickly artificial intelligence is developing, it's more possible that white-collar employment will be affected as well. Therefore, the question that needs to be answered is: For what kind of work should university students prepare (Gaur et al., 2020)?

It is necessary to contrast how AI and human judgment are used in situation analysis and decision-making. The comprehension of cause-and-effect relationship dynamics is often at the center of human cognitive processes when it comes to traditional situation analysis and decision-making. This technique entails figuring out which particular causes are most likely to result in a certain effect. Additionally, this evaluation is combined with a thorough examination of the possible expenses and damages related to the occurrence or non-occurrence of the expected effects—the development of informed and logical conclusions is the outcome of this complex and multifaceted process. It should be remembered, though, that while having well-organized cognitive processes, people are nonetheless prone to making mistakes and that decision-making can take a long time. These factors restrict the application of human judgment in intricate circumstances, particularly those necessitating instantaneous functioning (Su & Yang, 2023).

Artificial intelligence, on the other hand, mostly relies on statistical patterns to function. Even if historical evidence indicates a certain phenomenon has a high statistical probability of recurring, this does not ensure that it will happen again in the future. For instance, the great statistical likelihood of human responses in historical circumstances does not imply that those responses will be the same. Avoiding this ambiguity, or teaching individuals when to base judgments on cause-and-effect linkages, is the problem facing education. and when to use statistical trends to guide decision-making. Humans will play a vital role in preventing serious repercussions from AI's improper responses (Afaq et al., 2022).

Finding the ideal balance between AI-based autonomous and self-organizing systems and human planning and control will be a problem for future workers. Employees will need to show that they are intelligent (able to reason) beyond artificial intelligence (statistics) and that they can work with robots that are controlled by AI in situations where conditions, objectives, and goals change quickly. There's a chance that people will become so dependent on AI that they lose knowledge and become less capable of rational, logical thought. The more accurate and less mistake-prone AI becomes the greater the problem. However, a significant mistake made by AI that was missed by humans at the time could have disastrous effects. To meet this problem, educational systems need to be adequately redesigned to teach students how to collaborate with AI-based program robots.

6.3 Risks Associated With Generative AI in Education (Prasad et al., 2023)

Some of the potential risks associated are given below in points:

1. Quick and cursory education.
2. Impeding pupils' ability to acquire autonomous and critical thinking abilities, which may have long-term consequences.
3. A possible obstacle to the growth of creativity.
4. Giving insufficient details that cause a notion to be misunderstood.

5. Providing solutions that sound logical but are illogical, frequently leading to "fabricated" outcomes known as hallucinations.
6. Restrictions on how one may interpret quantitative data that is integrated into a text.
7. Frequently, there is no disclosure regarding the authorship or the source of the data used to support the results, which is also illegal.
8. Potentially detrimental consequences on the development of interpersonal skills, such as a compromise in peer and teacher communication and interaction.
9. Dishonest use of these tools is when the produced material is used without giving due credit, which is a kind of plagiarism. The unequal use and access to these technologies, especially the premium versions, by those who can afford them and those who cannot, may give rise to equity concerns.
10. The breach of confidentiality and privacy regarding data.
11. A rise in socioeconomic and racial discrimination as a result of data biases in the applications' training.
12. Possible harm to the environment as a result of the high processing power needed to get the desired outcomes.
13. Issues with cybersecurity.

However, there have also been concerns raised over the limitations of GenAI as well as problems with academic integrity, ethics, and plagiarism. The examination of AI-generated answers to academic writing prompts reveals that while the text output was largely unique and pertinent to the subjects, it also lacked human viewpoints and made improper allusions. These are two things that AI is typically not good at providing. It can be difficult for second language learners to create acceptable prompts because it takes a certain amount of language proficiency, and relying too much on GenAI tools can undermine students' sincere attempts to become proficient writers (Chan Hu, 2023).

7. REASONS WHY AI IS GOOD FOR EDUCATION

The advantages of GenAI highlight the technology's potential as an important learning aid for students. Also, more research must be there on the best ways to include GenAI into the teaching and learning process.

7.1 Artificial Intelligence in Education

How may artificial intelligence be used in education? Given the benefits and drawbacks of artificial intelligence, it is advised that teachers employ generative AI in the classroom only in situations where they believe their students will benefit. Generative AI should be carefully chosen and used to advance students' knowledge and skill development in the classroom today. Teachers incorporate GAIs into their lessons by utilizing them occasionally—not constantly—in a knowledgeable, modern, and highly selective way; by understanding the risks involved, pointing out the drawbacks to students, and ensuring that students do not lose critical skills from over-reliance on this recently developed technology.

First of all, AI gathers enormous volumes of data; in fact, we have never before been able to process data of this size. The second advantage is that by mining this data, AI can support and enhance human thought. The third benefit is that AI currently gives teachers and students access to comprehensive and more complex materials. Content production is helpful since GAI can create and transform multimodal, written, image-based, and musical texts. Fourthly, AI software can edit works, offer comprehension and solutions to complicated problems through texts, make suggestions for problem-solving strategies, and converse with users to exchange ideas. (E. A. M, 2023)

Moving on, it can also alter previously written texts, which is another advantage. Sixth, students can use AI to evaluate written writings generated by the system, collaborate with them, and develop them. The seventh advantage is that it can strengthen and expand educational institutions' capacity to adapt to changing circumstances, allowing them to avoid being restricted to a predetermined set of methods for assessment and content delivery. AI can also help with individualized learning for the needs of the individual, community, and school. The ability of AI to create immersive, dynamic, and adaptable learning experiences through recording—such as AI apps for language learning—is the eighth advantage. The multimodal experiences that support the development of skills and knowledge while fostering learners' engagement through a variety of simulated and interactive situations are an addition to this.

Another benefit of artificial intelligence (AI), albeit a drawback, is that it allows teachers to save a ton of time and effort by employing it to create curriculum, screenplays, tests, and other useful literature. As was previously said, AI can be used to create, assess, and teach assignments to students. The eleventh benefit is that AI makes schools and students more competitive globally and gets young people ready for the workforce. The last advantage—which I want to go into more depth about—is that AI gives users the freedom to experiment with, get around, and use the technology in novel ways that its creators never would have imagined (Afaq et al., 2023).

7.2 Proliferation of AI In Teaching Other Domain

Examining professional artists' creative and entertaining uses of AI is beneficial for instructors and students alike. Following this course, as described, can help educators foster creativity.

They are discussing the "co-construction" of art production through collaborative efforts with AI. The key concept here is the co-construction of art. It is not conventionally using AI, and by this means using programs like Dream or DALLE-E2. A studio where users can instruct AI to create a piece of art, and the AI will generate the artwork on demand. I am referring to the collaboration between artists and AI when both parties possess agency.

8. LIMITATIONS AND FUTURE

Even if AI offers the potential for large-scale improvements in education, there are still obstacles to overcome. Some sophisticated tools, for example, have technological limitations that lead to chatbots responding problematically. Examples of this include ChatGPT's hallucinations and Microsoft Bing's chatbot, which generates hostile and disparaging responses in response to ideas and information that at first glance appear to make sense but turn out to be nonsensical (Thierry-Aguilera et al., 2023). Apart from the technological impediments, educators encounter additional problems that hinder the efficient creation and incorporation of artificial intelligence systems in educational environments. Among these challenges are diminished learner agency and dubious data privacy. The black box, or the hidden and occasionally incomprehensible nature of AI processes, is another significant worry (Lim et al., 2023).

Future education will be AI-enabled at all levels, with GenAI playing important roles in all facets of the educational process. To solve the challenges, academics and educational institutions will creatively adopt new tools and technology. Despite some institutions' bans, learners will use AI-enabled technologies more and more. It is a good idea to reconsider and reshape research, writing, teaching, learning, and evaluation.

8.1 Issues That Generative AI Presents For Educational Systems Include but not limited to:

1. Teacher training in generative AI competencies;

2. All parties involved adapting to the constantly changing digital ecosystem resulting from generative AI
3. Practice-generating communities to exchange insights on the application of AI in education
4. The development of students' generative AI talents, with a focus on encouraging critical thinking abilities to comprehend its possibilities and constraints and to utilize these technologies ethically
5. Addressing resistance to change and reviewing, updating, and inventing curriculum content and teaching strategies that may have become outmoded to increase possibilities for students' reflection.
6. Investigation of alternatives and/or complementarities in assessment techniques, such as adding oral exams to written assignments, using open-ended assessments to promote creativity and originality, offering visual aids like diagrams or graphics, and stressing the value of the learning process rather than just the finished product
7. The creation of basic rules and ethical codes about generative AI, guaranteeing ethical and responsible methods in its application

According to research, bias detection technologies, continual AI model training, and routine monitoring can all assist reduce biases and guarantee inclusive and equitable results.

The requirement for excellent digital literacy and critical thinking abilities is a major hurdle when using GenAI. Whether today's pupils are "digital natives," or have grown up with technology, and are therefore more equipped to adjust to the use of new technologies, has been a contentious academic issue for more than 20 years (Ojha et al., 2024). This group includes a large number of university students, particularly those who were born in Western nations in the twenty-first century. They may already possess the necessary digital literacy skills to interact with GenAI tools in a way that will yield valuable results and allow them to be critically assessed. In several ways, the COVID-19 pandemic's remote learning experiences for students hastened the adoption of learning technologies and the development of the digital literacy needed to interact with them. It's crucial to remember, though, that a student may not be ready to use cutting-edge GenAI technologies in a formal learning setting just because they were raised in a technologically advanced home or had firsthand experience learning online during the COVID-19 pandemic. Even with strong general digital literacy, students may still feel confined in certain domains or adhere to routines they have formed over time rather than applying critical thinking to novel situations. For instance, preliminary research on ChatGPT indicates that individuals may be underestimating the degree to which the information affects their decisions.

8.2 How Can These Issues Be Avoided?

What tactics are available to us in this GAI era? 1. Since students are now utilizing GAIs, teachers can discuss with them the benefits and drawbacks of these fascinating tools, as well as their limits. Above all, we must be honest with students about this in the classroom. Encourage pupils to consider the broader effects of using AI. Students should be asked to assess GAIs critically. Give students criticism on AI texts and ask them to make them better. Teach pupils how to utilize them wisely and to conduct thorough research at all levels. It would be bad for students to fully rely on GAIs to generate all texts because research is a complex process, even though it is easy to conduct.

Emphasize to students the value of developing their abilities, which include obtaining information, solving problems, synthesizing ideas, and building them in an educated, analytical, and coherent way. Students should subsequently be able to communicate these talents through texts they create on their own, independent of technology. They develop into independent writers and thinkers by adhering to this method. This is quite important. Teach our teachers too, so that they can become skilled educators, learn about pedagogy, impart these skills to students, and create a curriculum that works. To do this, teacher education programs at colleges of education must provide adequate training in all facets of artificial intelligence.

People are becoming more concerned about the spread of bogus news, which has scared them. To identify growing censorship and fake news in K–12 and higher education, educators and students must learn how to gather data independently and impartially. All students and teachers should be able to recognize fake news, which is divided into three categories: 1) misinformation, which is the unintentional spread of false information; 2) disinformation, which is false information purposefully presented as true and accurate; and 3) weaponization, which is the defamation of legitimate, lawful, and thoroughly researched content by raising doubts about it and, in doing so, seeking to undermine experts (Jennings et al., 2011). The fourth tactic is to undermine carefully considered and well-written facts with unfounded, personal insults. The last method is the widespread usage of "deep fakes," which allow users of AI software to fabricate photos or videos of people and, in the case of moving images, allow them to say things that seem plausible.

There are numerous strategies to prevent censorship in education, such as teaching pupils to consider different points of view when gathering information. Additionally, students can utilize search engines other than Google, like DuckDuckGo, which prevents filter bubbles, avoids Internet microtargeting, guards against privacy mining, and does not deliver personalized search results. Additionally, GAI can readily assist in student plagiarism; yet, due to the imperfect and unreliable nature of the GAI systems, it remains difficult to detect plagiarism despite the existence of applications designed to do so, such as GPT Radar and Copy Leaks. Instructors should place a strong emphasis on procedure and in-class work to stop AI misuse. Throughout class—this enables teachers to keep a close eye on students' activities and the progress of their assignments. There is nothing wrong with teachers using handwritten assignments and assessments for students when they feel the need to. It has the potential to work now just as it did in the past. Maintaining students' interest in learning will depend on their ability to solve problems on their own, analyze ideas, write texts, and explain things to others (Gaur et al., 2023).

Teachers should work with GAIs, and if they discover any that are useful for instruction, they should make use of these tools. Being transparent is crucial for teachers to avoid problems as well; in fact, openness is the key word here. As with any use of technology, openness is essential.

In conclusion, technology ought to serve us rather than the other way around. If humans do not use artificial intelligence (AI) with discernment, prudence, and mindfulness, it could negatively impact children's learning abilities. We want to be in charge of the technology and make sure they are used for teaching and learning purposes. Research needs to be done on how AI models affect student learning outcomes, how to build algorithms to detect and reduce bias, how to integrate AI models with human expertise, and how to construct ethical frameworks for using AI in educational contexts. These initiatives have the potential to offer fact-based information that will inform policy choices and best practices for integrating GenAI into higher education. Technologies about generative artificial intelligence are also involved. There is huge transformative potential for ChatGPT in education, but must be done consistently with research & studies, while also giving additional insights.

REFERENCES

Afaq, A., Gaur, L., & Singh, G. (2022, April). A latent dirichlet allocation technique for opinion mining of online reviews of global chain hotels. In 2022 3rd International Conference on Intelligent Engineering and Management (ICIEM) (pp. 201-206). IEEE. DOI:10.1109/ICIEM54221.2022.9853114

Afaq, A., Gaur, L., & Singh, G. (2023). Social CRM: Linking the dots of customer service and customer loyalty during COVID-19 in the hotel industry. *International Journal of Contemporary Hospitality Management*, 35(3), 992–1009. DOI:10.1108/IJCHM-04-2022-0428

Afaq, A., Singh, G., Gaur, L., & Kapoor, S. (2023, November). Aspect-Based Opinion Mining of Customer Reviews in the Hospitality Industry: Leveraging Recursive Neural Tensor Network Algorithm. In 2023 3rd International Conference on Technological Advancements in Computational Sciences (ICTACS) (pp. 1392-1397). IEEE.

Baidoo-Anu, D., & Owusu Ansah, L. Education in the era of generative artificial intelligence (AI): Understanding the potential benefits of ChatGPT in promoting teaching and learning. *SSRN* 2023. [CrossRef] DOI:10.2139/ssrn.4337484

Barros, A., Prasad, A., & Śliwa, M. (2023). Generative artificial intelligence and academia: Implication for research, teaching and service. *Management Learning*, 54(5), 597–604. DOI:10.1177/13505076231201445

Chan, C. K. Y., & Hu, W. (2023). Students' voices on generative AI: Perceptions, benefits, and challenges in higher education. *International Journal of Educational Technology in Higher Education*, 20(1), 43. DOI:10.1186/s41239-023-00411-8

Chaudhary, M., Gaur, L., Singh, G., & Afaq, A. (2024). Introduction to Explainable AI (XAI) in E-Commerce. In *Role of Explainable Artificial Intelligence in E-Commerce* (pp. 1–15). Springer Nature Switzerland. DOI:10.1007/978-3-031-55615-9_1

Cherukuri, A. K., Jonnalagadda, A., & Murugesan, S. (2021, May). AI in education: Applications and impact. *Amplify*, 34(5), 26–33.

Cooper, G. (2023). Examining Science Education in ChatGPT: An Exploratory Study of Generative Artificial Intelligence. *Journal of Science Education and Technology*, 32(3), 444–452. DOI:10.1007/s10956-023-10039-y

Dyer, G. (2023). Yet Another Article on AI (p. 7). Winnipeg Free Press.

Elkhatat, A. M., Elsaid, K., & Almeer, S. (2023). Evaluating the efficacy of AI content detection tools in differentiating between human and AI-generated text. *International Journal for Educational Integrity*, 19(1), 17. DOI:10.1007/s40979-023-00140-5

Farrelly, T., & Baker, N. (2023, November 04). Generative Artificial Intelligence: Implications and Considerations for Higher Education Practice. *Education Sciences*, 13(11), 1109. DOI:10.3390/educsci13111109

Gaur, L., & Afaq, A. (2020). Metamorphosis of CRM: incorporation of social media to customer relationship management in the hospitality industry. In *Handbook of Research on Engineering Innovations and Technology Management in Organizations* (pp. 1–23). IGI Global. DOI:10.4018/978-1-7998-2772-6.ch001

Gaur, L., Afaq, A., Arora, G. K., & Khan, N. (2023). Artificial intelligence for carbon emissions using system of systems theory. *Ecological Informatics*, 76, 102165. DOI:10.1016/j.ecoinf.2023.102165

Gaur, L., Gaur, D., & Afaq, A. (2024). Ethical Considerations in the Use of the Metaverse for Healthcare. In Metaverse Applications for Intelligent Healthcare (pp. 248-273). IGI Global.

Gaur, L., Gaur, D., & Afaq, A. (2024). Demystifying Metaverse Applications for Intelligent Healthcare. In Metaverse Applications for Intelligent Healthcare (pp. 1-23). IGI Global.

Gordon, C. (2023). ChatGPT Is the Fastest Growing App in the History of Web Applications.

Jennings, P., & Greenberg, M. (2011). The Prosocial Classroom: Teacher Social and Emotional Competence about Student and Classroom Outcomes. Review of Educational Research -. *Review of Educational Research*, 79. Advance online publication. DOI:10.3102/0034654308325693

Lim, W. M., Gunasekara, A., Pallant, J. L., Pallant, J. I., & Pechenkina, E. Generative AI and the future of education: Ragnarök or reformation? A paradoxical perspective from management educators, The International Journal of Management Education, Volume 21, Issue 2, 2023, 100790, ISSN 1472-8117, DOI:10.1016/j.ijme.2023.100790

Michel-Villarreal, R., Vilalta-Perdomo, E., Salinas-Navarro, D. E., Thierry-Aguilera, R., & Gerardou, F. S. (2023). Challenges and Opportunities of Generative AI for Higher Education as Explained by ChatGPT. *Education Sciences*, 13(9), 856. DOI:10.3390/educsci13090856

Multidisciplinary perspectives on opportunities, challenges and implications of generative conversational AI for research, practice, and policy, International Journal of Information Management, Volume 71, 2023, 102642, ISSN 0268-4012

Ojha, Nitish & Pandita, Archana & Ramkumar, J. (2024). Cyber Security Challenges and Dark Side of AI: Review and Current Status. .DOI:10.4018/979-8-3693-0724-3.ch007

S. Murugesan and A. K. Cherukuri, "The Rise of Generative Artificial Intelligence and Its Impact on Education: The Promises and Perils," in Computer, vol. 56, no. 5, pp. 116-121, May 2023, Doi: .DOI:10.1109/MC.2023.3253292

Sarah, A. Chauncey, H. Patricia McKenna, A framework and exemplars for ethical and responsible use of AI Chatbot technology to support teaching and learning, Computers and Education: Artificial Intelligence, Volume 5, 2023, 100182, ISSN 2666-920X

Su, J., & Yang, W. (2023). Unlocking the power of ChatGPT: A framework for applying generative AI in education. *ECNU Review of Education*, 6(3), 355–366. DOI:10.1177/20965311231168423

Su, J., & Yang, W. (2023). Unlocking the power of ChatGPT: A framework for applying generative AI in education. *ECNU Review of Education*, 6(3), 355–366.

van Dis, E. A. M., Bollen, J., Zuidema, W., van Rooij, R., & Bockting, C. L. (2023, February). ChatGPT: Five priorities for research. *Nature*, 614(7947), 224–226. DOI:10.1038/d41586-023-00288-7 PMID:36737653

What ChatGPT means for universities: Perceptions of scholars and students. (2023). Journal of Applied Learning and Teaching, 6(1). DOI:10.37074/jalt.2023.6.1.22

Yeralan, S., & Lee, L. A. (2023). Generative AI: Challenges to higher education. *Sustainable Engineering and Innovation*, 5(2), 107–116.

Dwivedi, Y. K., Kshetri, N., Hughes, L., Slade, E. L., Jeyaraj, A., Kar, A. K., & Wright, R. (2023). Opinion Paper:"So what if ChatGPT wrote it?" Multidisciplinary perspectives on opportunities, challenges and implications of generative conversational AI for research, practice and policy. *International Journal of Information Management*, 71, 102642.

Zhai, X. ChatGPT user experience: Implications for education. *SSRN* 2022. [CrossRef] DOI:10.2139/ssrn.4312418

KEY TERMS AND DEFINITIONS

LLM: LLM stands for Large language model, which can understand and recognize text from large datasets. It is especially used in machine learning to generate new content.

Artificial Intelligence: A computer can think like humans and perform tasks intelligently by analyzing the cognitive process.

ChatGPT: ChatGPT stands for Chat Generative Pre-Trained Transformer, and is a chatbot powered, developed, and maintained by OpenAI.

Generative AI: An AI technology that can create content, text, audio, video, images.

Higher Education: Level of education that provides professional qualification.

COVID-19: An infectious disease caused by the SARS-COV-2 Virus.

DALL-E: An AI tool to create realistic images in any art form.

Midjourney: An AI chatbot to generate images in seconds.

Chapter 9
The Ethical Dilemma of Using (Generative) AI to Science and Research

Syed Ibad Ali

https://orcid.org/0000-0001-6312-6768

Parul Institute of Engineering and Technology, Parul University, Vadodara, India

Mohammad Shahnawaz Shaikh

https://orcid.org/0000-0002-1763-8989

Parul Institute of Engineering and Technology, Parul University, Vadodara, India

ABSTRACT

Using Artificial intelligence (AI) in research offers many important benefits for science and society but also creates novel and complex ethical issues. While these ethical issues do not necessitate changing established ethical norms of science, they require the Scientific community to develop new guidance for the appropriate use of AI. In this article, we briefly introduce AI and explain how it can be used in research, examine some of the ethical issues raised when using it, and offer recommendations for responsible use, including: Researchers are responsible for identifying, describing, reducing, and controlling AI-related biases and random errors; Researchers should disclose, describe, and explain their use of AI in research, including its limitations, in language that can be understood by non-experts; Researchers should engage with impacted communities, populations, and other stakeholders concerning the use of AI in research to obtain their advice and assistance and address their interests and concerns, such as issues related to bias.

DOI: 10.4018/979-8-3693-9173-0.ch009

1. INTRODUCTION

Exponential growth in the use of Artificial intelligence in scientific research in just a few years, Artificial intelligence (AI) has taken the world of scientific research by storm. AI tools have been used to perform or augment a variety of Scientific tasks, The applications of AI in Scientific research appears to be limitless, and in the next decade AI is likely to completely transform the process of Scientific discovery and innovation, Although using AI in scientific research has steadily grown, ethical guidance has lagged far behind. With the exception of using AI to draft or edit Scientific papers (see discussion in Sect. most codes and policies do not explicitly address ethical issues related to using AI in Scientific research (M. Wooldridge, 2021). For example, in the revision of the European Code of Conduct for Research Integrity briefly discusses the importance of transparency. The code stipulates that researchers should report "their results and methods including the use of external services or AI and automated tools" and considers "hiding the use of AI or automated tools in the creation of content or drafting of publications" as a violation of research integrity). One of the most thorough and up-to-date institutional documents, the National Institutes of Health Guidelines and Policies for the Conduct of Research provides guidance for using AI to write and edit manuscripts but not for other tasks. Codes of AI ethics, Ethics of Artificial Intelligence and the Office of Science and Technology Blueprint for an AI Bill of Rights, provide useful guidance for the development and use of AI in general without including specific guidance concerning the development and use of AI in Scientific research There is therefore a gap in ethical and policy guidance concerning AI use in Scientific research that needs to be filled to promote its appropriate use (Ali, 2024).

Moreover, the need for guidance is urgent because using AI raises novel epistemological and ethical issues related to objectivity, reproducibility, transparency, accountability, responsibility, and trust in science. In this chapter, we will examine important questions related to AI's impact on ethics of science. We will argue that while the use of AI does not require a radical change in the ethical norms of science, it will require the Scientific community to develop new guidance for the appropriate use of AI. To defend this thesis, we will provide an overview of AI and an account of ethical norms of science, and then we will discuss the implications of AI for ethical norms of science and offer recommendations for its appropriate use.

1.1 What is AI?

AI can be defined as "a technical and scientific field devoted to the engineered system that generates outputs such as content, forecasts, recommendations or decisions for a given set of human-defined objectives." AI is a subfield within the discipline of computer science. However, the term 'AI' is also commonly used to refer to technologies (or tools) that can perform human tasks that require intelligence, such as perception, judgment, reasoning, or decision-making. We will use both senses of 'AI' in this chapter, depending on the context While electronic calculators, cell phone apps, and programs that run on personal computers can perform functions associated with intelligence, they are not generally considered to be AI because they do not "learn" from the data. As discussed below, AI systems can learn from the data insofar as they can adapt their programming in response to input data (Stone, 2022). While applying the term 'learning' to a machine may seem misleadingly anthropomorphic, it does make sense to say that a machine can learn if learning is regarded as a change in response to information about the environment. Many different entities can learn in this sense of the term, including the immune system, which changes after being exposed to molecular information about pathogens, foreign objects, and other things that provoke an immune response, this paper will focus on what is commonly referred to as narrow (or weak) AI, which is already being extensively used in science (Tahiru, 2020). Narrow AI has been designed and developed to do a specific task, such as playing chess, modelling complex phenomena, or identifying possible brain tumours in diagnostic images. Other types of AI discussed in the literature include broad AI (also known as Artificial general intelligence or AGI), which is a machine than can perform multiple tasks requiring human-like intelligence, and Artificial consciousness (AC), which is a form of AGI with characteristics widely considered to be essential for consciousness. Because there are significant technical and conceptual obstacles to developing AGI and AC, it may be years before machines have this degree of human-like intelligence (Shaikh, 2024).

1.2 What is Machine Learning?

Machine learning (ML) can be defined as a branch of AI "that focuses on the using data and algorithms to enable AI to imitate the way that humans learn, gradually improving its accuracy." There are several types of ML, including support vector machines, decisions trees, and neural networks. In this chapter we will focus on ML that uses artificial neural networks (ANNs). An ANN is composed of artificial neurons, which are modelled after biological neurons. An artificial neuron receives a series of computational inputs, 5 applies a function, and produces an output. The inputs have different weightings. In most applications, a specific output is generated only when a certain threshold value for the inputs is reached. In the example below, an output of '1' would be produced if the threshold is reached, otherwise, the output would be '0'. A pair statements describing how a very simple artificial neuron processes inputs could be as follows:

If $[(x_1)(w_1) + (x_2)(w_2) + (x_3)(w_3) + (x_4)(w_4) > T]$, then output $U = 1$

If $[(x_1)(w_1) + (x_2)(w_2) + (x_3)(w_3) + (x_4)(w_4) \leq T]$, then output $U = 0$

Where x_1, x_2, x_3, and x_4 are inputs, w_1, w_2, w_3, and w_4 are weightings, T is a threshold value, and U is an output value (1 or 0). A single neuron may have dozens of inputs. An ANN may consist of thousands of interconnected neurons. In a deep learning ANN, there may be many hidden layers of neurons between the input and output layers. Training (or reinforcement) occurs when the weightings on inputs are changed in response to system's output. Changes in the weightings are based on their contribution to the neuron's error, which can be understood as the difference between the output value and the correct value as determined by the human trainers. Training can occur via supervised or unsupervised learning. In supervised learning, the ANN works with labelled data and becomes adept at correctly representing structures in the data recognized by human trainers (E. Alpaydin, 2016). In unsupervised learning, the ANN works with unlabelled data and discovers structures inherent in the data that might not have been recognized by humans. For example, to use supervised learning to train an ANN to recognize dogs, human beings could present the system with various images and evaluate the accuracy of its output accordingly. If the ANN labels an image a "dog" that human beings recognize as a dog, then its output would be correct, otherwise, it would be incorrect. In unsupervised learning, the ANN would be presented with images and would be reinforced for accurately modelling structures inherent in the data, which may or may not correspond to patterns, properties, or relationships that humans would recognize or conceive of. For an example of the disconnect between ML and human processing of information, consider research conducted by Roberts et al. In this study, researchers trained an ML system on radiologic images from hospital patients so that it would learn to identify patients with COVID-19 and predict the course of their illness (Preeti, 2024). Since the patients who were sicker tended to laying down when their images were taken, the ML system identified laying down as a diagnostic criterion and disease predictor. However, laying down is a confounding factor that has nothing to do with the likelihood of having COVID-19 or getting very sick from it. The error occurred because the ML system did not account for this fundamental fact of clinical medicine (J. D. Kelleher, 2019).

Despite problems like the one discovered by Roberts et al. the fact that ML systems process and analyse data differently from human beings can be a great benefit to science and society because these systems may be able to identify useful and innovative structures, properties, patterns, and relationships that human beings would not recognize. For example, ML systems have been able to design novel compounds and materials that human beings might not be able to conceive. That said, the disconnect between AI/ML and human information processing can also make it difficult to anticipate, understand, control, and reduce errors produced by ML systems (J. D. Kelleher, 2019).

Training ANNs is a resource-intensive activity that involves gigabytes of data, thousands of computers, and hundreds of thousands of hours of human labour. A system can continue to learn after the initial training period as it processes new data. ML systems can be applied to any dataset that has been properly prepared for manipulation by computer algorithms, including digital images, audio and video recordings, natural language, medical records, chemical formulas, electromagnetic radiation, business transactions, stock prices, and games. One of the most impressive feats accomplished by ML systems is their contribution to solving the protein folding problem. A protein is composed of one or more long chains of amino acids known as polypeptides. The three-dimensional (3-D) structure of the protein is produced by folding of the polypeptide(s), which is caused by the interplay of hydrogen bonds, Van der Waals attractive forces, and conformational entropy between different parts of the polypeptide. Molecular biologists and biochemists have been trying to develop rules for predicting the 3-D structures of proteins from amino acid sequences since the 1960s, but this is, computationally speaking, a very hard problem, due to the immense number of possible ways that polypeptides can fold. Tremendous progress on the protein-folding problem was made in 2022, when scientists demonstrated that an ML system, DeepMind's AlphaFold, can predict 3-D structures from amino acid sequences with 92.4% accuracy. AlphaFold, which built upon available knowledge of protein chemistry, was trained on thousands of amino acids sequences and their corresponding 3-D structures. Although human researchers still needed to test and refine AlphaFold's output to ensure that the proposed structure is 100% accurate, the ML system greatly improves the efficiency of protein chemistry research. Recently developed ML systems can generate new proteins by going in the opposite direction and predicting amino acids sequences from 3-D protein structures. Since proteins play a key role in the structure and function of all living things, these advances in protein science are likely to have important applications in different areas of biology and medicine (E. Alpaydin, 2016).

1.3 What is Generative AI?

Not only can ML image processing systems recognize patterns in the data that correspond to objects (e.g., cat, dog, car), when coupled with appropriate algorithms they can also generate images in response to visual or linguistic prompts. The term 'generative AI' refers to "deep learning models that can generate high-quality text, images, and other content based on the data they were trained on" (M. White, 2023). Perhaps the most well-known types of generative AI are those that are based on large language models (LLMs), such as chatbots like OpenAI's ChatGPT and Google's Gemini, which analyse, paraphrase, edit, translate, and generate text, images and other types of content. LLMs are statistical algorithms trained on huge sets of natural language data, such as text from the internet, books, journal articles, and magazines. By processing this data, LLMs can learn to estimate probabilities associated with possible responses to text and can rank responses according to the probability that they will be judged to be correct by human beings

and magazines (T. Wang, 2023). By processing this data, LLMs can learn to estimate probabilities associated with possible responses to text and can rank responses according to the probability that they will be judged to be correct by human beings In just a few years, some types of generative AI, such as ChatGPT, have become astonishingly proficient at responding to text data. ChatGPT has passed licensing exams for medicine and law and scored in the 93rd percentile on the Scholastic Aptitude Test reading exam and in the 89th percentile on the math exam. Some researchers have used ChatGPT to write Scientific papers and have even named them as authors. Some LLMs are so adept at mimicking the type of discourse associated with conscious thought that computer scientists, philosophers, and cognitive psychologists are updating the Turing test to more reliably distinguish between humans and machines (J. Hutson, 2023).

2. LITERATURE SURVEY

Our search included both academic research and gray literature containing principles and guidelines for ethical AI. Documents addressing the following aspects were selected: artificial intelligence, healthcare, ethics, and guidelines. The literature search was conducted through an automatic search in each search engine listed, using the keywords and synonyms. In recent years, AI has developed rapidly and it now affects people's lives in many fields, including healthcare, intelligent transportation, and education. For instance, genetic algorithms are used to predict outcomes in critically ill patients in healthcare, sensing algorithms are applied for self-driving vehicles in smart transportation, and natural language processing (NLP) is combined with machine learning (ML) to facilitate online learning in education (M. White, 2023). As one of the core technologies of AI, ML has brought the development of AI to an advanced level. ML employs algorithmic methods that enable machines to solve problems without explicit computer programming (T. Wang, 2023). Deep learning (DL) is a subset of ML based on multi-layered artificial neural networks, which can be further utilized to solve complex problems using unstructured data, much like the human brain. Healthcare is one of the most promising application domains for ML and DL (J. D. Kelleher, 2019). AI techniques and their applications can help to detect cancer faster and earlier than before, make more accurate medical diagnoses, care for and monitor the elderly using robots, etc. ML techniques can process massive amounts of data and make increasingly accurate assessments and predictions. Although ML has advanced the development of AI. It also brings up ethical issues, especially in the healthcare domain. Ethics has been identified as a priority for developing and deploying AI across sectors (Sheikh, 2024). Ethical decision making by AI systems refers to the computational process of evaluating and selecting alternatives in a manner compliant with social, ethical, and legal requirements. The resulting ethical issues affect the further development and acceptance of AI, especially in healthcare, where technology must comply with the law, regulations, and privacy principles to ensure the maintenance of the common good (S. Lucci, 2022). The use of sophisticated ML algorithms employing DL and other complex techniques leads to black-box models, which may have low transparency and explainability. Blackbox models make it difficult even for their developers to explain how an AI system makes decisions. Meanwhile, users are confronted with decisions without an explanation for these decisions. The "black-box" nature of ML often clashes with legislation in high-stakes domains, where stakeholders can experience severe consequences if a bad decision is made (Mungale, 2024). Particularly in the healthcare domain, where lives are at stake, the actual adoption of AI in everyday practice is limited by numerous factors, including accuracy, explainability, transparency, and compatibility. This makes it important to promote the explainability of AI algorithms. Explainability is essential to responsible AI and can build trust in and engagement with AI (Mungale, 2024). Algorithmic transparency and explainability have been requested by several societal bodies, such as the government, the media, and the legal community. The research community has embraced this notion over the last few years, and numerous efforts have been made to design explainable AI systems Nevertheless, aside from explainability, multiple ethical concerns still exist when using AI-enabled solutions in the healthcare field, and they are gradually becoming the dominant factors influencing the adoption of AI. Policymakers and related professionals have been looking for approaches to cope with the ethical risks associated with AI development (Shaikh, 2024). Examples of rules and regulations are the "Ethics Guidelines for Trustworthy AI" from the European Commission, "Report on the Future of Artificial Intelligence" from the US, and the "Beijing AI Principles" from the Chinese government. Among these governing bodies, the European Union (EU) has been acknowledged as a leader in establishing a framework for ethical regulations and rules for AI. Unlike the other two sets of guidelines, the fundamental principle of the EU guidelines is to promote a "human-centred" approach that respects European values and regulations. The ethical challenges addressed by the EU framework are globally relevant. As they are based on a fundamental-rights approach, the relevance and importance of these guidelines can be considered universal. The authority and obligations underlying these guidelines form the framework for most of the United Nations' (UN) (C. Ebert, 2023).

Sustainable Development Goals (SDGs). This also affects the development strategies in low and middle-income countries outside the EU. These guidelines apply to all industrial sectors, and none of them are specifically and directly related to AI's ethical and legal aspects in healthcare. In addition to the ethical regulations and policies mentioned above, many academic publications discuss general ethical issues related to AI (Google, 2023). Examples are "The global landscape of AI ethics guidelines" by Jobin's group, which presents an overview of existing ethical guidelines and strategies "The Ethics of AI Ethics: An Evaluation of Guidelines" by Rangeomorph, which analyses 22 ethical guidelines for AI, and providing recommendations for overcoming their relative ineffectiveness and "Artificial intelligence ethics guidelines for developers and users: clarifying their content and normative implications" by Ryan and Carsten Stahl, which provides a elaborative explanation of 11 normative implications of current AI ethical guidelines directed to AI developers and organizational users. Although these three documents present very useful discussions of ethical AI issues in a general domain, none of them specifically address the ethics of AI in healthcare Academic publications discussing the ethical issues concerning AI in healthcare do exist, such as "The ethics of AI in health care: A mapping review" by Morley's research group, "Ethical and legal challenges of artificial intelligence-driven healthcare" by Gerke's group, and "A governance model for the application of AI in healthcare" by Reddy's group. Morley's group focused on mapping the ethical issues based one epistemic, normative, and overarching perspectives. Gerke's group explored ethical issues from the perspective of legal challenges, but did not present a systematic review of how AI can influence them in healthcare applications. Reddy's group addressed the introduction and implementation of a proposed governance model in healthcare. In short, governmental policy and academic research have often addressed the ethics of AI in a general sense, but have devote much less attention to the specific field of healthcare (Shaikh, 2024).

3. ETHICAL NORMS OF SCIENCE

With this overview of AI in mind, we can now consider how using AI in research impacts the ethical norms of science. But first, we need to describe these norms. Ethical norms of science are principles, values, or virtues that are essential for conducting good research. These norms apply to various practices, including research design, experimentation and testing, modelling, concept formation, data collection and storage, data analysis and interpretation, data sharing, publication, peer review, hypothesis theory formulation and acceptance, communication with the public, as well as mentoring and education (C.L. Bockting, 2023). Many of these norms are expressed in codes of conduct, professional guidelines, institutional or journal policies, or books and papers on Scientific methodology. Others, like collegiality, might not be codified but are implicit in the practice of science. Some norms, such as testability, rigor, and reproducibility, are primarily epistemic, while others, such as fair attribution of credit, protection of research subjects, and social responsibility, are primarily moral (when enshrined in law, like instance of fraud, these norms become legal but here we only focus on ethical norms). There are also some like honesty, openness, and transparency, which have both epistemic and moral dimensions. Scholars from different fields, including philosophy, sociology, history, logic, decision theory, and statistics have studied ethical norms of science (C. Huang, 2024). Sociologists such as Merton and Shapin, tend to view ethical norms as generalizations that accurately describe the practice of science, while philosophers, such as Kitcher and Haack, conceive of these norms as prescriptive standards that scientists ought to follow. These approaches need not be mutually exclusive, and both can offer useful insights about ethical norms of science. Clearly, the study of norms must take the practice of science as its starting point, otherwise our understanding of norms would have no factual basis. However, one cannot simply infer the ethical norms of science from the practice of science because scientists may endorse and defend norms without always following them. For example, most scientists would agree that they should report data honestly, disclose significant conflicting interests, and keep good research records, but evidence indicates that they sometimes fail to do so. One way of bridging the gap between descriptive and prescriptive accounts of ethical norms of science is to reflect on the social and epistemological founda-tions (or justifications) of these norms. Ethical norms of science can be justified in at least three ways. First, these norms help the Scientific community achieve its epistemic and practical goals, such as understanding, predicting, and controlling nature. It is nearly impossible to understand how a natural or social process works or make accurate predictions about it without standards pertaining to honesty, logical consistency, empirical support, and reproducibility of data and results (UNESCO, 2022). These and other epistemic standards distinguish science form superstition, pseudoscience, and sophistry

Table 1. Norms of science

Honesty - Accountability
Testability - Freedom of inquiry
Rigor - Fair sharing of credit

continued on following page

Table 1. Continued

Empiricism - Confidentiality of peer review
Scepticism - Collegiality
Explanatory power - Non-discrimination
Objectivity - Respect for intellectual property
Realism - Protection of human subjects
Precision - Protection of animal subjects
Openness - Safety (physical, biological, psychosocial)
Transparency - Stewardship of resources

Second, ethical norms promote trust among scientists, which is essential for collaboration, peer review, publication, sharing of data and resources, mentoring, education, and other Scientific activities. Scientists need to be able to trust that the data and results reported in papers have not been fabricated, falsified or manipulated, that reviewers for journals and funding agencies will maintain confidentiality that colleagues or mentors will not steal their ideas and other forms of intellectual property, and that credit for collaborative work will be distributed fairly. Third, ethical norms are important for fostering public support for science. The public is not likely to financially, legally, or socially support research that is perceived as corrupt, incompetent, untrustworthy, or unethical. Taken together, these three modes of justification link ethical norms to science's social foundations, that is, ethical norms are standards that govern the Scientific community, which itself operates within and interacts with a larger community, namely society. Although vital for conducting science, ethical norms are not rigid rules. Norms sometimes conflict, and when they do, scientists must make decisions concerning epistemic or moral priorities (Shaikh M. S., 2024). For example, model-building in science may involve trade of among various epistemic norms, including generality, precision, realism, simplicity, and explanatory power. Research with human subjects often involves trade of between rigor and protection of participants. For example, placebo control groups are not used in clinical trials when receiving a placebo instead of an effective treatment would cause serious harm to the participant. Although the norms can be understood as guidelines, some have higher priority than others. For example, honesty is the hallmark of good science, and there are very few situations in which scientists are justified in deviating from this norm. Openness, on the other hand, can be deemphasized to protect research participants' privacy, intellectual property, classified information, or unpublished research. Finally, science's ethical norms have changed over time, and they are likely to continue to evolve. While norms such as empiricism, objectivity, and consistency originated in ancient Greek science, others, such as reproducibility and openness, developed during the 1500s, and many, such as protection of research subjects and social responsibility, did not emerge as formalized norms until the twentieth century. This evolution is in response to changes in science's social, institutional, economic, and political environment and advancements in scientific instruments, tools, and methods. For example, the funding of science by private companies and their requirements concerning data access and release policies have led to changes in norms related to open sharing of data and materials. The increased presence of women and racial and ethnic minorities in science has led to the development of policies for preventing sexual and other forms of harassment. The use of computer software to analyse large sets of complex data has challenged traditional views about norms related to hypothesis testing.

4. AI AND THE ETHICAL NORMS OF SCIENCE WITH CONCLUSION

We will divide our discussion of AI and the ethics of science into six topics corresponding to the problems and issues previously identified in this paper and seventh topic related to scientific education. While these topics may seem somewhat disconnected, they all involve ethical issues that scientists who use AI in research are currently dealing with AI biases and the ethical norms of science Bias can undermine the quality and trustworthiness of science and its social impacts. While reducing and managing bias are widely recognized as essential to good scientific methodology and practice, they become crucial when AI is employed in research because AI can reproduce and amplify biases inherent in the data and generate results that lend support to policies that are discriminatory, unfair, harmful, or ineffective. More-over, by taking machines' disinterestedness in findings as a necessary and sufficient condition of objectivity, users of AI in research may overestimate the objectivity of their findings (A. Spirling, 2023). AI biases in medical research have generated considerable concern, since biases related to race, ethnicity, gender, sexuality, age, nationality, and socioeconomic status in health-related datasets can perpetuate health disparities by supporting biased hypotheses, models, theories, and policies. Biases also negatively impact areas of science outside the health sphere, including ecology, forestry, urban planning, economics, wildlife management, geography, and agriculture. OpenAI, Google, and other generative AI developers have been using filters that prevent their systems from generating text that is outright racist, sexist, homophobic, pornographic, offensive, or dangerous. While bias reduction is a necessary step to make AI safe for human use, there are reasons to be sceptical of the idea that AI can be appropriately sanitized. First, the biases inherent in data are so pervasive that no amount of filtering can remove all of them. Second, AI systems may also have political and social biases that are difficult to identify or control. Even in the case of generative AI models where some filtering has happened, changing the inputted prompt may simply confuse and of research design, data analysis and interpretation, but also address issues related to data diversity, sampling, and representativeness. They must also realize that they are ultimately accountable for AI biases, both to other scientists and to members of the public. As such, they should only use AI in contexts where their expertise and judgement are sufficient to identify and remove biases. This is important because given the accessibility of AI systems and the fact that they can exploit our cognitive shortcomings, they are creating an illusion of understanding. Furthermore, to build public trust in AI and promote transparency and accountability, scientists who use AI should engage with impacted populations, communities and other stakeholders to address their needs and concerns and seek their assistance in identifying and reducing potential biases.18 During the engagement process, researchers should help populations and communities understand how their AI system works, why they are using it, and how it may produce bias. To address the problem of AI bias, the Biden Administration recently signed an executive order that directs federal agencies to identify and reduce bias and protect the public from algorithmic discrimination.

AI random errors and the ethical norms of science Like bias, random errors can undermine the validity and reliability of scientific knowledge and have disastrous consequences for public health, safety, and social policy. For example, random errors in the processing of radiologic images in a clinical trial of a new cancer drug could harm patients in the trial and future patients who take an approved drug, and errors related to the modelling of the transmission of an infectious disease could undermine efforts to control an epidemic. Although some random errors are unavoidable in science, an excessive amount when using AI could be considered carelessness or recklessness when using AI. Reduction of random errors, like reduction of bias, is widely recognized as essential to good scientific methodology and practice. Although some random errors are unavoidable in research, scientists have obligations to identify, describe, reduce, and correct them because they are ultimately accountable for both human and AI errors. Scientists who use AI in their research should disclose and discuss potential limitations and (known) AI-related errors. push a system to generate biased content anyway. Third, by removing, reducing and controlling some biases, AI developers may create other biases, which are difficult to anticipate, identify or describe at this point. If we try to weed undesirable features of this data, we will eliminate parts of our language and culture, and ultimately. If we want to use LLMs to make sound moral and political judgments, sanitizing their data processing and output may hinder their ability to excel at this task, because the ability to make sound moral judgements or anticipate harm may depend, in part, on some familiarity with immoral choices and the darker side of humanity. It is only by understanding evil that we can freely and rationally choose the good. We admit this last point is highly speculative, but it is worth considering. Clearly, the effects of LLM bias management bear watching. While the problem of AI bias does not require a radical revision of scientific norms, it does imply that scientists who use AI systems in research have special obligations to identify, describe, reduce, and control bias. To fulfil these obligations, scientists must not only attend to matters.

5. CONCLUSION

Using AI in research benefit science and society but also creates some novel and complex ethical issues that affect accountability, responsibility, transparency, trustworthiness, reproducibility, fairness, and objectivity, and other important values in research. Although scientists do not need to radically revise their ethical norms to deal with these issues, they do need new guidance for the appropriate use of AI in research. Since AI continues to advance rapidly, scientists, academic institutions, funding agencies and publishers, should continue to discuss AI's impact on research and update their knowledge, ethical guidelines and policies accordingly. Guidance should be periodically revised as AI becomes woven into the fabric of scientific practice (or normalized) and researchers learn about it, adapt to it, and use it in novel ways. Since science has significant impacts on society, public engagement in such discussions is crucial for responsible the use, development, and AI in research. In closing, we will observe that many scholars, including ourselves, assume that today's AI systems lack the capacities necessary for moral agency. This assumption has played a key role in our analysis of ethical uses of AI in research and has informed our recommendations. We realize that a day may arrive, possibly sooner than many would like to believe, when AI will advance to the point that this assumption will need to be revised, and that society will need to come to terms with the moral rights and responsibilities of some types of AI systems. Perhaps AI systems will one day participate in science as full partners in discovery and innovation. Although we do not view this as a matter that now demands immediate attention, we remain open to further discussion of this issue in the future.

REFERENCES

Ali, S. I., Kale, G. P., Shaikh, M. S., Ponnusamy, S., & Chouhan, P. S. (2024). AI Applications and Digital Twin Technology Have the Ability to Completely Transform the Future. In S. Ponnusamy, M. Assaf, J. Antari, S. Singh, & S. Kalyanaraman (Eds.), Harnessing AI and Digital Twin Technologies in Businesses (pp. 26-39). IGI Global. https://doi.org/. J. Denning et al., "Computing as a Discipline," Computer, Vol. 22, No. 2, 1989, pp. 63–70, DOI:10.4018/979-8-3693-3234-4.ch003

Alpaydin, E. (2016). *Machine Learning: The New AI*. MIT Press.

Bockting, C. L., van Dis, E. A. M., van Rooij, R., Zuidema, W., & Bollen, J. (2023, October 19). Living guidelines for generative AI—Why scientists must oversee its use. *Nature*, 622(7984), 03266–1. https://www.nature.com/articles/d41586-023-. DOI:10.1038/d41586-023-03266-1 PMID:37857895

Chopkar, P., Wanjari, M., Jumle, P., Chandankhede, P., Mungale, S., & Shaikh, M. S. (2024), A Comprehensive Review on Cotton Leaf Disease Detection using Machine Learning Method, Grenze International Journal of Engineering and Technology, June Issue, Grenze ID: 01.GIJET.10.2.537, Grenze Scientific Society, 2024

Ebert, C., & Lourida, P. (2023). Generative AI for Software Practitioners, (2023). *IEEE Software*, 40(4), 30–38. DOI:10.1109/MS.2023.3265877

Google. "Responsible AI practices": https://ai.google/responsibility/responsible-ai-practices/, accessed November 5, 2023.

Huang, C., Zhang, Z., Mao, B., & Yao, X. "An Overview of Artificial Intelligence Ethics," IEEE Transactions on Artificial Intelligence (2024), Vol. 4, Issue 4, pp. 799-819, 2022.

Hutson, J., & Harper-Nichols, M. "Generative AI and Algorithmic Art: Disrupting the Framing of Meaning and Rethinking the Subject- Object Dilemma," Global Journal of Computer Science and Technology (2023): https://digitalcommons.lindenwood.edu/faculty-researchpapers/461

Kelleher, J. D. (2019). *Deep Learning*. MIT Press. DOI:10.7551/mitpress/11171.001.0001

Lucci, S., Musa, S. M., & Kopec, D. (2022). *Artificial Intelligence in the 21st Century* (3rd ed.). Mercury Learning and Information. DOI:10.1515/9781683922520

Mungale, S. G., Mungale, N. G., Shaikh, M. S., Mungale, S. G., Wazalwar, S. S., Wanjari, M. M., & Jichkar, R. A. (2024). Safeguard Wrist: Empowering Women's Safety. In Ponnusamy, S., Bora, V., Daigavane, P., & Wazalwar, S. (Eds.), *Wearable Devices, Surveillance Systems, and AI for Women's Wellbeing* (pp. 192–205). IGI Global., DOI:10.4018/979-8-3693-3406-5.ch012

Shaikh, M. S., Ali, S. I., Deshmukh, A. R., Chandankhede, P. H., Titarmare, A. S., & Nagrale, N. K. (2024). AI Business Boost Approach for Small Business and Shopkeepers: Advanced Approach for Business. In Ponnusamy, S., Assaf, M., Antari, J., Singh, S., & Kalyanaraman, S. (Eds.), *Digital Twin Technology and AI Implementations in Future-Focused Businesses* (pp. 27–48). IGI Global., DOI:10.4018/979-8-3693-1818-8.ch003

Shaikh, M. S., Chandrawat, U. B., Choudhary, S. M., Ali, S. I., Ponnusamy, S., Khan, R. A., & Sheikh, A. G. (2024). Harnessing Logistic Industries and Warehouses With Autonomous Carebot for Security and Protection: A Smart Protection Approach. In Ponnusamy, S., Assaf, M., Antari, J., Singh, S., & Kalyanaraman, S. (Eds.), *Harnessing AI and Digital Twin Technologies in Businesses* (pp. 239–257). IGI Global., DOI:10.4018/979-8-3693-3234-4.ch017

Shaikh, M. S., Chandrawat, U. B., Choudhary, S. M., Ali, S. I., Ponnusamy, S., Khan, R. A., & Sheikh, A. G. (2024). Harnessing Logistic Industries and Warehouses With Autonomous Carebot for Security and Protection: A Smart Protection Approach. In Ponnusamy, S., Assaf, M., Antari, J., Singh, S., & Kalyanaraman, S. (Eds.), *Harnessing AI and Digital Twin Technologies in Businesses* (pp. 239–257). IGI Global., DOI:10.4018/979-8-3693-3234-4.ch017

Shaikh, M. S., Ponnusamy, S., Ali, S. I., Wanjari, M., Mungale, S. G., Ali, A., & Baig, I. (2024). AI-Based Advanced Surveillance Approach for Women's Safety. In Ponnusamy, S., Bora, V., Daigavane, P., & Wazalwar, S. (Eds.), *Wearable Devices, Surveillance Systems, and AI for Women's Wellbeing* (pp. 13–25). IGI Global., DOI:10.4018/979-8-3693-3406-5.ch002

Sheikh, M. S.. (2024). Harnessing Logistic Industries Using Autonomous Carebot for Smart Surveillence, Protection and Security. In Al-Turjman, F. (Ed.), *The Smart IoT Blueprint: Engineering a Connected Future. AIoTSS 2024. Advances in Science, Technology & Innovation.* Springer., DOI:10.1007/978-3-031-63103-0_20

Spirling, A. (2023, April). Why open-source generative AI models are an ethical way forward for science. *Nature*, 616(7957), 413. DOI:10.1038/d41586-023-01295-4 PMID:37072520

Stone, P. Brooks, et al. (2022) Artificial Intelligence and Life in 2030: The One Hundred Year Study on Artificial Intelligence. Rep. 2015–2016 Study Panel. Available online: https://ai100.stanford.edu

Tahiru, F. (2008). Pannu (2020), A. Artificial Intelligence and Its Application in Different Areas. *Int. J. Eng. Innov. Technol.*, 4, 79–84.

UNESCO. "Ethics of Artificial Intelligence" (2022): https://www.unesco.org/en/artificialintelligence / recommendation-ethics.

Wang, T. "Navigating Generative AI (ChatGPT) in Higher Education: Opportunities and Challenges," (2023) in C. Anutariya, D. Liu, Kinshuk, A. Tlili, J. Yang, M. Chang (eds.), Smart Learning for A Sustainable Society, ICSLE 2023, Lecture Notes in Educational Technology, Singapore: Springer: https://doi.org/DOI:10.1007/978-981-99-5961-7_28

White, M. "A Brief History of Generative AI," Medium (2023): https://matthewdwhite.medium.com/a-brief-historyof-generative-ai-cb1837e67106

Wooldridge, M. (2021). *A Brief History of Artificial Intelligence: What It Is, Where We Are, and Where We Are Going.* Macmillan Publishers.

Chapter 10
Ethical Leadership in the Age of AI

Sunil Kumar
https://orcid.org/0000-0002-2362-1972
Shoolini University, India

ABSTRACT

This chapter examines leaders' perspectives on the ethical use of AI, drawing insights from 17 respondents across various industries. Six open-ended questions were crafted to explore ethical AI usage, with responses analyzed through sentiment and topic analysis. Most leaders expressed positive views on the ethical dimensions of AI and leadership. Findings highlighted the significance of ethical and moral principles in ensuring data privacy and security within AI-augmented systems. Organizational management is actively developing strategies to address the risks and challenges posed by AI. The study identified five key themes: the need for AI systems to be human-centric, the importance of training and development, ethical AI development, data privacy, and the integration of AI into organizational mainframes. Overall, the research underscores the necessity for leaders to cultivate a responsible approach to AI usage.

INTRODUCTION

The integration of Artificial intelligence in our daily lives has made a lot of things easier and more convenient for all us in all sorts of fields. But this is not without its fair share of problems and on the major problem is the ethical use of AI. So, in a landscape where more often than not, AI is prone to being misused, this study aims to navigate this landscape with the integration of ethical conduct with AI. This

DOI: 10.4018/979-8-3693-9173-0.ch010

will set the tone to both acts ethically and inspire others to follow suit and that best describes ethical leadership in the age of AI.

Ethics are essentially the standards or requirements that help us decide what's right and what is incorrect (Rae, 2018). They are like a moral compass guiding our conduct and choices. Ethics inform us how we need to deal with others fairly, be sincere, and do what's excellent for everyone concerned. They are like rules we as humans evolved to believe under righteousness (Holmes, 2007).

Morality is sort of a personal code of behavior that tells us what is right and wrong (Rae, 2018). It is shaped by way of our upbringing, lifestyle, and reports. Morality can vary from individual to individual, and will be different considering upbringing, tradition, and private ideals. For human beings, morality is frequently influenced by social norms, spiritual teachings, and character experiences. What one person considers morally perfect might not be same for someone else. In fact, morality is such a subjective concept that there are various philosophies about morality given by different theorists and what these philosophies preach is at times even contradictory to each other (Haidt, 2008).

According to 'The Future of Artificial Intelligence and Cybernetics' inception of artificial intelligence as a field can be traced to the mid-20th century, an era characterized by pioneering theories about computational intelligence (Warwick, 2013). Notably, Alan Turing proposed the notion of machines capable of cognitive processes, a profound step that would shape the future discourse on AI (Bishop, 2021; Cappuccio, 2016). In parallel, John McCarthy emerged as an influential figure, famously introducing the term 'artificial intelligence,' thereby solidifying its identity as a distinct sub field in computer science (O'Regan, 2008).

In the ensuing years spanning from the 1950s through to the 1970s, researchers concentrated on symbolic AI (Augusto, 2021). It is a method involving the manipulation of symbols and the application of logic to tackle complex problems. It was during these decades that innovative endeavors, exemplified by the groundbreaking Logic Theorist, provided early demonstrations of what would later be known as artificial intelligence (Luger, 1998).

However, the trailblazing momentum experienced a downturn during the latter part of the 1970s continuing into the 1980s. This period, often referred to as the "AI winter," was marked by a notable slowdown in AI advancements (Mitchell, 2021). Predicated on the inherent constraints of the era's algorithms and a subsequent reduction in research financing, the interest in AI diminished significantly during these years.

The resurgence of artificial intelligence began to take shape in the 1990s, carrying through into the 2000s, fueled by significant strides in machine learning and the neural network domain (Deng, 2018). This revival of interest was underscored by achievements such as IBM's Deep Blue chess program, which captured the global imagination by besting world champion Garry Kasparov (Sharples, 2021).

The last decade has been a golden era for AI, propelled by the maturation of deep learning techniques and an unprecedented access to vast datasets. Leading tech enterprises, including but not limited to Google, Facebook, and Microsoft, have channeled extensive resources into AI development (Franco et al. 2023; Rikap, 2023). Such investments have yielded transformative results across a multitude of domains, firmly establishing AI as an integral component of the modern technological landscape (Corea, 2017).

The machine learning and deep learning in data science are driving the progression of generative AI (Deng, 2018). Foundational and large-scale language models, such as OpenAI's GPT-4 and Google's BERT, play critical roles in transforming natural language processing applications (Myers, 2024). Multimodal AI frameworks, like OpenAI's CLIP and DALL-E, integrate various data types—text, images, audio, and video—to enable automated video synthesis and enhance content generation (Lin, 2024). In addition to its commercial and entertainment uses, generative AI fosters human creativity in fields such as music, visual arts and literature. Technological innovations such as diffusion models, Neural Radiance Fields (NeRFs), and advanced Generative Adversarial Networks (GANs) are crucial in expanding the capabilities of generative AI (Po et al., 2024).

Leadership

The concept of leadership has grown increasingly complex, reflecting human progress and organizational challenges. Yukl (2013) notes that 'leadership' has shifted from common to technical vocabulary in scientific discourse, with definitions influenced by researchers' perspectives. Robbins, Judge, and Vohra (2013) define leadership as the ability to influence a group towards a vision or goals, arising from social processes or formally appointed authorities. Drafke (2009) links formal leadership to organizational authority, responsibility, and accountability. Jain and Bhargav (2010) highlight leadership as a dynamic force crucial for organizational success. Rowe and Guerrero (2010) describe it as a transactional process where leaders and followers mutually influence each other. Greenberg (2011) identifies non-coercive influence, goal-directed influence, and the necessity of followers as key aspects, emphasizing leadership's reciprocal nature. Thus, leadership, both formal and informal, is essential at all organizational levels for modern organizations' survival and maintenance.

Leadership is the ability to be able to influence the masses and drive them towards a cause in way that both furthers the cause the holistic development of the followers. Leadership is an essential factor of effective management, related to the guidance and direction of individuals or organizations toward shared goals (Wood, 2005). It encompasses inspiring, influencing, and guiding others via interpersonal abilities, choice-making skills, and strategic vision. effective leaders show empathy, robust communication abilities, integrity, and adaptability.

Formal leadership arises from appointed or elected roles, such as CEOs and managers, who have authority based on their positions and are responsible for directing their teams (Brown, 1999). In contrast, informal leadership stems from personal characteristics and influence, where respected individuals guide through their expertise and charisma, despite lacking formal titles. Their influence is significant but can be overridden by formal leaders. Autocratic leadership dismisses input from followers, leading to poor decisions and dissatisfaction, while democratic leadership values collective opinions, enhancing employee satisfaction. Free-rein leadership provides maximum freedom to employees, requiring high self-motivation to maintain productivity (Nagarathinam, 2020).

Skilled leaders excel in interpersonal abilities, connecting with coworkers to understand their perspectives and foster an inclusive environment. Their authentic integrity builds a culture of trust, and they embrace proactive accountability, confidently facing challenges. With considerate decision-making, they balance fairness and insight for inclusive outcomes. Adaptability defines them, as they blend resilience with flexibility to adjust strategies based on new information. Their inspirational presence instills purpose and enthusiasm, while their visionary direction strategically charts a course for lasting success. Transparent communication ensures their directives are clear, aligning team efforts with organizational goals.

Ethics and Leadership

The literature on organizational behavior is increasingly examining the connection between ethics and leadership (Fulmer, 2004; Hollander, 1995; Rost, 1995). Ethical leaders prioritize serving their followers ethically, avoiding any actions that would benefit themselves at the expense of their followers (Howell & Avolio, 1992). Such leaders create an environment where followers feel comfortable sharing their concerns and providing honest feedback (Walumbwa & Schaubroeck, 2009). This emphasis on values and ethical leadership highlights the relevance of leadership in contemporary organizations. Scholars in the field are particularly focused on the ethical use of power by leaders who inspire followers to selflessly pursue common or group goals. This trend is reflected in the growing body of conceptual and empirical research on ethical leadership within organizational settings (Greenleaf, 2002; Hale

& Fields, 2007; Neubert, Kacmar, Carlson, Chonko, & Roberts, 2008; Spears, 2004; Van Dierendonck, 2011; Walumbwa, Hartnell, & Oke, 2010).

Leadership is the foundation of responsible management, fostering fairness and impartiality to create a respectful and effective environment for all team members (Cameron, 2012). Ethical leaders cultivate deep trust within their teams, enhancing a robust and collaborative atmosphere that improves teamwork and cohesion (Mercader et al. 2021). By wielding power judiciously, these leaders encourage a culture of thoughtful decision-making throughout the organization. Their dedication to ethical standards enables them to foresee and mitigate potential risks, protecting the organization from legal and ethical issues. Prioritizing integrity, these leaders set a high moral benchmark, inspiring their teams to embrace similar values and boosting overall employee morale. Ultimately, ethical decision-making ensures enduring success, valuing long-term prosperity over short-term gains and paving the way for continued growth and a lasting positive reputation (Stainer, 2003).

Ethics, AI and Leadership

In the realm of AI and ethical leadership, it's crucial to explore traditional moral theories like deontological ethics, consequentialism, and virtue ethics, adapting them to guide AI development and usage. Deontological ethics, rooted in duty-based principles, advocates for AI systems programmed to adhere to moral rules aligned with human rights and dignity (Tseng and Wang, 2021). Ethical leadership in AI from this perspective focuses on creating algorithms that inherently prioritize what is morally right, regardless of outcomes. Consequentialism evaluates AI systems based on their outcomes, striving to maximize benefits and minimize harms, particularly in public policy decisions (Card and Smith, 2020). Virtue ethics shifts focus to the character of AI developers and operators, emphasizing virtues like empathy and fairness reflected in AI design (Farina et al. 2022). Integrating these theories, ethical leadership in AI must balance rule adherence, positive outcomes, and moral character throughout the AI lifecycle, promoting technologies that align with human values and ethical principles across diverse scenarios for robust AI governance.

Man, and machine are significantly contributing in revival and survival of hospitality sector (Kumar et al. 2024). The occupational stress is major issue in hospitality, and leaders are helping employees psychologically and socially to cope with occupation stress (Kumar and Yadav, 2024). AI empowers leaders across various sectors: in health institutions, it enables precise interpretation of medical images; in administrative tasks, it streamlines processes, allowing focus on strategic thinking; in customer interactions, it personalizes experiences; in hiring, it enhances recruitment efficiency; in banking, it detects fraud swiftly; in market analysis, it predicts trends; in global communication, it overcomes language barriers; in manufacturing, it ensures efficiency and precision; in product innovation, it enables personalized solutions; and in promoting environmental sustainability, it optimizes energy usage.

Leaders must prioritize data privacy and security as AI increasingly relies on personal data for decision-making, necessitating vigilant protection of individual privacy (Bankins & Formosa, 2023). Instances like Facebook's use of algorithms for content curation highlight the urgency of discussions about user data privacy and its implications for personal freedom. Addressing bias in AI decision-making is crucial to prevent exacerbating existing prejudices, exemplified by Amazon's discarded AI recruitment software due to gender discrimination (Bankins & Formosa, 2023). The accountability issue is evident in cases like the Uber autonomous vehicle incident, raising questions about culpability when humans aren't directly controlling the technology. With AI reshaping the workforce, leaders must ethically manage employment transitions, offering skill development opportunities to mitigate job displacement (Bankins & Formosa, 2023). Transparency in AI decision-making, particularly in fields like finance, is imperative, sparking debates about the necessity for transparency to those affected by AI decisions.

The ethical and regulatory issues raised by generative AI are critical in discussions about data protection, consent, and misuse (Wach et al., 2023). AI models require large datasets, which raises serious concerns about data privacy and personal information security (Shahriar, 2023) Informed consent is essential for ensuring that individuals understand how their data is used and have the opportunity to withdraw consent at any time (Jones et al., 2018). Misuse of AI includes propagating misinformation using deepfakes, violating intellectual property rights, and perpetuating biases, all of which raise ethical concerns in areas such as surveillance and predictive police (Gosain et al. 2024). To mitigate these concerns, strong laws, ethical AI design standards, and extensive public education are required to promote digital literacy and responsible AI usage. Addressing these concerns enables stakeholders to prudently balance the benefits of generative AI with essential ethical and regulatory safeguards.

Leader's Perspective on AI

In 2023, CEOs reported that the primary AI focus within organizations was hiring talent with AI skills, with 41 percent stating this action was completed. The most common uninitiated AI effort was forming an AI task force reporting directly to the C-suite (EY, 2023). A survey of 1,325 CEOs revealed that 82 percent were concerned AI could introduce new attack methods for adversaries, and nearly 60 percent cited ethical challenges as their main concern in implementing generative AI (KPMG, 2024). Most commercial leaders indicated that their organizations used generative AI infrequently and machine learning occasionally in 2023, with 20 percent noting frequent use of both technologies (McKinsey & Company, 2023).

These leaders believed their organizations should regularly or almost always employ machine learning and generative AI, with no interest in rare or infrequent use (McKinsey & Company, 2023). By 2023, the majority of senior AI executives in the US and Europe were responsible for their companies' AI strategies, except in the consumer industry, where many reported no designated AI strategy owner, and about five percent in healthcare and biotech were unsure (Heidrick & Struggles, 2023).

REVIEW OF LITERATURE

In a high-tech world, leaders grapple with ethical challenges, especially in decision-making. True leadership means navigating these complexities, promoting fairness, and fostering accountability. How leaders handle technology impacts their teams directly, highlighting the need for fairness in automated decisions. It's essential for leaders to cultivate an ethical culture and advocate for fairness throughout the company. The researcher emphasizes the crucial role of ethical integrity in building trust and fairness in technology systems (Uddin, 2023).

In AI-integrated organizations, ethical leadership is crucial for productivity within ethical boundaries. As AI becomes more prevalent, ensuring its ethical alignment is vital. This research highlights the importance of cultivating ethical AI governance, emphasizing accountability and transparency. It showcases real-life examples, demonstrating how ethical leadership in AI-driven organizations acts as a beacon for integrity and responsibility, shaping a future focused on social well-being (Mohav, 2023).

In the realm of business ethics, the allure of AI as a savior is strong. However, De Cremer and Narayanan (2023) caution against overestimating AI's ability to make ethical decisions. They argue that despite AI's intelligence, humans still hold the reins of ethical decision-making. Instead of relying excessively on AI, they advocate for strengthening the ethical capabilities of companies and their leaders. AI serves as a mirror, revealing our biases and mistakes, prompting us to improve our decision-making. While AI excels in data analysis and scenario forecasting, the real power lies in combining human judgment with AI's capabilities. This partnership, they believe, is key to enhancing ethical standards in preparation for a digital future that is not only smart but also genuinely ethical.

Nicolae and Nicolae (2018) discuss the challenges confronting higher education amidst rapid global changes, particularly with the rise of artificial intelligence. They highlight the unprecedented pace of AI development, posing difficulties for education systems to stay abreast. Leaders in higher education must navigate concerns like job displacement and data control while ensuring institutions remain current with AI advancements. The article explores how education leaders are tackling these challenges and adapting to AI's fast-paced changes, stressing the importance of proactive leadership in integrating AI into education and preparing for its societal effects.

Milton and Al-Busaidi (2023) explore the necessity for education leaders to adjust to the evolving landscape shaped by artificial intelligence (AI). They emphasize the increasing competitiveness in higher education and the vital role of digital technology in this context. As AI becomes a central component of digital transformation, leaders must be prepared to adapt, stay technologically current, and comprehend AI's implications. The article delves into how AI could redefine leadership expectations, posing crucial questions about its impact on leadership roles, the requisite skills for leaders in an AI-driven environment, and the potential for AI to assume leadership responsibilities. By addressing these questions, the paper aims to shed light on the future of educational leadership amid AI integration.

Abasaheb and Subashini (2023) highlight the transformative impact of digitalization on business, particularly with the rise of AI. They underscore how AI is poised to reshape future workforces and empower leadership. In today's intricate business landscape, ethical leadership is paramount. The article delineates how ethical leaders set the tone for organizational culture, guide decision-making, and establish norms for team behavior. It emphasizes attributes like transparency, honesty, accountability, and integrity in ethical leaders, who serve as role models for others. Research consistently showcases the positive outcomes of ethical leadership, including building trust, credibility, and fostering ethical behavior. The article advocates for cultivating ethical leaders through targeted training, aimed at enhancing their ability to make ethical decisions and navigate complex ethical issues within their organizations. It examines proactive measures organizations should take to promote ethical conduct and uphold a culture of integrity (Taj, 2023).

Vaja (2017) underscores the unique ethical challenges arising from rapid technological advancement in organizations. Ethical leadership is pivotal in navigating these challenges, fostering an ethical culture, and ensuring responsible technology use. Leaders set the tone for ethical behavior, create an environment conducive to ethical decision-making, and consider the ethical implications of new technology in today's digital age.

De Cremer (2022) sheds light on the increasing influence of AI on organizational dynamics and stakeholder expectations. The emergence of AI ethics underscores the importance of human-centric leadership in contrast to a "machines first" approach. Leaders must prioritize a "humans first" mindset, embracing purpose-driven and inclusive leadership to counteract the tendency to rely solely on machines.

As technology evolves, AI becomes more integrated into organizations, serving as more than just a tool. However, AI leadership can evoke negative feelings among employees, impacting job satisfaction and ethical decisions. Leadership research highlights the complex interplay between leaders, followers, and context, emphasizing the importance of considering various factors in understanding organizational outcomes (Tiago, 2021; Silva et al., 2024).

As AI becomes increasingly prominent in various aspects of life, we're also witnessing the emergence of AI-equipped bots that facilitate human interaction. However, along with the benefits, challenges related to ethics and leadership arise in the development and use of AI and robotics. As creators of AI, we bear ethical responsibility towards its development and its impact on human leadership. Recognizing these challenges, leaders must address them with care and foresight to shape a responsible future (Kodish, 2023).

While much research focuses on digital transformation, the human element is often overlooked, particularly in understanding a leader's role in driving organizational change in the digital age. A human-centric approach to leadership is crucial for successful digital transformation. In today's fast-paced world, leadership requires more than traditional management skills; foresight and adaptability are paramount. AI can assist leaders in managing more effectively, but success ultimately hinges on a leader's ability to integrate AI's analytical capabilities with their own intuition (Schiuma et al., 2024; Teixeira & Pacione, 2024).

The literature exploring ethical leadership and the integration of artificial intelligence (AI), several key themes emerge. Scholars discuss the challenges faced by leaders in tech-driven contexts and the importance of ethical leadership in fostering accountability and fairness (Uddin, 2023; Mohav, 2023). They explore how AI is reshaping leadership responsibilities in various sectors, including higher education and business organizations, emphasizing the need for technological literacy and agility (Nicolae & Nicolae, 2018; Milton & Al-Busaidi, 2023). Additionally, researchers highlight the foundational role of ethical leadership in shaping organizational culture and decision-making processes, particularly in navigating the ethical challenges associated with AI and robotics (Taj, 2023; Vaja, 2017). Ultimately, the literature underscores the importance of human-centric leadership in driving organizational transformation in the digital age, balancing AI integration with ethical considerations and maintaining a focus on the well-being of individuals within organizations (De Cremer, 2021; Rodrigues, 2021; Schiuma et al., 2024; Teixeira & Pacione, 2024).

Objectives of the study

- To explore the sentiments of leaders toward ethical use of AI
- To identify the topics associated with ethical use of AI by leaders.

METHOD

A research methodology is a structured and scientific approach used to collect, analyze, and interpret qualitative or quantitative data to answer research questions or test hypothesis (**Choy, 2014**). It's like a plan for carrying out research and helps keep researchers on track by limiting the scope of the research. The "Ethical Leadership in the Age of AI" research examines how AI has been shaping professional experiences, zooming in on the ethical dimensions that have surfaced. A qualitative research approach was used, and 6 open ended questions were prepared and given below.

- **Item_1:** How do you perceive your role in navigating the ethical challenges posed by the integration of AI technologies in your industry?
- **Item_2:** How can you help to foster an ethical AI culture within your organizations?
- **Item_3:** Can you share any example of AI implementation and ethical issues during your job?
- **Item_4:** How you as a leader adapting to the AI driven changes?
- **Item_5:** What strategies as a leader you employ to ensure that AI development and deployment align with ethical standards in your job?
- **Item_6:** What do you believe the most pressing ethical dilemmas that you will face in the near future regarding ethical use of AI?

The items were validated by two academic experts for face validity. This study engaged with 17 individuals, chosen because they could offer in-depth and illustrative takes on the topic. The data collection started through an online questionnaire, circulated among professionals via emails and industry networks, ensuring that respondents were well-positioned to provide valuable insights. The structured nature of Google Forms made it an ideal platform, as it accommodated a mixture of straightforward, choice-based queries alongside more expansive, open-ended ones, but this time we edged towards the latter as we it was a descriptive interview rather than a choice based one. Out of 17 individual 9 were male and 8 were female. The average age of respondent was 30 years with an average experience of 4 years.

To ensure clarity and consistency in the findings, the responses underwent a detailed cleaning process with the assistance of Python's programming tools, such as Pandas and NLTK. Additionally, sentiment analysis was performed. This step is akin to shifting through the subtleties of human emotions conveyed in text, categorizing them into positive, negative, or neutral sentiments with the help of the aforementioned libraries in Python.

RESULTS

Objective 1: To Explore the Sentiments of Leaders Toward Ethical Use of AI

Text Analysis-Sentiment Analysis: Sentiment analysis was used to analyze the collected information, revealing key themes. Sentiment analysis is a natural language processing used to determine whether the data is positive, negative, or neutral **(Stine, 2019).** It is often performed on textual data to help businesses monitor brand and product sentiment in customer feedback and understand customer needs. This approach provided subtle insights about what experienced professionals feel about AI culture and its implementation in the organizations.

Table 1. Sentiment analysis scores

Respondent	Item_1 (Compound score)	Item_2 (Compound score)	Item_3 (Compound score)	Item_4 (Compound score)	Item_5 (Compound score)	Item_6 (Compound score)
Resp_1	0.945 (Positive)	0.796 (Positive)	0.402 (Positive)	0 (Neutral)	0.709 (Positive)	0.226 (Positive)
Resp_2	-0.296 (Negative)	0.572 (Positive)	0.340 (Positive)	0.153 (Positive)	-0.402 (Negative)	0.402 (Positive)
Resp_3	0 (Neutral)	-0.151 (Negative)	-0.296 (Negative)	-0.296 (Negative)	-0.296 (Negative)	0 (Neutral)
Resp_4	0.875 (Positive)	0.637 (Positive)	0.434 (Positive)	0.896 (Positive)	0.852 (Positive)	0.944 (Positive)
Resp_5	-0.0772 (Negative)	0 (Neutral)	0 (Neutral)	0.493 (Positive)	0 (Neutral)	0 (Neutral)
Resp_6	0.724 (Positive)	0 (Neutral)	0 (Neutral)	0 (Neutral)	-0.510 (Negative)	-0.422 (Negative)
Resp_7	0.984 (Positive)	0.990 (Positive)	0.920 (Positive)	0.920 (Positive)	0.986 (Positive)	0.963 (Positive)
Resp_8	0.9828 (Positive)	0.993 (Positive)	0.965 (Positive)	0.920 (Positive)	0.987 (Positive)	0.963 (Positive)
Resp_9	0.9382 (Positive)	0.883 (Positive)	0.679 (Positive)	0.807 (Positive)	0.964 (Positive)	-0.318 (Negative)
Resp_10	-0.296 (Negative)	0 (Neutral)	0.382 (Positive)	0.805 (Positive)	0.758 (Positive)	-0.296 (Negative)
Resp_11	0 (Neutral)	0.878 (Positive)	0 (Neutral)	0.659 (Positive)	0.671 (Positive)	0.153 (Positive)
Resp_12	0.128 (Positive)	0.512 (Positive)	-0.318 (Negative)	-0.052 (Negative)	-0.119 (Negative)	0 (Neutral)

continued on following page

Table 1. Continued

Respondent	Item_1 (Compound score)	Item _2 (Compound score)	Item_3 (Compound score)	Item_4 (Compound score)	Item_5 (Compound score)	Item_6 (Compound score)
Resp_13	0.9678 (Positive)	0.856 (Positive)	0.459 (Positive)	0 (Neutral)	0.934 (Positive)	0.950 (Positive)
Resp_14	0.4019 (Positive)	0.511 (Positive)	0.512 (Positive)	0.318 (Positive)	0.765 (Positive)	0.440 (Positive)
Resp_15	0.645 (Positive)	0.742 (Positive)	0 (Neutral)	0.338 (Positive)	0.732 (Positive)	-0.402 (Negative)
Resp_16	0.7269 (Positive)	0.402 (Positive)	0 (Neutral)	0 (Neutral)	0 (Neutral)	0 (Neutral)
Resp_17	0.7227 (Positive)	0.361 (Positive)	0 (Neutral)	0.892 (Positive)	0 (Neutral)	-0.153 (Negative)

Source: Primary data

Item_1: How do you perceive your role in navigating the ethical challenges posed by the integration of AI technologies in your industry?

The perception of respondents regarding their role in addressing the ethical challenges posed by AI integration in their industry is predominantly positive. Out of 17 responses, 12 expressed a favorable view of their role, with high compound scores like 0.9828 and 0.9678 indicating strong positive feelings. A few respondents (2) maintained a neutral outlook, with compound scores of 0 reflecting a lack of strong feelings either way. Only 3 responses showed negative sentiment, and even these had low compound scores (-0.296), signifying minor concerns rather than strong opposition. The key themes extracted from their responses include ethical challenges, AI integration, responsible usage, human well-being, ethical principles, transparency, and accountability.

Item_2: How can you help to foster an ethical AI culture within your organization?

Most respondents have a proactive stance on fostering an ethical AI culture within their organizations. The majority of responses were positive, with high compound scores (0.9902, 0.9928, 0.8834) indicating strong conviction towards promoting ethical AI practices. A few neutral responses (3) suggest some respondents are undecided or take an analytical approach, possibly due to evolving regulations and inadequate training. There was only one negative response, which had a mild compound score of -0.1511. The main themes from their responses include AI governance frameworks, ethical guidelines, education, transparency, continuous monitoring, and responsible AI practices.

Item_3: Can you share any example of AI implementation and ethical issues during your job?

When asked about examples of AI implementation and associated ethical issues, most respondents shared positive experiences. The majority of positive responses had compound scores below 0.5, indicating well-considered views with a neutral component. This question elicited the highest number of neutral replies (6), suggesting that real-life AI implementation is complex and nuanced. There were 2 negative responses with moderate compound scores (-0.296, -0.3182), indicating some challenges but nothing severe. Key themes included AI implementation, ethical issues, privacy, bias, transparency, and accountability.

Item_4: How are you as a leader adapting to the AI-driven changes?

Respondents generally expressed positive sentiments about their adaptability to AI-driven changes. The majority felt confident and positive, with compound scores ranging from 0.1531 to 0.9201, reflecting varying levels of confidence. A significant portion (23%) had a neutral stance, suggesting a cautious and analytical approach to AI adoption in their organizations. Only 2 responses were negative, with mild compound scores (-0.296, -0.0516), indicating minimal challenges. The main themes included continuous learning, responsible implementation, collaboration, transparency, strategic planning, and innovation.

Item_5: What strategies as a leader do you employ to ensure that AI development and deployment align with ethical standards in your job?

More than half of the respondents (59%) shared positive sentiments about their strategies to align AI development with ethical standards. Their responses had compound scores ranging from 0.6705 to 0.9869, indicating thoughtful and well-considered strategies. Three neutral responses suggest some uncertainty about the effectiveness of their strategies. Four responses were negative, with compound scores ranging from -0.5106 to -0.296, indicating moderate to small challenges in formulating these strategies. The key themes were ethical guidelines, transparency, accountability, interdisciplinary collaboration, continuous learning, and ethical implications.

Item_6: What do you believe are the most pressing ethical dilemmas you will face in the near future regarding the ethical use of AI?

Respondents expressed the lowest positive sentiment for this question, with only 8 out of 17 (47%) having an optimistic outlook on dealing with future ethical dilemmas. The compound scores for positive responses varied widely (0.1531 to 0.9633). Four neutral responses suggest a degree of uncertainty and caution. Five responses were negative, with compound scores ranging from -0.4215 to -0.1531, indicating varying levels of concern. The main themes were deep fakes, misinformation, bias, fairness, transparency, accountability, human control, socio-economic inequalities, and responsible AI use.

Objective 2: To Identify the Topics Associated With Ethical Use of AI by Leaders

Text analysis- Topic Modelling: Text mining is a process of deriving meaningful information from text. Qualitative studies benefit from the topic modeling analysis approach. To apply topic models to text data, one needs a quality metric for matching human judgment (Nikolenko et al., 2017). Topic models are based on natural language processing, machine learning, and are generative in nature. Latent Dirichlet Allocation (LDA) is used as a generative technique in the present study. The LDA method observes the unobserved text patterns in the data (Tong and Zhang, 2016). The R package is used to apply LDA topic modeling. The outcome of the analysis identified 5 topics with relative probabilities. Figure 1 presents the 5 topics with their terms and beta values. The following are the topics identified using LDA on the text analysis of ethics, leadership, and AI-based question responses.

- *AI users concern in organization:* This topic seems to focus on issues related to privacy, data security, and user concerns, possibly in the context of algorithms and user consent. The top terms are privacy, data, fair, concern, user, algorithm, challenging, account, consent, clear.
- *AI integration and implementation:* This topic appears to center around technological advancement, integration, impact, and the importance of staying informed and engaged with industrial developments. The top terms are technology, integration, impact, help, inform, industrial, crucial, stay, engage, advancements.
- *Ethics in development of AI:* This topic seems to deal with ethics, development, transparency, promotion, and the implementation of research and development processes. The top terms are ethic, develop, can, transparency, promote, make, implement, research, foster, deploy.

- **AI in human centered system:** This topic focuses on systems, bias, human factors, addressing issues, responses, and implications, suggesting a focus on fairness and protection in systems. The top terms are system, bias, human, address, response, protect, implication.

- **AI training and development:** This topic seems to be about organizational practices, ensuring proper use, education, cultural values, and processes. The top terms are use, work, organization, ensure, education, practice, culture, value, process, sure.

Figure 1. Topic modelling: ethics, leadership, and AI

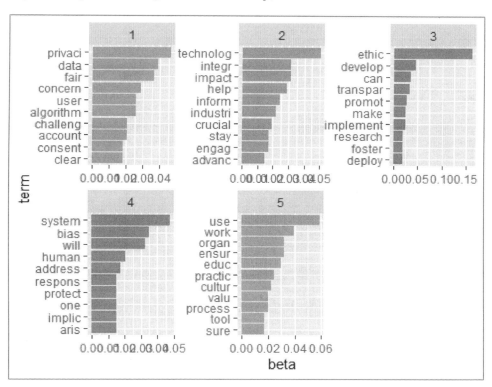

The LDA topic modeling analysis identifies various aspects of ethical considerations in AI. These encompass immediate practical issues such as promoting ethical practices and monitoring AI systems, alongside strategic concerns like ensuring algorithmic integrity and mitigating bias. It examines real-world examples and challenges in integrating AI, while it also focuses on how organizations adapt to AI-driven transformations. Topics in model anticipates future ethical dilemmas related to privacy and accountability. This structured analysis enables organizations to better understand and prioritize different dimensions of ethical AI integration, spanning from implementation challenges to long-term strategies and emerging ethical issues, thereby aiding informed decision-making in AI development and deployment.

Table 2. Relative importance of topics

Items/Questions	Topic 1 Relative Importance	Topic 2 Relative Importance
How do you perceive your role in navigating the ethical challenges posed by the integration of AI technologies in your industry?	0.53	0.47
How can you help to foster an ethical AI culture within your organizations?	0.58	0.42
Can you share any example of AI implementation and ethical issues during your job?	0.77	0.23
How you as a leader adapting to the AI driven changes?	0.79	0.21
What strategies as a leader you employ to ensure that AI development and deployment align with ethical standards in your job?	0.52	0.48
What do you believe the most pressing ethical dilemmas that you will face in the near future regarding ethical use of AI?	0.68	0.32

Source: Primary data

The first topic (Topic 1) focuses on practical and immediate ethical issues of AI implementation, including real-world examples, adapting to changes, and future dilemmas. The second topic (Topic 2) emphasizes strategic and procedural aspects, such as navigating challenges and ensuring ethical standards alignment.

DISCUSSION

The 62 percent of responses were positive, showing an optimistic outlook on ethical leadership in AI. These responses had high compound scores, indicating strong positivity. This sentiment suggests a favorable view of integrating ethics into AI in leadership roles, which can be leveraged to encourage further engagement and proactive AI ethics strategies. 21% of responses were neutral, indicating a cautious "wait and see" approach to AI's integration with ethics. This suggests that while most people are positive, a significant number remain uncertain and prefer to observe how changes unfold before making irreversible decisions. The 17 negative responses reflect minor grievances about AI or its ethical implementation, indicated by their low compound scores. While these issues are minor, they should be addressed to prevent escalation. Transparency and inclusive discussions can help mitigate negativity and enhance positive sentiment. The sentiment analysis shows a predominantly positive view of AI and its ethical considerations, with most respondents feeling confident and proactive. However, there were also significant neutral and negative responses, highlighting areas needing further engagement. Addressing these concerns can strengthen ethical AI strategies. Maddula (2018) explores the impact of AI and reciprocal symmetry on leadership and organizational culture, emphasizing data-driven decision-making and inclusivity, and the need for ethical and regulatory frameworks for responsible AI use. The responses indicate that while many are optimistic about handling ethical dilemmas in AI, others remain cautious or skeptical. Overall, the sentiment distribution reflects a strong engagement with AI and a commitment to ethical standards, despite the mixed views on the challenges involved.

The leader perceived their role positive in navigating the ethical challenges posed by the integration of AI technologies. A positive outlook toward integrating AI technologies into organizational culture is perceived by the leaders in organizations. Leaders are more human-centric in adoption and implementation of digital technologies in contemporary organizations (Teixeira & Pacione, 2024).

In topic modeling, AI users' concerns about data security and privacy are the foremost topics discussed in the literature, and this study confirms the same. According to Carmody et al. (2021), advanced AI models are more capable of extracting sensitive personal information. The execution of AI at the organizational level is quite a complex task; communication and engagement at different levels are required to ensure the success of various AI-based technologies in organizations. Leaders perceive this as a challenge in the implementation and adoption of AI in organizations. The role of moral consciousness seems to be important in the ethical implementation of AI tools and models. Leaders perceive ethical transparency in adoption and implementation as key factors for AI integration in organizations. Viability, feasibility, and usability are the core ideas around human-centric implementation of AI. User capability and capacity are core components of AI model development and implementation stages. Lastly, leaders advocate the role of training and development in the deployment of AI models in organizations.

CONCLUSION

The findings resulted in a rich combination of insights displaying the nuanced ethical predicament faced in professional environments when AI is concerned. The study spotlighted these issues and attempted to suggest ways to combat these dilemmas. While the study is candid about its constraints, such as the concentrated sample size and the subjective nature of the narrative-based answers, the hope is that this research contributes a meaningful layer to the dialogue around ethics in the AI sphere. It is an invitation for organizations to recalibrate and prioritize responsible direction in the rapidly evolving landscape of artificial intelligence. Overall, Study confirms a positive stance of leaders toward ethical use of AI in organizational contexts.

REFERENCES

Abasaheb, S. A., & Subashini, R. (2023). Maneuvering of Digital Transformation: Role of Artificial Intelligence in Empowering Leadership-An Empirical Overview. *International Journal of Professional Business Review: Int.J. Prof. Bus. Rev.*, 8(5), 20.

Augusto, L. M. (2021). From symbols to knowledge systems: A. Newell and HA Simon's contribution to symbolic. *AI*, 2(1), 29–62.

Bankins, S., & Formosa, P. (2023). The ethical implications of artificial intelligence (AI) for meaningful work. *Journal of Business Ethics*, 185(4), 725–740. DOI:10.1007/s10551-023-05339-7

Bishop, J. M. (2021). Artificial intelligence is stupid and causal reasoning will not fix it. *Frontiers in Psychology*, 11, 513474. DOI:10.3389/fpsyg.2020.513474 PMID:33584394

Brown, C. V. (1999). Horizontal mechanisms under differing IS organization contexts. *Management Information Systems Quarterly*, 23(3), 421–454. DOI:10.2307/249470

Cameron, K. (2012). Responsible leadership as virtuous leadership. *Responsible leadership*, 25-35.

Cappuccio, M. L. (2016). The Seminal Speculation of a Precursor: Elements of Embodied Cognition and Situated AI in Alan Turing. *Fundamental Issues of Artificial Intelligence*, 479-496.

Card, D., & Smith, N. A. (2020). On consequentialism and fairness. *Frontiers in Artificial Intelligence*, 3, 34. DOI:10.3389/frai.2020.00034 PMID:33733152

Carmody, J., Shringarpure, S., & Van de Venter, G. (2021). AI and privacy concerns: A smart meter case study. *Journal of Information. Communication and Ethics in Society*, 19(4), 492–505. DOI:10.1108/JICES-04-2021-0042

Choy, L. T. (2014). The strengths and weaknesses of research methodology: Comparison and complimentary between qualitative and quantitative approaches. *IOSR journal of humanities and social science, 19*(4), 99-104.

Corea, F. (2017). *Artificial intelligence and exponential technologies: Business models evolution and new investment opportunities*. Springer.

De Cremer, D. (2022). With AI entering organizations, responsible leadership may slip! *AI and Ethics*, 2(1), 49–51. DOI:10.1007/s43681-021-00094-9

De Cremer, D., & Narayanan, D. (2023). How AI tools can—and cannot—help organizations become more ethical. *Frontiers in Artificial Intelligence*, 6, 1093712. DOI:10.3389/frai.2023.1093712 PMID:37426304

Deng, L. (2018). Artificial intelligence in the rising wave of deep learning: The historical path and future outlook [perspectives]. *IEEE Signal Processing Magazine*, 35(1), 180–177. DOI:10.1109/MSP.2017.2762725

Drafke, M. (2009). *Human side of organizations* (10th ed.). PHI Learning Private Ltd.

EY. (October 24, 2023). CEO perspectives on the current state of artificial intelligence (AI) at their organization worldwide in 2023 [Graph]. In *Statista*. Retrieved June 16, 2024, from https://www.statista.com/statistics/1445865/ceo-perspectives-on-ai-usage-at-their-organization/

Farina, M., Zhdanov, P., Karimov, A., & Lavazza, A. (2022). AI and society: A virtue ethics approach. *AI & Society*, 1–14.

Franco, S. F., Graña, J. M., Flacher, D., & Rikap, C. (2023). Producing and using artificial intelligence: What can Europe learn from Siemens's experience? *Competition & Change*, 27(2), 302–331. DOI:10.1177/10245294221097066

Fulmer, R. M. (2004). The challenge of ethical leadership. *Organizational Dynamics*, 33(3), 307–317. DOI:10.1016/j.orgdyn.2004.06.007

Gosain, M. T., Bhatia, M. K., Sharma, M. R., Bhanvra, M. S., Shaw, A., Singh, T., & Kashyap, B. H. (2024). Artificial Intelligence and the Privacy Paradox: Challenges and Opportunities in Legal Adaptations. *Educational Administration: Theory and Practice*, 30(5), 10384–10394.

Greenberg, J. (2011). *Behaviour in organizations* (10th ed.). PHI Learning Private Ltd.

Greenleaf, R. K. (2002). *Servant leadership: A journey into the nature of legitimate power and greatness*. Paulist Press.

Haidt, J. (2008). Morality. *Perspectives on Psychological Science*, 3(1), 65–72. DOI:10.1111/j.1745-6916.2008.00063.x PMID:26158671

Hale, J. R., & Fields, D. L. (2007). Exploring servant leadership across cultures: A study of followers in Ghana and the USA. *Leadership*, 3(4), 397–417. DOI:10.1177/1742715007082964

Heidrick & Struggles. (October 17, 2023). American and European companies' responsible person for the artificial intelligence (AI) strategy in 2023, by industry [Graph]. In *Statista*. Retrieved June 16, 2024, from https://www.statista.com/statistics/1455283/ai-strategy-leaders-by-industry/

Hollander, E. P. (1995). Ethical challenges in the leader-follower relationship. *Business Ethics Quarterly*, 5(1), 55–65. DOI:10.2307/3857272

Holmes, A. F. (2007). *Ethics: Approaching moral decisions*. InterVarsity Press.

Howell, J. M., & Avolio, B. J. (1992). The ethics of charismatic leadership: Submission or liberation? *The Academy of Management Perspectives*, 6(2), 43–54. DOI:10.5465/ame.1992.4274395

Jain, S., & Bhargav, S. (2010). *Human resource management*. Knowledge Book Distributors.

Jones, M. L., Kaufman, E., & Edenberg, E. (2018). AI and the ethics of automating consent. *IEEE Security and Privacy*, 16(3), 64–72. DOI:10.1109/MSP.2018.2701155

Kodish, S. (2020). The Age of Artificial Intelligence and Robotics: Challenges and Issues. *Journal of Leadership, Accountability and Ethics*, 17(5), 42–52. https://articlearchives.co/index.php/JLAE/article/view/3954

KPMG. (January 1, 2024). CEO perspectives on generative artificial intelligence (AI) in their workplace worldwide in 2023 [Graph]. In *Statista*. Retrieved June 16, 2024, from https://www.statista.com/statistics/1445762/ceo-perspectives-on-generative-ai-worldwide/

Kumar, S., & Yadav, M. (2024). Occupation Stress and Leadership in the Hospitality Industry. In Valeri, M., & Sousa, B. (Eds.), *Human Relations Management in Tourism* (pp. 163–187). IGI Global., DOI:10.4018/979-8-3693-1322-0.ch008

Kumar, S., Yadav, M., & Kumar, D. (2024). Viability of Man and Machine as Co-Workers in the Hotel Industry. In Nozari, H. (Ed.), *Building Smart and Sustainable Businesses With Transformative Technologies* (pp. 189–204). IGI Global., DOI:10.4018/979-8-3693-0210-1.ch011

Lin, L., Gupta, N., Zhang, Y., Ren, H., Liu, C. H., Ding, F., . . . Hu, S. (2024). Detecting multimedia generated by large ai models: A survey. arXiv preprint arXiv:2402.00045. DOI:10.36227/techrxiv.170723324.44685515/v1

Luger, G. F. (1998). *Artificial Intelligence: Structures and Strategies for Complex Problem Solving, 5/e*. Pearson Education India.

Maddula, S. S. (2018). The Impact of AI and Reciprocal Symmetry on Organizational Culture and Leadership in the Digital Economy. *Engineering International*, 6(2), 201–210. DOI:10.18034/ei.v6i2.703

McKinsey & Company. (May 11, 2023). Extent of which commercial leaders feel their organizations are using machine learning (ML) or generative artificial intelligence (AI) in 2023 [Graph]. In *Statista*. Retrieved June 16, 2024, from https://www.statista.com/statistics/1411472/commerical-leader-use-of-gen-ai/

McKinsey & Company. (May 11, 2023). Extent of which commericial leaders feel their organizations should be using machine learning (ML) or generative artificial intelligence (AI) in 2023 [Graph]. In *Statista*. Retrieved June 16, 2024, from https://www.statista.com/statistics/1411478/desired-commerical-leaders-use-of-generative-ai/

Mercader, V., Galván-Vela, E., Ravina-Ripoll, R., & Popescu, C. R. G. (2021). A focus on ethical value under the vision of leadership, teamwork, effective communication and productivity. *Journal of Risk and Financial Management*, 14(11), 522. DOI:10.3390/jrfm14110522

Milton, J., & Al-Busaidi, A. (2023). New role of leadership in AI era: Educational sector. *In SHS Web of Conferences* (Vol. 156, p. 09005). EDP Sciences.

Mitchell, M. (2021). Why AI is harder than we think. *arXiv preprint arXiv:2104.12871*. DOI:10.1145/3449639.3465421

Mohav, Y. (2023). Ethical Leadership in AI-driven Organizations: Balancing Innovation and Responsibility. *International Journal of Advanced Engineering Technologies and Innovations*, 1(04), 204–220.

Myers, D., Mohawesh, R., Chellaboina, V. I., Sathvik, A. L., Venkatesh, P., Ho, Y. H., Henshaw, H., Alhawawreh, M., Berdik, D., & Jararweh, Y. (2024). Foundation and large language models: Fundamentals, challenges, opportunities, and social impacts. *Cluster Computing*, 27(1), 1–26. DOI:10.1007/s10586-023-04203-7

Nagarathinam, D. (2020). Leadership styles, qualities, and characteristics of the world great leaders with constitutional and judicial flavors. *Issue 4 Int'l JL Mgmt. &. Human.*, 3, 1428.

Neubert, M. J., Kacmar, K. M., Carlson, D. S., Chonko, L. B., & Roberts, J. A. (2008). Regulatory focus as a mediator of the influence of initiating structure and servant leadership on employee behavior. *The Journal of Applied Psychology*, 93(6), 1220–1233. DOI:10.1037/a0012695 PMID:19025244

Nicolae, M., & Nicolae, E. E. (2018). Leadership in Higher Education–coping with AI and the turbulence of our times. *InProceedings of the International Conference on Business Excellence* (Vol. 12, No. 1, pp. 683-694). DOI:10.2478/picbe-2018-0061

Nikolenko, S. I., Koltcov, S., & Koltsova, O. (2017). Topic modelling for qualitative studies. *Journal of Information Science*, 43(1), 88–102. DOI:10.1177/0165551515617393

O'Regan, G. (2008). Artificial Intelligence and Expert Systems. In *A Brief History of Computing* (pp. 149–177). Springer London. DOI:10.1007/978-1-84800-084-1_5

Po, R., Yifan, W., Golyanik, V., Aberman, K., Barron, J. T., Bermano, A., Chan, E., Dekel, T., Holynski, A., Kanazawa, A., Liu, C. K., Liu, L., Mildenhall, B., Nießner, M., Ommer, B., Theobalt, C., Wonka, P., & Wetzstein, G. (2024, May). State of the art on diffusion models for visual computing. *Computer Graphics Forum*, 43(2), e15063. DOI:10.1111/cgf.15063

Rae, S. (2018). *Moral choices: An introduction to ethics.* Zondervan Academic.

Rikap, C. (2023). The expansionary strategies of intellectual monopolies: Google and the digitalization of healthcare. *Economy and Society*, 52(1), 110–136. DOI:10.1080/03085147.2022.2131271

Robbins, S. P., Judge, T. A., & Vohra, N. (2013). *Organizational behaviour* (15th ed.). Pearson India.

Rost, J. C. (1995). Leadership: A discussion about ethics. *Business Ethics Quarterly*, 5(1), 129–142. DOI:10.2307/3857276

Rowe, W. G., & Guerrero, L. (2010). *Cases in leadership (2nd* ed.). New Delhi. Sage Publication India Pvt. Ltd.

Schiuma, G., Santarsiero, F., Carlucci, D., & Jarrar, Y. (2024). Transformative leadership competencies for organizational digital transformation. *Business Horizons*, 67(4), 425–437. Advance online publication. DOI:10.1016/j.bushor.2024.04.004

Shahriar, S., Allana, S., Hazratifard, S. M., & Dara, R. (2023). A survey of privacy risks and mitigation strategies in the Artificial Intelligence life cycle. *IEEE Access : Practical Innovations, Open Solutions*, 11, 61829–61854. DOI:10.1109/ACCESS.2023.3287195

Sharples, J. (2021). 'The machine being set in motion': The automaton chess-player in urban and literary culture, 1839–1851. *Sport in History*, 41(2), 181–212. DOI:10.1080/17460263.2021.1906310

Silva, F. A. P., Velez, M. J. P., & Borrego, P. J. V. B. M. (2024). Perceived leadership effectiveness and turnover intention in remote work: The mediating role of communication. *Caderno Pedagógico*, 21(3), e3200–e3200. DOI:10.54033/cadpedv21n3-093

Spears, L. C. (2004). Practicing servant-leadership. *Leader to Leader*, 2004(34), 7–11. DOI:10.1002/ltl.94

Stainer, L. (2003). Ethical dimensions of management decision-making: a stakeholder values approach to performance and strategy. https://doi.org/DOI:10.18745/th.14114

Stine, R. A. (2019). Sentiment analysis. *Annual Review of Statistics and Its Application*, 6(1), 287–308. DOI:10.1146/annurev-statistics-030718-105242

Taj, Y. (2023). Ethical Dilemmas in Leadership: Navigating Difficult Situations with Integrity. *International Journal For Multidisciplinary Research*, 5(4), 4222. DOI:10.36948/ijfmr.2023.v05i04.4222

Teixeira, N., & Pacione, M. (2024). *Implications of Artificial Intelligence on Leadership in Complex Organizations: An Exploration of the Near Future.* (Master's Thesis)

Tiago, I. R. (2021). The influence of artificial intelligence leadership on employee ethical decision-making (Doctoral dissertation).

Tong, Z., & Zhang, H. (2016, May). A text mining research based on LDA topic modelling. In *International conference on computer science, engineering and information technology* (pp. 201-210). DOI:10.5121/csit.2016.60616

Tseng, P. E., & Wang, Y. H. (2021). Deontological or utilitarian? An eternal ethical dilemma in outbreak. *International Journal of Environmental Research and Public Health*, 18(16), 8565. DOI:10.3390/ijerph18168565 PMID:34444311

Uddin, A. S. M. (2023). The Era of AI: Upholding ethical leadership. *Open Journal of Leadership*, 12(04), 400–417. DOI:10.4236/ojl.2023.124019

Vaja, J. R. (2017). Ethical Leadership in the Digital Age: Assessing the Role of Leaders in Nurturing Ethical Behavior in Technology-Driven Organizations. *International Journal of Management and Development Studies*, 6(10), 118–126. DOI:10.53983/ijmds.v6n10.013

Van Dierendonck, D. (2011). Servant leadership: A review and synthesis. *Journal of Management*, 37(4), 1228–1261. DOI:10.1177/0149206310380462

Wach, K., Duong, C. D., Ejdys, J., Kazlauskaitė, R., Korzynski, P., Mazurek, G., Paliszkiewicz, J., & Ziemba, E. (2023). The dark side of generative artificial intelligence: A critical analysis of controversies and risks of ChatGPT. *Entrepreneurial Business and Economics Review*, 11(2), 7–30. DOI:10.15678/EBER.2023.110201

Walumbwa, F. O., Hartnell, C. A., & Oke, A. (2010). Servant leadership, procedural justice climate, service climate, employee attitudes, and organizational citizenship behavior: A cross-level investigation. *The Journal of Applied Psychology*, 95(3), 517–529. DOI:10.1037/a0018867 PMID:20476830

Walumbwa, F. O., & Schaubroeck, J. (2009). Leader personality traits and employee voice behavior: Mediating roles of ethical leadership and work group psychological safety. *The Journal of Applied Psychology*, 94(5), 1275–1286. DOI:10.1037/a0015848 PMID:19702370

Warwick, K. (2013). *The future of artificial intelligence and cybernetics. There's a future: visions for a better world. BBVA Open Mind*. TF Editores.

Wood, M. S. (2005). Determinants of shared leadership in management teams. *International Journal of Leadership Studies*, 1(1), 64–85.

Yukl, G. A. (2013). *Leadership in organizations*. Pearson Education India.

ADDITIONAL READINGS

Afaq, A., Gaur, L., & Singh, G. (2022, April). A latent dirichlet allocation technique for opinion mining of online reviews of global chain hotels. In 2022 3rd International Conference on Intelligent Engineering and Management (ICIEM) (pp. 201-206). IEEE. DOI:10.1109/ICIEM54221.2022.9853114

Chaudhary, M., Gaur, L., Singh, G., & Afaq, A. (2024). Introduction to Explainable AI (XAI) in E-Commerce. In *Role of Explainable Artificial Intelligence in E-Commerce* (pp. 1–15). Springer Nature Switzerland. DOI:10.1007/978-3-031-55615-9_1

Gaur, L., Afaq, A., Arora, G. K., & Khan, N. (2023). Artificial intelligence for carbon emissions using system of systems theory. *Ecological Informatics*, 76, 102165. DOI:10.1016/j.ecoinf.2023.102165

Gaur, L., Afaq, A., Singh, G., & Dwivedi, Y. K. (2021). Role of artificial intelligence and robotics to foster the touchless travel during a pandemic: A review and research agenda. *International Journal of Contemporary Hospitality Management*, 33(11), 4079–4098. DOI:10.1108/IJCHM-11-2020-1246

Gaur, L., Gaur, D., & Afaq, A. (2024). Demystifying Metaverse Applications for Intelligent Healthcare. In Metaverse Applications for Intelligent Healthcare (pp. 1-23). IGI Global.

Gaur, L., Gaur, D., & Afaq, A. (2024). Ethical Considerations in the Use of the Metaverse for Healthcare. In Metaverse Applications for Intelligent Healthcare (pp. 248-273). IGI Global.

KEY TERMS AND DEFINITIONS

Artificial Intelligence (AI): Is a man-made technology where machines or systems start thinking and acting like humans.

Ethical Leadership: Deals with the greater good of people, the means to achieve the outcome, and the consequences of leaders' actions.

Ethics: Are moral principles of human behavior under given conditions.

Leader: Is an individual who leads followers toward desired common goals.

Leadership: Is a process in which leaders use non-coercive force and motivate followers toward desired common goals.

Chapter 11
Exploring Ethical Dimensions of AI Assistants and Chatbots

Shweta Bhattacharjee Porna
https://orcid.org/0009-0003-9612-2079
Ahsanullah University of Science and Technology, Bangladesh

Munir Ahmad
https://orcid.org/0000-0003-4836-6151
Survey of Pakistan, Pakistan

Rubén González Vallejo
https://orcid.org/0000-0002-9697-6942
Universidad de Málaga, Spain

Iram Shahzadi
PMAS Arid Agricultural University, Pakistan

Md. Asifur Rahman
https://orcid.org/0009-0004-3310-0769
University of Dhaka, Bangladesh

ABSTRACT

This chapter explores the ethical implications of chatbots and AI assistants, focusing on privacy and data security, bias and fairness, transparency and accountability, user autonomy, employment impacts, manipulation and persuasion, user well-being, cultural sensitivity, and sustainability. Privacy and data security are paramount, emphasizing user consent and trust in data handling. Bias in AI systems can perpetuate discrimination, necessitating proactive measures for fairness. Transparency

DOI: 10.4018/979-8-3693-9173-0.ch011

and accountability are critical for user trust, requiring clear communication and regulatory frameworks. User autonomy must be respected, allowing control over AI interactions. The impact on employment highlights the need for strategies to manage job displacement and skill development. AI's potential for manipulation necessitates ethical guidelines to prevent misuse. Ensuring user well-being involves addressing mental health risks and misinformation. Cultural sensitivity in AI design promotes inclusivity. Sustainability considerations address the environmental impact of AI systems.

INTRODUCTION

AI assistants, generally considered virtual assistants or intelligent virtual agents, constitute software applications that allow the accomplishment of tasks and render services to users (Maedche et al., 2019). This type of responsibility can vary from simple queries and data retrieval to more advanced acting processes like making appointments, booking the recreations, or managing the smart home devices are among them. What is particularly different is their application of artificial intelligence (AI) technologies to recognize the natural language input from users, process the information, and then take the right actions and make the proper response. Instead, they frequently use machine learning principles to provide such facilities that can intelligently understand and meet customer's needs whilst becoming more advanced in the near future. These assistants are mostly composed of smartphones, smart speakers, and other connected devices. There are numerous well-known AI assistants, like Siri from Apple, Alexa from Amazon, Google Assistant, and Cortana from Microsoft. These can be deployed to run efficiently, generate more work, and give the clients a comfortable platform, where they can give instructions or commands verbally or through text features (Kamoonpuri & Sengar, 2023; Maedche et al., 2019; Tulshan & Dhage, 2019).

Chatbots are a kind of specialized AI instruments, which are trained to imitate the natural human-like conversations with users by means of text or voice. It is the AI algorithms, anywhere precision and natural language processing (NLP) in forces that are running the organizations, and also that is used to understand and reply to user questions or commands in real-time. Chatbots operate in the chat format, facilitating exchanges, giving help, answering neither definite nor indefinite questions, or doing pre-commanded tasks if there is a relative context (Ahmad et al., 2024; Assayed et al., 2024; Luo et al., 2022). AI Assistants are often pre-installed into platforms or devices but Chatbots are mostly stand-alone applications or Internet sites, messaging channels, or customer care stations. They cover a variety of functions, including customer service processes, sales bots, information retrieval, and entertainment, in different domains. Chatbots replicate humanistic communication using different modalities such as rules encoding systems, machine learning models, and NLG techniques. Some of the sophisticated chatbots utilize deep learning structures like recurrent neural networks (RNNs) and transformers to move further in the field of realistic communications and accommodating diverse user inputs.

The extensive application areas of AI are an important issue especially in natural language processing technology and AI assistants and chatbots. These augmented reality systems are melded within smartphones, smart speakers, websites, customer service platforms and so on, equipping people with tools for communication, information retrieval, automation, and decision-making. With the growing popularity of these technologies which have been mainstreamed into our way of life, it's very important to take guidance from these facts from those who have the power to influence people as well as businesses and then society as a whole (R. S. T. Lee, 2020; S. Lee et al., 2023).

The development of personal AI assistants and chatbots not only concerns itself with many ethical dilemmas and risks but resonates throughout our society. Within this front, a number of areas are involved, namely user privacy, data security, algorithmic bias, transparency, accountability, social impact, and user control (Garcia Valencia et al., 2023; Kooli, 2023; Yeung et al., 2023). For example, the gathering and manipulation of user data by AI systems will bring up issues relating to data privacy, privacy consent, and data misuse without permission or unauthorized access. Apart from this, machine learning capabilities of AI assistants and chatbots can reproduce or amplify the disparity and unequal conditions in the community which will ultimately lead to the unfair limitation of opportunities. Faced with the manifold aspects that are associated with the ethical challenges of AI assistants and chatbots, there is now an urgent demand to undertake a exploration of their ethical dimensions.

The chapter delves into various ethical dimensions relevant to AI assistants and chatbots, including transparency and explainability, fairness and bias, privacy and data security, user interaction and consent, social and psychological impact, accountability, and responsibility. The chapter aims to contribute to the ongoing discourse on AI ethics by offering valuable insights, critical analysis, and actionable recommendations for promoting ethical practices and responsible innovation in the design, development, and deployment of AI assistants and chatbots. By fostering ethical awareness and accountability, we can strive towards harnessing the potential of these technologies for the benefit of individuals and society while mitigating potential risks and harms.

EVOLUTION OF AI ASSISTANTS AND CHATBOTS

The development of AI Assistants and Chatbots has seen significant milestones that is elaborated below along with key technologies.

Early Beginnings

1966: ELIZA

ELIZA was one of the pioneering chatbots and marked a significant milestone in the history of artificial intelligence and natural language processing. Created by MIT computer scientist Joseph Weizenbaum, ELIZA was designed to simulate a conversation between a human and a computer. It mimicked a Rogerian psychotherapist, meaning it would reflect questions back to the user, creating the illusion of understanding and empathy. ELIZA used a simple pattern matching and substitution methodology. It identified keywords in the user's input and applied pre-defined transformation rules to generate responses. The responses were generated using scripts written in a language called MAD-SLIP (Meta Language SLIP). The most famous script was the DOCTOR script, which made ELIZA simulate the behavior of a psychotherapist. Despite its simplicity, ELIZA was remarkably effective at convincing users that they were interacting with an intelligent entity. This early experiment demonstrated the potential for machines to engage in natural language conversation, laying the groundwork for future developments in AI-driven communication (Bassett, 2019; Ciesla, 2024; Rajaraman, 2023; Weizenbaum, 1966; ZEMČÍK, 2019).

1972: PARRY

PARRY, developed by psychiatrist Kenneth Colby at Stanford University, was designed to simulate a person with paranoid schizophrenia. It was a significant advancement over ELIZA in terms of complexity and realism. PARRY included a model of a paranoid patient, incorporating various mental states and thought processes typical of paranoid schizophrenia. It used heuristic pattern matching to generate responses, making its conversations more contextually relevant and psychologically complex than those of ELIZA. PARRY demonstrated the potential for AI to simulate complex human psychological states, contributing to both AI research and the understanding of psychiatric conditions. It was subjected to the Turing Test, where psychiatrists were challenged to distinguish between PARRY and real patients. The test showed that PARRY could convincingly simulate a human patient, raising questions about the nature of intelligence and the capabilities of AI (Ciesla, 2024; Shum et al., 2018; ZEMČÍK, 2019).

1980s to 1990s

1986: Jabberwacky

Jabberwacky was developed to simulate natural human chat in an entertaining and engaging manner. It marked a shift towards more interactive and conversational AI experiences aimed at entertainment and companionship. Unlike rule-based systems, Jabberwacky learned from interactions with users, continuously improving its conversational abilities over time. It aimed to provide humorous and engaging conversations, making it popular among users looking for entertainment rather than functional assistance. Jabberwacky's approach to learning from user interactions foreshadowed the adaptive learning capabilities seen in modern AI. It demonstrated the potential for chatbots to evolve and improve through continued use, paving the way for more interactive and engaging AI systems (Bors et al., 2020; Ciesla, 2024; Rajaraman, 2023).

1995: ALICE (Artificial Linguistic Internet Computer Entity)

ALICE was a significant step forward in chatbot development, utilizing AIML (Artificial Intelligence Markup Language) to create more sophisticated and contextually aware conversations. ALICE's conversations were driven by AIML, which allowed for a more structured and flexible approach to generating responses. AIML enabled the creation of complex conversational patterns and context management. ALICE improved upon earlier pattern matching techniques, allowing for more accurate and meaningful interactions. ALICE won the Loebner Prize Turing Test multiple times, demonstrating its ability to engage in convincing human-like conversations. The development and use of AIML influenced subsequent chatbot development, providing a robust framework for creating interactive and intelligent conversational agents (Ciesla, 2024; Rajaraman, 2023; Wallace, 2009).

2000s: Emergence of Virtual Assistants

2001: SmarterChild

SmarterChild was deployed on popular instant messaging platforms like AOL Instant Messenger and MSN Messenger, marking the transition of chatbots from academic projects to widely used consumer applications. SmarterChild provided users with information retrieval services, such as weather updates, sports scores, and trivia, through natural language interaction. It used a combination of scripted responses and database queries to generate its replies. SmarterChild's integration into mainstream messaging platforms demonstrated the practicality and appeal of chatbots for everyday use. It was a precursor to the virtual assistants we use today, showing the viability of chatbots in providing real-time information and enhancing user experiences in digital communication (Kumaraguru Diderot et al., 2024; Okonkwo & Ade-Ibijola, 2021).

2006: IBM Watson

IBM Watson was initially developed to compete on the quiz show Jeopardy!, where it famously defeated human champions. Watson's success showcased the potential of AI in processing and understanding natural language on a massive scale. Watson used advanced natural language processing, machine learning, and data retrieval techniques to analyze and answer questions posed in natural language. It could parse complex queries, search vast amounts of data, and provide accurate answers. Watson's victory on Jeopardy! highlighted the capabilities of AI in understanding and processing natural language, inspiring further developments in AI assistants. It demonstrated the potential for AI to handle complex information tasks, leading to its application in various industries, including healthcare and finance (Alexsoft, 2021; Chow et al., 2023; Oliveira et al., 2019).

2010s: Rise of Mainstream AI Assistants

2011: Siri

Siri was the first widely adopted AI assistant integrated into smartphones, revolutionizing the way users interacted with their devices. It brought voice-activated AI to the masses, making it a standard feature in mobile technology. Siri combined voice recognition, natural language processing, and contextual understanding to perform tasks, answer questions, and provide recommendations. It leveraged Apple's extensive ecosystem to deliver a seamless user experience. Siri set a new standard for AI assistants, influencing the development of similar technologies by other tech giants. Its success demonstrated the demand for voice-activated AI and its potential to transform user interaction with technology (Crowder, 2023; Shum et al., 2018; Tulshan & Dhage, 2019).

2012: Google Now

Google Now introduced predictive information delivery, providing users with relevant information based on their habits and preferences before they even asked for it. This proactive approach set it apart from earlier AI assistants. Google Now used machine learning algorithms to analyze user data, such as search history and location, to anticipate their needs and provide timely information. It integrated deeply with Google's services to offer a comprehensive experience. Google Now's predictive capabilities influenced the design of future AI assistants, emphasizing the importance of context-aware AI that could anticipate user needs. It showcased the potential of AI to enhance productivity and convenience through proactive assistance (Case, 2022; Goksel Canbek & Mutlu, 2016).

2014: Amazon Alexa

Alexa, integrated with Amazon Echo devices, popularized the use of voice-controlled assistants in smart home environments. It became a central hub for controlling smart devices, accessing information, and performing tasks through voice commands. Alexa used advanced voice recognition and natural language understanding to process user requests. It is connected with a wide range of smart devices and services, creating a versatile and extensible ecosystem. Alexa's widespread adoption demonstrated the growing demand for voice-activated AI in everyday life. It highlighted the potential for AI assistants to become integral parts of smart homes, leading to the development of similar products by other companies and the expansion of the smart home market (Bogdan et al., 2021; Tulshan & Dhage, 2019; Yang et al., 2021).

2016: Google Assistant

Google Assistant enhanced the capabilities of Google Now with more advanced natural language processing and machine learning. It offered a more conversational and interactive experience, capable of understanding and responding to complex queries. Google Assistant leveraged Google's vast knowledge graph and machine learning models to provide accurate and contextually relevant responses. It integrated with various Google services and third-party applications to perform a wide range of tasks. Google Assistant's success further cemented the role of AI assistants in everyday life, influencing the development of more advanced and capable AI systems. Its ability to engage in natural, flowing conversations set a new benchmark for AI interaction quality (Google, 2019; Hong et al., 2021; Tulshan & Dhage, 2019).

Late 2010s to 2020s: Advanced AI and Deep Learning

2018: BERT (Bidirectional Encoder Representations from Transformers)

BERT revolutionized natural language processing by introducing bidirectional training of transformers, significantly improving the understanding of context in language models. This advancement enabled AI systems to grasp the nuances of human language better. BERT uses transformers to read text bidirectionally, understanding the context of words based on their surrounding words. This approach allows for a deeper and more accurate understanding of language. BERT's introduction led to significant improvements in AI's ability to comprehend and generate human-like text, enhancing the capabilities of chatbots and virtual assistants. It influenced the development of more sophisticated language models and applications in various AI-driven services (Bello et al., 2023; Sayeed et al., 2023).

2019: GPT (Generative Pre-Trained Transformer)

GPT, with its multibillion parameters, demonstrated unprecedented capabilities in generating human-like text, pushing the boundaries of conversational AI. Its size and training data allowed it to perform a wide range of language tasks with high accuracy. GPT uses a transformer architecture to generate text based on input prompts. Its pre-training on diverse datasets enables it to produce coherent and contextually appropriate responses across various topics. GPT's capabilities showcased the potential of large-scale language models in creating more advanced and versatile chatbots and virtual assistants. It influenced ongoing research in AI, highlighting the importance (Kolade et al., 2024; Wu et al., 2023). ChatGPT-3.5 and ChatGPT-4 represent the latest advancements in the GPT model series, featuring enhanced capabilities. These versions have been trained on more extensive datasets, allowing them to produce more nuanced and sophisticated responses. ChatGPT-4, in particular, marks a significant milestone in chatbot development, with its ability to generate responses that closely mimic human conversation (Doshi et al., 2023; Kolade et al., 2024).

METHODOLOGY

To comprehensively explore the ethical dimensions of AI assistants and chatbots, this chapter conducted a thorough review of scholarly literature. By drawing from a diverse range of academic sources, including research articles, books, and journals, this review synthesized existing knowledge and insights on the topic. Examining the findings and arguments presented in scholarly literature, this chapter aimed to provide a comprehensive understanding of the ethical considerations surrounding AI assistants and chatbots.

RESULTS AND DISCUSSION

This section presents the results of the chapter. Figure 1 shows the results in visual form.

Privacy and Data Security

Privacy and data security imply the notion of preserving the secrecy hitherto and unauthorized access to the personal information that AI assistants collect from the users (Bolton et al., 2021; Edu et al., 2021). Privacy makes sure that personal details can be accessed and used with the permission of individuals. AI applications such as virtual assistants like Siri, Alexa, or chatbots used in customer interactions, collect information about users, including their preferences, behaviors, and even if someone is sensitive regarding their personal data, they may disclose it to these applications.

There must be trust between the user and the company that they will be taking care of user information appropriately and safely. If apprehensions in regard to a user's privacy or data given into danger of breach ensue, AI helper may not be used, or information provided to it. Transparency, which is the key factor that enables the establishment of and keeping trust, should be the major element in all the processes. Companies should really make it obvious to the users the data that their AI assistants are gathering, how it is going to be used, and which parties will get authority over it. It ensures that customers are aware of the AI's functionality and will be able to choose whether or not they wish to use the AI assistant and share their personal data.

Personal data should belong to individuals so that they be able to decide which information they want to share and with whom they want to share it. It will work for the online environment where people may select not to contribute their data, delete their own data, or adjust their privacy settings. Ensuring that users take control over their own data puts in their hands the kind of protection and exposure they get and promotes privacy and respect for their ideas and values (Hasal et al., 2021; Vimalkumar et al., 2021).

Bias and Fairness

AI systems learn from historical data, and if this data contains biases, the AI system can perpetuate or even amplify these biases (Chen et al., 2023; Devillers et al., 2021). For example, if a hiring AI is trained on historical hiring data that favored certain demographics, such as gender or race, the AI might inadvertently prioritize candidates from those demographics, perpetuating discrimination.

Biased AI systems can lead to unfair outcomes, where certain individuals or groups are systematically disadvantaged or discriminated against (Ferrara, 2024). For instance, biased AI algorithms used in lending decisions might unfairly deny loans to certain demographic groups, perpetuating economic inequality. It's essential to proactively identify biases in AI systems. This involves thorough testing and evaluation to uncover any systematic patterns of unfairness or discrimination. Various techniques, such as fairness audits and bias detection algorithms, can help in identifying biases.

Once biases are identified, steps should be taken to mitigate them. This might involve retraining the AI system with more diverse and representative data, adjusting algorithmic parameters, or implementing fairness-aware algorithms designed to mitigate biases in decision-making. Certain domains, such as hiring, lending, and criminal justice, are particularly sensitive to biases and discrimination. Extra care should be taken to ensure that AI systems used in these areas are fair and unbiased. This might involve regulatory oversight, industry standards, or independent auditing to ensure compliance with fairness principles (Ferrara, 2024; Makanadar et al., 2024; Manyika et al., 2019).

Transparency and Accountability

It is imperative for users to be conscious of the time when they are communicating with AI technologies and when interacting with humans. Informed communication should be provided to users from the earliest, which would state whether they are transacting with a bot or any other human agent. This translucency produces trust and controls the expectations of users, whether AI systems manage to perform some tasks or should not perform them at all (Elahi et al., 2019; Vogel et al., 2021). AI systems are to be smarter in what they can and cannot do. The audience should be advised concerning the AI's capabilities in terms of the AI's strong points, shortcomings, and lengths of bias. Users get guidance for following through when and how to use AI systems which allows them to make proper decisions.

AI software developers, system manufacturers, and their users should be treated just like compliance overseers for the AI systems' actions. This accountability entails putting in place structures for supervising and appraising AI-system performance but has mechanisms for handling problems and receiving complaints too. Human society wants AI systems, but they have to be defect-free and follow all ethical rules. AI systems that cause harm or violate ethical principles cannot be permitted, and there should be clear redress channels and accountability.

The implementation of regulatory frameworks and performance standards to guarantee the reliability and predictability of AI systems is the key to transparent and accountable AI systems. The regulatory bodies can institute rules which, in turn, define the guidelines, as well as the overall requirements for transparency, accountability and responsible AI utilization both during the development and implementation period. Implementation of such regulations puts the power of checking into AI developers' hand while guaranteeing the ethics standards are met.

Developers must adhere to ethical principles and the best practices in AI technology innovation and R&D. One of these guidelines, for example, is the IEEE Ethically Aligned Design, which can help to determine proper standards and deploying AI systems with some ethical concerns (DataEthics, 2021; Weng & Hirata, 2018). These measures can provide assurance that AI technology makes transparency an integral part, holds accountability, and the users' trust is its core value.

User Autonomy and Agency

Adopting AI systems should be serving the purpose of enabling users by letting them enjoy different options that would be provided to them as well as giving them autonomy when their AI interactions are being carried out. Therefore, the interfaces need to be developed in such a way that they appear structureless and simple enough to facilitate ease while users use AI systems (Hu et al., 2021). Respecting user autonomy implies that their will to publicly announce their opinions and make decisions is theirs to decide (Huh et al., 2023). AI systems should not ever try to subject users into making any decisions or performing any actions without their permission. The opposite way they should respect users' opinion and choices and provide them with an opportunity either to exclude or sign up for certain features or interactions.

Users must keep their interface with the brains of an AI system to the standard of their selections. It not only means that people are able to control the way the exchange occurs by, for example, starting, pausing, or terminating the interaction when desired, but also match the communication to their individual preferences and values. Users ought to be given the alternative to deactivate specific actions or AI systems' characteristic attributes if they are unwilling to accept them. Moreover, AI systems should enable users to have a great level of customization, providing them with the possibility to define their own experiences in a manner that is in harmony with their own emotions and values.

AI systems should be tailored to the goals and values of individuals as well as to the general human welfare (Laitinen & Sahlgren, 2021). This is achievable through research, which looks at the user and their preferences using different research methods acquiring feedback from them, and subsequently integrating the insights into the design and development process. Transparency is one vital factor that can be used to empower users about the possible consequences of their actions in a given system (Felzmann et al., 2020). Transparency is essential in terms of AI systems, and it should be clear in terms of the limits, capabilities, and effects on users. Also, letting users learn about AI by way of thorough education and information can help them grasp the AI systems and free will of choosing.

Impact on Employment

The use of AI assistants and chatbots can take away the role of humans as they can perform multiple jobs that were initially performed by humans (Maedche et al., 2019). As a result, certain jobs will be lost in specific industries because of automation (Georgieff & Hyee, 2022). It is crucial for the country to act in advance and identify from which professions a higher risk of automation comes and have strategies to control them. Some jobs may become fully automated also some job expectations may change, and workers' abilities are to be redefined (Yin et al., 2024). On the other hand, all the workers may not have resources, or the chance required to reskill or be more attractive. This could just make the gap bigger and grate the jobs for the scholars between different industries.

Similarly, to automation which is AI technology-led doing the same job where it could only follow commands, the production and the efficiency may rise. However, only some segments of society may be economically benefited. The case is that specific workers or certain types of jobs might be more likely to adopt AI, in contrast to other groups of workers that may experience wage stagnation or job loss as a result of automation (Georgieff & Hyee, 2022). Encouraging the diffusion of AI-related skills among the population is the key factor in making sure that these benefits of AI are benefiting the whole society more equally than they do now.

AI might be an instrument of job replacement as some of that work is automatically traditional, however, it can be the most possible source of emerging industries and occupations. Nevertheless, two things may stand in their way: (1) the new jobs may necessitate the learning of different skill sets, and (2) it might prove to be problematic for workers to transition or adapt to the new requirements. AI based job displacement may lead to extensive social and economic impact, this may include high unemployment rate, income disparities, ill-being, as well as social disturbances (Hornung & Smolnik, 2022; Nazareno & Schiff, 2021). We need to come up with solutions and government policies to control these unwanted effects and feed the vulnerable population.

Organizations that use AI-based assistants and chatbots for customer services must think through the effect their technology may have on job prospects and take measures to avoid the undesirable effects. In this case, the government can prioritize workforce development and resume preparation programs, and create a more inclusive work environment for the needs of the displaced.

Manipulation and Persuasion

An AI system can be configured in a manner which may produce undesirable changes in one's behavior or the user may be unaware of it. AI programs can gather large volumes about users, based on which it can help to select proper content and interactions that could have an impact on people's behavior. In other words, it is not uncommon to come across personalized ads, suggestion systems, and specially designed messages that might bring out particular emotions in users (Ahn et al., 2021; Ienca, 2023; Ischen et al., 2022). While users may not be conscious of the mechanisms by which AI systems are shaping their conduct, behaviors changing in such a fashion may not always feel right, even though the way in which the systems interact with people may still be beyond understanding. When people do not clearly understand what data is being collected about them, how it's being used, and who has access to it, they can feel less in control and start to question whether they are being taken advantage of or even manipulated.

Furthermore, AI systems are potent political machinery, such as controlled messaging and biased opinion making. This leads to questions about the possibility of AI cutting off democratic processes and spreading information of a complementing nature. The reproducers of AI systems, as well as the organizations that allocate the systems to the public, have the duty to make sure that their systems do not emasculate the customers or the people for profiteering purposes or political interests. Such an endeavor comprises setting up ethical norms and safeguards, to avoid AI's excessive use, which may otherwise pose potential adverse outcomes. It is crucial for transparency that we don't lose sight of the issues with persuasion and manipulation of AI systems. Users must be told about the way their data are being used and the impacts of managing human behavior by AI systems. Besides, mandate mechanisms must be put in role to assess whether developers and organizations leverage AI to fulfill ethical obligations.

User Well-Being

AI systems should put human well-being and a sense of security as their highest priorities. It is important for the IT community to foresee a possible declining effect of AI systems on mental welfare, for example, addiction, isolation, and misinformation, and should take measures to prevent them (Dhimolea et al., 2022; Kettle & Lee, 2023). AI systems, in the context of belonging to the entertainment or socialization bodily class of AI, contain the ability that can lead to addictive conduct (Maedche et al., 2019). This could happen, e.g., if the algorithms used by social media were geared toward engagement and not the duration of use could make the user develop dependency and unhealthy patterns. To this effect, AI developers should consider the risks related to unhealthy behavior because of the use of the devices and provide users with suitable tools and means to limit their screen time.

What is more, the AI systems that are created in this way can link people better and help them establish relationships but at the same time, it can be responsible for the social isolation of users who make greater use of virtual rather than real communication. AI developers must be aware that the aim should be to enhance, not substitute, offline real-life conversations and create a desire for personal contacts (Dhimolea et al., 2022; Kang & Shao, 2023; Maedche et al., 2019).

AI systems can also manipulate and accelerate the growth and spread of propaganda and false news, which can negatively affect others' health by spreading incorrect medical information as well as fuel further conspiracy theories (Blauth et al., 2022; Kertysova, 2018; Whyte, 2020). Developers should install means that would search out and promote the mitigation of the spread of misinformation like fact-checking algorithms, and moderation policies. AI tools like mental health support bots touch people's lives and could have a significant impact on the health of users (Boucher et al., 2021; Chin et al., 2023; Pham et al., 2022). Nevertheless, the use of AI in the field of mental health also involves some risks, like missing diagnoses or deteriorating mental state if not properly supervised. Designers of AI models for mental health assistance should make sure that they follow ethical rules and best practices in gathering and the assembly of data like providing correct details, referring users to professionals when needed, and above all ensuring users' privacy and confidentiality.

Cultural Sensitivity and Diversity

AI systems need to be built around the humane approach to not only reproduce but to also accept and respect the diversity of cultures and perspectives of people (Seaborn et al., 2024). AI-based systems should be open, for cultural diversity and biological perspectives. It further suggests the importance of understanding the cultural norms, values, and traditions of different communities worldwide. Developers should take the impedance given to the AI system to systematically foster detrimental stereotypes and culturally biased categories in consideration carefully. AI algorithms that are data-driven and trained on biased and unrepresentative data may inadvertently give the impression to society of accentuating or promoting existing unfair stereotypes and discrimination.

AI systems should be created to convert as many languages as possible to accommodate people with different cultural backgrounds (Chin et al., 2023; Zhai & Wibowo, 2022). This could be achieved by taking into account factors, which are related to language variety, the accessibility for disabled people as well as cultural sensitiveness while you are building user interface and creating content. It is incumbent on the developers to grow intercultural cohesion within their teams in order to secure a broad dispersion of views and experiences within their team. This may help correct and bring cultural sensitivities and prejudices to the forefront at the beginning of the making of AI systems.

Sustainability

The environmental impact of AI systems, particularly in terms of energy consumption and electronic waste, should be considered (Al-Sharafi et al., 2023; Dauvergne, 2020; Nishant et al., 2020). While AI systems can have a positive contribution to reducing pollution, the environmental effects of AI systems, especially in terms of energy consumption and electronic waste, must be addressed. Developers aim at unveiling green AI systems; the use of non-energy-sapping AI systems is a desideratum. AI systems, especially those that are characterized by deep learning and neural networks, are typical in nature in terms of computational power, i.e. they are power-hungry. A majority of AI systems' energy consumption is produced as their output of carbon emissions and environmental degradations, chiefly coming from non-renewable energy sources such as fossil fuels.

New and advanced technologies in the IT area and AI field are designed at a pace. Such a development of modern technologies reduces significantly the life of devices used to operate AI systems. It is e-waste generation that comes as an outcome of the discarding of outdated or obsolete units and the replacement of newer models (Albreem et al., 2021). The environmental impact of AI systems is not limited to overusing power and e-waste generation, instead, it is much more far-reaching (energy usage and discarded electronics). To illustrate, the extraction and refining of raw required for manufacturing hardware components may be ecologically disastrous. The outcome can be anything related: to the destruction of habitats, pollution including water and air pollution, depletion of resources, and so on.

The AI developers ought to focus on the development of systems that are friendly to the environment and that are energy efficient. It likely includes efforts such as improving and refining algorithms and software that will require little or no computational resources, architecting hardware that utilizes energy-efficient designs, and using clean energy sources as a power source for AI infrastructure. Sustainability implications must be present in the lifecycle of AI systems naturally, from the point of design and development to deployment and disposal.

Long-Term Implications

AI technologies have the capability to exert power over democratic processes in different facets. Some include the shaping of the public opinion through tailored contents and applications aiming to influence their decisions; as well as the prospect of disinformation and misinformation spreading rapidly through AI tools. Therefore, it is key to the supervision of AI influences on democratic processes through which AI promotes democratic values instead of dismantling them, including democracy as transparency, accountability, and civic engagement. AI systems create mimetic interactions, and framing of personality and communication can alter the social structure and established cultural norms. As an example, AI-generated recommendation systems may create echo chambers of views and ideologies which only strengthen the existing opinions and thus could possibly result in polarization and the heightening of the social gap.

The purpose of AI technology ought to focus on combining both human superiority with existing technology to create an acceptable human quality of life and condition. With this, a balanced decision should take into account the existence as well as the ethical and societal implications of AI development and deployment. The AI systems must be designed to support human abilities, defend human dignity and moral values, and follow local values and human rights. Having the ability to anticipate, predict, and handle likely ethical, societal, and economic implications of AI technologies is realized through the implementation of ethical foresight. This requires involving diverse groups of actors, such as ethicists, policymakers, and technologists, as well as the people impacted in a given community, to map out potential risks and avenues that may arise from AI innovation and the eventual deployment of AI. Ethical cogency has a chance of mediating the accountable generation, regulation, and governance of AI technologies, through which risks can be mitigated, and benefits for society are maximized.

Figure 1. Ethical dimensions of AI assistants and Chatbots

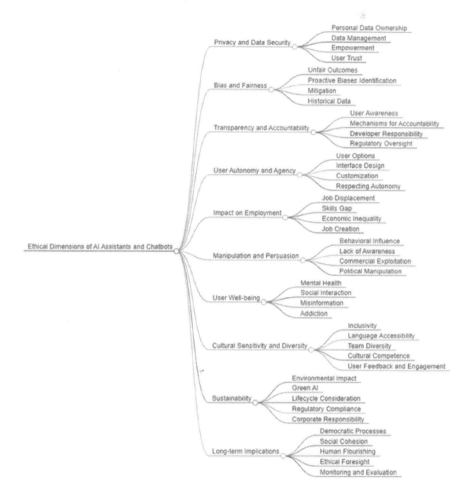

CONCLUSION

Alongside the development of the range of AI assistants and their increasing capability, the context of ethical issues may lead to new ethical questions. For instance, these might involve considerations on the influence of user choice behavior, the "gray area" between human and machine interplay, and the commitment to confidentiality in unique AI experiences. Besides that, as an AI system become more and more involved in forming important decisions it is possible to raise the issues of responsibility, transparency, and even unintentional consequences.

AI technologies firmament is quite fast with natural language processing, computer vision, as well as reinforcement learning, give rise the ethical questions about AI assistants and chatbots. For instance, the preabuse of deepfake technology is fraught with the invasion of audio and video tools resulting in the spread of misinformation and deception. Likewise, AI-autonomous systems driven by artificial intelligence game forth the concern about safety, accountability, also the ethical aspect of allocating decision-making algorithms in a high-stakes scenario.

While the reason for AI system is posed by emerging AI technologies, it is also a challenge that is also a chance for us to promote ethical AI development and application. As an example, the progress in explainable techniques could enhance AI systems transparency and accountability in the decision-making process, improving user understanding of how decisions are made and detecting possible hidden biases and errors. Also, using ethical standards in the design process, for example, fairness, transparency, and human-first is a risk mitigation move as well as ensures trust among AI assistants and chatbots. Furthermore, multidisciplinary cooperation among ethicists, technologists, policymakers, and other orthodox entities could be a great step in finding solutions to ethical issues and developing ethical standards and guidelines for AI development and deployment.

REFERENCES

Ahmad, M., Naeem, M. K. H., Mobo, F. D., Tahir, M. W., & Akram, M. (2024). Navigating the Journey: How Chatbots Are Transforming Tourism and Hospitality. In Darwish, D. (Ed.), *Design and Development of Emerging Chatbot Technology* (pp. 236–255). IGI Global., DOI:10.4018/979-8-3693-1830-0.ch014

Ahn, J., Kim, J., & Sung, Y. (2021). AI-powered recommendations: The roles of perceived similarity and psychological distance on persuasion. *International Journal of Advertising*, 40(8), 1366–1384. Advance online publication. DOI:10.1080/026 50487.2021.1982529

Al-Sharafi, M. A., Al-Emran, M., Arpaci, I., Iahad, N. A., AlQudah, A. A., Iranmanesh, M., & Al-Qaysi, N. (2023). Generation Z use of artificial intelligence products and its impact on environmental sustainability: A cross-cultural comparison. *Computers in Human Behavior*, 143, 107708. Advance online publication. DOI:10.1016/j.chb.2023.107708

Albreem, M. A., Sheikh, A. M., Alsharif, M. H., Jusoh, M., & Mohd Yasin, M. N. (2021). Green Internet of Things (GIoT): Applications, Practices, Awareness, and Challenges. *IEEE Access : Practical Innovations, Open Solutions*, 9, 38833–38858. Advance online publication. DOI:10.1109/ACCESS.2021.3061697

Alexsoft. (2021). *Comparing Machine Learning as a Service: Amazon, Microsoft Azure, Google Cloud AI, IBM Watson*. Alexsoft.

Assayed, S. K., Woods, D., Alkahtib, M., & Shaalan, K. (2024). Effective Human-Chatbot Interaction: A High School Student Perspective. *International Journal of Computing and Digital Systems*, 15(1).

Bassett, C. (2019). The computational therapeutic: Exploring Weizenbaum's ELIZA as a history of the present. *AI & Society*, 34(4), 803–812. Advance online publication. DOI:10.1007/s00146-018-0825-9

Bello, A., Ng, S. C., & Leung, M. F. (2023). A BERT Framework to Sentiment Analysis of Tweets. *Sensors (Basel)*, 23(1), 506. Advance online publication. DOI:10.3390/s23010506 PMID:36617101

Blauth, T. F., Gstrein, O. J., & Zwitter, A. (2022). Artificial Intelligence Crime: An Overview of Malicious Use and Abuse of AI. *IEEE Access : Practical Innovations, Open Solutions*, 10, 77110–77122. Advance online publication. DOI:10.1109/ ACCESS.2022.3191790

Bogdan, R., Tatu, A., Crisan-Vida, M. M., Popa, M., & Stoicu-Tivadar, L. (2021). A practical experience on the amazon alexa integration in smart offices. *Sensors (Basel)*, 21(3), 734. Advance online publication. DOI:10.3390/s21030734 PMID:33499092

Bolton, T., Dargahi, T., Belguith, S., Al-Rakhami, M. S., & Sodhro, A. H. (2021). On the security and privacy challenges of virtual assistants. *Sensors (Basel)*, 21(7), 2312. Advance online publication. DOI:10.3390/s21072312 PMID:33810212

Bors, L., Samajdwer, A., & van Oosterhout, M. (2020). Introduction to Oracle Digital Assistant. In *Oracle Digital Assistant*. DOI:10.1007/978-1-4842-5422-6_1

Boucher, E. M., Harake, N. R., Ward, H. E., Stoeckl, S. E., Vargas, J., Minkel, J., Parks, A. C., & Zilca, R. (2021). Artificially intelligent chatbots in digital mental health interventions: a review. *Expert Review of Medical Devices, 18*(sup1). DOI: 10.1080/17434440.2021.2013200

Case, M. (2022). Google, Big Data, & Antitrust. *Delaware Journal of Corporate Law, 46*(2).

Chen, P., Wu, L., & Wang, L. (2023). AI Fairness in Data Management and Analytics: A Review on Challenges, Methodologies and Applications. In *Applied Sciences (Switzerland)* (Vol. 13, Issue 18). DOI:10.3390/app131810258

Chin, H., Song, H., Baek, G., Shin, M., Jung, C., Cha, M., Choi, J., & Cha, C. (2023). The Potential of Chatbots for Emotional Support and Promoting Mental Well-Being in Different Cultures: Mixed Methods Study. *Journal of Medical Internet Research*, 25, e51712. Advance online publication. DOI:10.2196/51712 PMID:37862063

Chow, J. C. L., Wong, V., Sanders, L., & Li, K. (2023). Developing an AI-Assisted Educational Chatbot for Radiotherapy Using the IBM Watson Assistant Platform. *Healthcare (Basel)*, 11(17), 2417. Advance online publication. DOI:10.3390/healthcare11172417 PMID:37685452

Ciesla, R. (2024). The Book of Chatbots. In *The Book of Chatbots*. DOI:10.1007/978-3-031-51004-5

Crowder, J. (2023). *AI chatbots: The good, the bad, and the ugly*. Springer Nature.

DataEthics. (2021). *Addressing Ethical Dilemmas in AI: Listening to Engineers*. Association of Nordic Engineers.

Dauvergne, P. (2020). *AI in the Wild: Sustainability in the Age of Artificial Intelligence*. MIT Press. DOI:10.7551/mitpress/12350.001.0001

Devillers, L., Fogelman-Soulié, F., & Baeza-Yates, R. (2021). AI & Human Values: Inequalities, Biases, Fairness, Nudge, and Feedback Loops. In *Lecture Notes in Computer Science (including subseries Lecture Notes in Artificial Intelligence and Lecture Notes in Bioinformatics): Vol. 12600 LNCS*. DOI:10.1007/978-3-030-69128-8_6

Dhimolea, T. K., Kaplan-Rakowski, R., & Lin, L. (2022). *Supporting Social and Emotional Well-Being with Artificial Intelligence*. DOI:10.1007/978-3-030-84729-6_8

Doshi, R., Amin, K., Khosla, P., Bajaj, S., Chheang, S., & Forman, H. P. (2023). Utilizing Large Language Models to Simplify Radiology Reports: a comparative analysis of ChatGPT3.5, ChatGPT4.0, Google Bard, and Microsoft Bing. *MedRxiv*. DOI:10.1101/2023.06.04.23290786

Edu, J. S., Such, J. M., & Suarez-Tangil, G. (2021). Smart Home Personal Assistants: A Security and Privacy Review. In *ACM Computing Surveys* (Vol. 53, Issue 6). DOI:10.1145/3412383

Elahi, H., Wang, G., Peng, T., & Chen, J. (2019). On transparency and accountability of smart assistants in smart cities. *Applied Sciences (Basel, Switzerland)*, 9(24), 5344. Advance online publication. DOI:10.3390/app9245344

Felzmann, H., Fosch-Villaronga, E., Lutz, C., & Tamò-Larrieux, A. (2020). Towards Transparency by Design for Artificial Intelligence. *Science and Engineering Ethics*, 26(6), 3333–3361. Advance online publication. DOI:10.1007/s11948-020-00276-4 PMID:33196975

Ferrara, E. (2024). Fairness and Bias in Artificial Intelligence: A Brief Survey of Sources, Impacts, and Mitigation Strategies. In *Sci* (Vol. 6, Issue 1). DOI:10.3390/sci6010003

Garcia Valencia, O. A., Suppadungsuk, S., Thongprayoon, C., Miao, J., Tangpanithandee, S., Craici, I. M., & Cheungpasitporn, W. (2023). Ethical Implications of Chatbot Utilization in Nephrology. In *Journal of Personalized Medicine* (Vol. 13, Issue 9). DOI:10.3390/jpm13091363

Georgieff, A., & Hyee, R. (2022). Artificial Intelligence and Employment: New Cross-Country Evidence. *Frontiers in Artificial Intelligence*, 5, 832736. Advance online publication. DOI:10.3389/frai.2022.832736 PMID:35620279

Goksel Canbek, N., & Mutlu, M. E. (2016). On the track of Artificial Intelligence: Learning with Intelligent Personal Assistants. *Uluslararas Insan Bilimleri Dergisi*, 13(1), 592. Advance online publication. DOI:10.14687/ijhs.v13i1.3549

Google. (2019). Google Assistant | Your own personal Google. In *Google*.

Hasal, M., Nowaková, J., Ahmed Saghair, K., Abdulla, H., Snášel, V., & Ogiela, L. (2021). Chatbots: Security, privacy, data protection, and social aspects. *Concurrency and Computation*, 33(19), e6426. Advance online publication. DOI:10.1002/cpe.6426

Hong, G., Folcarelli, A., Less, J., Wang, C., Erbasi, N., & Lin, S. (2021). Voice assistants and cancer screening: A comparison of alexa, siri, google assistant, and cortana. *Annals of Family Medicine*, 19(5), 447–449. Advance online publication. DOI:10.1370/afm.2713 PMID:34546951

Hornung, O., & Smolnik, S. (2022). AI invading the workplace: Negative emotions towards the organizational use of personal virtual assistants. *Electronic Markets*, 32(1), 123–138. Advance online publication. DOI:10.1007/s12525-021-00493-0

Hu, Q., Lu, Y., Pan, Z., Gong, Y., & Yang, Z. (2021). Can AI artifacts influence human cognition? The effects of artificial autonomy in intelligent personal assistants. *International Journal of Information Management*, 56, 102250. Advance online publication. DOI:10.1016/j.ijinfomgt.2020.102250

Huh, J., Whang, C., & Kim, H. Y. (2023). Building trust with voice assistants for apparel shopping: The effects of social role and user autonomy. *Journal of Global Fashion Marketing*, 14(1), 5–19. Advance online publication. DOI:10.1080/2093 2685.2022.2085603

Ienca, M. (2023). On Artificial Intelligence and Manipulation. *Topoi*, 42(3), 833–842. Advance online publication. DOI:10.1007/s11245-023-09940-3

Ischen, C., Araujo, T. B., Voorveld, H. A. M., Van Noort, G., & Smit, E. G. (2022). Is voice really persuasive? The influence of modality in virtual assistant interactions and two alternative explanations. *Internet Research*, 32(7), 402–425. Advance online publication. DOI:10.1108/INTR-03-2022-0160

Kamoonpuri, S. Z., & Sengar, A. (2023). Hi, May AI help you? An analysis of the barriers impeding the implementation and use of artificial intelligence-enabled virtual assistants in retail. *Journal of Retailing and Consumer Services*, 72, 103258. Advance online publication. DOI:10.1016/j.jretconser.2023.103258

Kang, W., & Shao, B. (2023). The impact of voice assistants' intelligent attributes on consumer well-being: Findings from PLS-SEM and fsQCA. *Journal of Retailing and Consumer Services*, 70, 103130. Advance online publication. DOI:10.1016/j. jretconser.2022.103130

Kertysova, K. (2018). Artificial Intelligence and Disinformation: How AI Changes the Way Disinformation is Produced, Disseminated, and Can Be Countered. *Security and Human Rights, 29*(1–4).

Kettle, L., & Lee, Y. C. (2023). User Experiences of Well-Being Chatbots. *Human Factors*. Advance online publication. DOI:10.1177/00187208231162453 PMID:36916743

Kolade, O., Owoseni, A., & Egbetokun, A. (2024). Is AI changing learning and assessment as we know it? Evidence from a ChatGPT experiment and a conceptual framework. *Heliyon*, 10(4), e25953. Advance online publication. DOI:10.1016/j.heliyon.2024.e25953 PMID:38379960

Kooli, C. (2023). Chatbots in Education and Research: A Critical Examination of Ethical Implications and Solutions. *Sustainability (Basel)*, 15(7), 5614. Advance online publication. DOI:10.3390/su15075614

Kumaraguru Diderot, P., Sakthidasan Sankaran, K., Jawarneh, M., Pallathadka, H., Arias-Gonzáles, J. L., & Sanchez, D. T. (2024). Evaluation of Chabot Text Classification Using Machine Learning. In *Conversational Artificial Intelligence*. DOI:10.1002/9781394200801.ch13

Laitinen, A., & Sahlgren, O. (2021). AI Systems and Respect for Human Autonomy. *Frontiers in Artificial Intelligence*, 4, 705164. Advance online publication. DOI:10.3389/frai.2021.705164 PMID:34765969

Lee, R. S. T. (2020). Artificial Intelligence in Daily Life. In *Artificial Intelligence in Daily Life*. DOI:10.1007/978-981-15-7695-9

Lee, S., Lee, M., & Lee, S. (2023). What If Artificial Intelligence Become Completely Ambient in Our Daily Lives? Exploring Future Human-AI Interaction through High Fidelity Illustrations. *International Journal of Human-Computer Interaction*, 39(7), 1371–1389. Advance online publication. DOI:10.1080/10447318.2022.2080155

Luo, B., Lau, R. Y. K., Li, C., & Si, Y. W. (2022). A critical review of state-of-the-art chatbot designs and applications. In *Wiley Interdisciplinary Reviews: Data Mining and Knowledge Discovery* (Vol. 12, Issue 1). DOI:10.1002/widm.1434

Maedche, A., Legner, C., Benlian, A., Berger, B., Gimpel, H., Hess, T., Hinz, O., Morana, S., & Söllner, M. (2019). AI-Based Digital Assistants: Opportunities, Threats, and Research Perspectives. *Business & Information Systems Engineering*, 61(4), 535–544. Advance online publication. DOI:10.1007/s12599-019-00600-8

Makanadar, A., Shahane, S., & Patil, U. (2024). Tracing Algorithmic and AI-biased Data. *Economic and Political Weekly*, 59(7).

Manyika, J., Silberg, J., & Presten, B. (2019). What do we do about the biases in AI? In *Harvard Business Review*.

Nazareno, L., & Schiff, D. S. (2021). The impact of automation and artificial intelligence on worker well-being. *Technology in Society*, 67, 101679. Advance online publication. DOI:10.1016/j.techsoc.2021.101679

Nishant, R., Kennedy, M., & Corbett, J. (2020). Artificial intelligence for sustainability: Challenges, opportunities, and a research agenda. *International Journal of Information Management*, 53, 102104. Advance online publication. DOI:10.1016/j.ijinfomgt.2020.102104

Okonkwo, C. W., & Ade-Ibijola, A. (2021). Chatbots applications in education: A systematic review. In *Computers and Education* (Vol. 2). Artificial Intelligence., DOI:10.1016/j.caeai.2021.100033

Oliveira, J. D. S., Espindola, D. B., Barwaldt, R., Ribeiro, L. M. I., & Pias, M. (2019). IBM Watson Application as FAQ Assistant about Moodle. *Proceedings - Frontiers in Education Conference, FIE, 2019-October*. DOI:10.1109/FIE43999.2019.9028667

Pham, K. T., Nabizadeh, A., & Selek, S. (2022). Artificial Intelligence and Chatbots in Psychiatry. In *Psychiatric Quarterly* (Vol. 93, Issue 1). DOI:10.1007/s11126-022-09973-8

Rajaraman, V. (2023). From ELIZA to ChatGPT: History of Human-Computer Conversation. *Resonance*, 28(6), 889–905. Advance online publication. DOI:10.1007/s12045-023-1620-6

Sayeed, M. S., Mohan, V., & Muthu, K. S. (2023). BERT: A Review of Applications in Sentiment Analysis. In *HighTech and Innovation Journal* (Vol. 4, Issue 2). DOI:10.28991/HIJ-2023-04-02-015

Seaborn, K., Sawa, Y., & Watanabe, M. (2024). Coimagining the Future of Voice Assistants with Cultural Sensitivity. *Human Behavior and Emerging Technologies*, 2024, 2024. DOI:10.1155/2024/3238737

Shum, H. yeung, He, X. dong, & Li, D. (2018). From Eliza to XiaoIce: challenges and opportunities with social chatbots. In *Frontiers of Information Technology and Electronic Engineering* (Vol. 19, Issue 1). DOI:10.1631/FITEE.1700826

Tulshan, A. S., & Dhage, S. N. (2019). Survey on virtual assistant: Google assistant, Siri, Cortana, Alexa. *Communications in Computer and Information Science*, 968, 190–201. Advance online publication. DOI:10.1007/978-981-13-5758-9_17

Vimalkumar, M., Sharma, S. K., Singh, J. B., & Dwivedi, Y. K. (2021). 'Okay google, what about my privacy?': User's privacy perceptions and acceptance of voice based digital assistants. *Computers in Human Behavior*, 120, 106763. Advance online publication. DOI:10.1016/j.chb.2021.106763

Vogel, K. M., Reid, G., Kampe, C., & Jones, P. (2021). The impact of AI on intelligence analysis: Tackling issues of collaboration, algorithmic transparency, accountability, and management. *Intelligence and National Security*, 36(6), 827–848. Advance online publication. DOI:10.1080/02684527.2021.1946952

Wallace, R. S. (2009). *The anatomy of ALICE*. Springer.

Weizenbaum, J. (1966). ELIZA-A computer program for the study of natural language communication between man and machine. *Communications of the ACM*, 9(1), 36–45. Advance online publication. DOI:10.1145/365153.365168

Weng, Y. H., & Hirata, Y. (2018). Ethically Aligned Design for Assistive Robotics. *2018 International Conference on Intelligence and Safety for Robotics, ISR 2018*. DOI:10.1109/IISR.2018.8535889

Whyte, C. (2020). Deepfake news: AI-enabled disinformation as a multi-level public policy challenge. *Journal of Cyber Policy*, 5(2), 199–217. Advance online publication. DOI:10.1080/23738871.2020.1797135

Wu, T., He, S., Liu, J., Sun, S., Liu, K., Han, Q. L., & Tang, Y. (2023). A Brief Overview of ChatGPT: The History, Status Quo and Potential Future Development. *IEEE/CAA Journal of Automatica Sinica, 10*(5). DOI:10.1109/JAS.2023.123618

Yang, S., Lee, J., Sezgin, E., Bridge, J., & Lin, S. (2021). Clinical advice by voice assistants on postpartum depression: Cross-sectional investigation using apple siri, amazon Alexa, google assistant, and microsoft cortana. *JMIR mHealth and uHealth*, 9(1), e24045. Advance online publication. DOI:10.2196/24045 PMID:33427680

Yeung, L. K. C., Tam, C. S. Y., Lau, S. S. S., & Ko, M. M. (2023). Living with AI personal assistant: An ethical appraisal. *AI & Society*. Advance online publication. DOI:10.1007/s00146-023-01776-0

Yin, M., Jiang, S., & Niu, X. (2024). Can AI really help? The double-edged sword effect of AI assistant on employees' innovation behavior. *Computers in Human Behavior*, 150, 107987. Advance online publication. DOI:10.1016/j.chb.2023.107987

Zemčík, M. (2019). A brief history of chatbots. DEStech Transactions on Computer Science and Engineering, 10.

Zhai, C., & Wibowo, S. (2022). A systematic review on cross-culture, humor and empathy dimensions in conversational chatbots: The case of second language acquisition. *Heliyon*, 8(12), e12056. Advance online publication. DOI:10.1016/j.heliyon.2022.e12056 PMID:36531630

Chapter 12
Transformative Potential and Ethical Challenges of Generative AI in E-Commerce:
Data Bias, Algorithm Bias

Shilpa Narula

Asian School of Business, India

Anam Afaq

https://orcid.org/0000-0003-3181-7630

Asian School of Business, India

Shikha Nagar

Asian School of Business, India

Meenu Chaudhary

https://orcid.org/0000-0003-3727-7460

Noida Institute of Engineering and Technology, Greater Noida, India

ABSTRACT

This chapter depicts how generative AI can redefine the e-commerce industry through enhanced personalised experience, dynamic pricing, and operational efficiency. However, the chapter also illustrates how generative AI can be accompanied by problems in terms of biases and ethical challenges. This chapter discusses ways to uncover, tackle, and prevent bias in AI systems to ensure that no customers are discriminated against. The chapter highlights the necessity of accountability, transparency, and robust privacy standards in AI decision-making to win over users and uphold

DOI: 10.4018/979-8-3693-9173-0.ch012

ethical standards. Generative AI development is expected to become increasingly important when used with decision-making accountability. In this way, Generative AI can promote justice, uphold trust, and achieve sustainable digital economy growth as e-commerce develops. As the e-commerce industry grows, generative AIs can ensure fairness, restore confidence, and drive sustainable development.

1. INTRODUCTION

E-commerce breakthroughs that are revolutionising the business, such as real-time pricing algorithms and product suggestion engines, are primarily attributed to generative AI. Driven by machine learning, these cutting-edge technologies analyse vast amounts of data to forecast consumer intent and adjust pricing tactics or advertising campaigns with previously unheard-of precision (Afaq and Gaur, 2021). For competing e-commerce enterprises, this boosts client satisfaction, enabling operational efficiency and boosting corporate growth. Although generative AI offers revolutionary advantages for e-commerce, there are equally different drawbacks, most notably the potential for bias and fairness-related problems when using this technology (Gupta et al., 2024). Systematic mistakes or biases that lead to the unjust treatment of a certain group and individual in AI systems are known as biases. In e-commerce, these biases can manifest in a number of ways, such as distorted product recommendations (Afaq et al., 2022).

It thoroughly examines the various contexts in which bias arises and how it can impact multiple facets of the technology industry, both upstream and downstream. Examples of how bias is applied are given for the classification of products using Amazon's product suggestion generation algorithms. Generative adversarial networks are designed to reveal and improve upon problems; by proactively managing these risks, companies may fully utilise AI's potential to guarantee fair and level playing fields for all parties (Sohn et al., 2024).

2. UNDERSTANDING BIAS IN E-COMMERCE AI

One way to conceptualise AI bias is as a systemic inaccuracy that, when it occurs, results in the mistreatment of a group or an individual. Biases in e-commerce AI can manifest themselves in a variety of ways, including skewing customer marketing campaigns or impacting price and product suggestions (Afaq et al., 2023a). Understanding bias's many forms and origins is essential to keep it from influencing our choices and ensure that all stakeholders receive fair coverage (Ghaffari et al., 2024). This is a direct outcome of algorithmic bias, which arises during the phases of training, development, and design of algorithms. These prejudices may originate from a variety of sources, such as:

- *Feature Selection and Representation*: AI models can be trained using datasets that one is unaware of or are not adequately represented by various demographic groups within an intended market. In this case, the AI may become over-focused on appealing to a specific demographic (e.g., wealthy customers), learning from historical data primarily featuring purchasing behaviours of that type and tuning recommendation or pricing strategies based around them but potentially at the expense of other groups.
- Training Data Imbalances: Biases can also emerge from imbalanced training data, representing certain groups more frequently (or less) than others. The AI can wobble by confusing its consumer preferences and behaviours, so it only provides tilted recommendations or prices (Khan, 2023).
- Algorithmic Choices: The decisions made in designing AI algorithms can also lead to biases. Suppose algorithms are designed and optimised to maximise short-term profits without reference to long-term fairness or any other social good. In that case, they may also end up discriminating in their pricing or promotional offers against certain sub-groups of customers.

2.1 Data Bias

At its core, AI bias is simply a systematic error that results in unfair treatment of some group or individual. The biases around e-commerce AI can manifest in many ways, from skewing product recommendations or pricing strategies to confounding customer marketing efforts. Awareness of the variations and origins of bias is significant for pursuing news that we consider legitimate on all sides.

2.2 Algorithmic Bias

This is precisely the consequence of algorithmic bias, which occurs during all stages - design, development and training. Bias - These biases can stem from several different sectors, including -

- *Feature Selection and Representation:* To train any AI model, one would need a feature set which is unknown or not well represented (by all demographic groups within the market being targeted). Here, the AI may veer towards media targeted at a specific audience (e.g. affluent customers), training on historical data and optimising recommendation or pricing strategies around its behaviours but possibly doing so to the detriment of other demographics (Afaq et al., 2023b).

- *Imbalances in Training Data*: This bias can also be present because of imbalanced training data, where certain groups are over-represented (or under). This concept can also make the AI a bit unstable, where it starts clouding consumer preferences and behaviours, but all that AI does is give biased recommendations or prices (Afaq et al., 2023c).

- *Algorithmic decisions*: The choices we make about how to design AI algorithms can also introduce bias. More crucially, should the algorithms be designed and optimised towards maximising short-term profit without any consideration for long-term fairness (amongst other social goods), they could also potentially price or make promotional offers that discriminate against specific sub-groups of customers.

3. MANIFESTATIONS OF BIAS IN E-COMMERCE AI

Bias can show up in a number of critical e-commerce dimensions that determine both the experiences consumers will have and how these companies perform in their operations. These manifestations include:

Product Recommendations: For AI-driven product recommendation systems in e-commerce platforms, the lists can be biased towards well-established brands or popular categories. That bias could lead to minorities or specific demographics being underrepresented in products. For example, if recommendations are based on historical sales data and one of the brands that sell more is a mainstream / bigger known brand, this can end up, for instance, in AI algorithms also pushing these other items than products from a small or unknown (by Cloudshopping) producers; opening potentially less repeating sale possibilities.

Pricing Algorithms: E-commerce dynamic pricing algorithms can introduce unintended bias in the form of price discrimination according to demographic profile or shopping history. This process, known as price discrimination, can lead to citizens being victims of unjust treatment whenever the pricing changes over time if a specific group constantly suffers from being offered higher prices or unattractive discounts relative to others. For instance, if an AI system looks at patterns of purchase and prices its products on the basis that those buyers are willing to pay more but ignore social-economic conditions, then it will discriminate against other customers (Afaq et al., 2022).

Marketing Strategies: AI-driven personalised marketing strategies, on the other hand, can unwittingly be discriminatory by intentionally or unintentionally excluding and discriminating against certain demographics (Gaur et al., 2023). AI leverages user data around web browsing history, purchase behaviour and demographic information to tailor its marketing campaigns based on individual preferences. However, only if the algorithms were based on diversity is it also possible that the algorithm itself was biased in certain campaign targeting or content creation. This may result in culturally oriented and default images/icons/messages that are just attractive to a sub-part of the possible users away from other outsider (Gaur et al., 2024a).

4. BIAS IN E-COMMERCE AI

Biases in sampling - if the data used to train the model comes from a specific, non-representative demographic or part of customers, you may be training your AI on skewed and biased input, leading it to not generalising accurately across all customer bases. This happens because the AI has picked up on specific patterns and behaviours unique to one demographic sample size, which could overshadow or wrongly express those of other customer clusters. For instance, an e-commerce platform which captures data mainly from young urban girls might lead to a biased AI model with incorrect predictions when applied to older rural customers (Sharma et al., 2022).

Poor Labelling: Wrongfully labelled or biased training data can skew the thought process and decision-making of AI to a great extent. Examples include incorrectly listing products or mis-categorising customer interactions, which can ultimately lead the AI to make wrong recommendations and decisions (Gaur and Afaq. 2020). Photo by ThisisEngineering RAEng on Unsplash)This problem occurs when data points are wrongly labelled due to human annotators or automatically assigned labels, introducing bias into our training dataset. Those biases can get carried from the input level through to later parts of any AI system, making it more likely that those inaccuracies are preserved and could harm particular demographics. Training algorithms and model architectures

Professional Biases: It is also observed how unconscious biases weaved their way in and out of the perceptions on AI development deployment among e-commerce professionals - e.g., marketers, data scientists or product managers. Those biases are based on personal beliefs, experiences or cultural norms that may not be consciously understood or recognised. When biased results are presented to those making decisions about AI development, they may inadvertently introduce biases into the model's design, training data selection or algorithmic choices. Let's say a product manager thinks that specific demographics have a higher propensity to buy high-end products due to some anecdotal evidence (Gaur et al., 2021a). If that is the case, they could unconsciously guide the AI to reinforce this bias. That means it runs the risk of telling those demographics to buy more high-end products, ignoring other customer segments or even fueling stereotypes.

Historical and Societal Contexts: Historically, purchasing data is tainted by societal biases and has been influenced by the preferences consumers have developed over time. These biases can also concern gender, race, and cultural inheritance. However, if AI systems are trained using historical data without considering such biases, they can unknowingly reinforce discrimination in e-commerce contexts. To take a prominent example, if historical purchasing patterns indicate that one gender is over-represented for purchasing particular items as society views them or some unconscious marketing bias influences AI use, this can lead to it recommending these types of items more than others. It propagates disparities and creates a scenario where customers of certain demographic properties might be treated differently (Gaur et al., 2021b).

4.1 Consequences of Biased Generative AI in E-Commerce

- *Discrimination towards demographic groups:* A prejudiced AI will lead to discrimination as there is an influx in broader differences among varying groups. If AI algorithms are trained on biased data or do not fully represent all people, they may fail to serve the needs and preferences of other groups accurately. So, suppose an e-commerce AI is trained to be biased based on the products it recommends. In that case, this can lead to a problematic introduction of selection bias (Which leads directly into model injustice) where one gender or age group disproportionately buys from other groups. This kind of disparate treatment funnels into our everyday lives and further compounds social inequalities by solidifying stereotypes and denying entire sects equitable access to products or services that all United Statesians ought to be afforded. Unfair artificial intelligence (AI) in credit or insurance algorithms, for instance, may result in people losing out on opportunities they deserve because their demographic profile indicates they cannot afford a particular product unless it is changed before hand (Ooi et al., 20230.

- *Social and Economic Outacts on Consumers and Businesses:* AI discrimination in the sector directly impacts consumer trust development processes and can lead to large social and economic outcomes for consumers and enterprises. It can be stated that as consumers realise that AI systems work unfairly to them (even though it is a known behaviour of the processor), they may lose trust in the platform or brand no matter how one decides and what decisions are taken. This lack of confidence can ultimately reduce customer satisfaction and repeat business, causing e-commerce to lose its customer base. There is an even greater reputational risk that biased AI systems will be exposed for being unfair or producing discriminatory outcomes (Harreis et al., 2023). It might damage the image of a company or its brand reputation and may impact building long-term relationships with customers and stakeholders. Biased AI hurts us more than just disrupting individual transactions; we should start thinking about it from a societal perspective. They can reinforce systemic biases, deepen social chasms and prevent progression towards a fairer society. Therefore, addressing bias in AI is a matter of fairness, trust and broader considerations such as privacy and inclusive sustainable economic development (Chaudhary et al., 2024).

4.2 Ethical and Societal Implications

Bias & Fairness: In order to guarantee equitable treatment for all consumer groups, independent of the many variables impacting their choices, the AI algorithms used in e-commerce must be impartial. Specific demographics already disadvantaged and discriminated against in terms of service availability, pricing, and product recommendations may see even more differences due to these biased algorithms. This requires robust data collection methods, transparency about what goes into the algorithms, and ongoing observation to ensure discrimination is not being facilitated (Chakraborty et al.,2023).

—Privacy and Data Protection: E-commerce AI extensively utilises consumer data to experience personalisation and business operations optimisation. However, ethical issues are associated with collecting and storing individuals' data, such as their gender, age, and other demographic details. These have to do a lot with transparency, where consent is greatly affected. The mechanisms needed to safeguard consumer privacy are strong data security protocols, transparency about the usage of such practices and adherence with - or arguably adaptation against - existing privacy statutes, which will preserve the trust and buy-in of consumers in AI-driven e-commerce platforms.

5. TECHNIQUES FOR IDENTIFYING BIAS IN E-COMMERCE AI

5.1 Methods for Detecting Bias in E-Commerce Datasets

Statistical Analysis of Consumer Data

Statistical techniques like descriptive statistics and distribution analysis can identify biases in e-commerce datasets. Analysts can identify such biases by comparing metrics like mean, median, variance and skewness calculated across different demographic cohorts or product offerings. If, for example, some demographic groups are over-/under-represented in the data, that may hint towards particular bias during sampling, which could dampen AI model generalizability (Gaur et al., 2024b).

Fairness Metrics

As mentioned earlier, fairness metrics pertaining to e-commerce outcomes are instrumental in understanding AI systems' ethical and equitable performance. The metrics offer methodologies to determine whether AI-driven decisions and consequences are fair across the board for different demographic groups.

- Demographic Parity: The metric of demographic parity looks into whether the outcomes — e.g. product recommendations or promotional offers toggled at deployment time happen proportionally across demographic groups. For instance, in an e-commerce setting, demographic parity would mean near identical group percentages as targets vs all groups with respect to ages, genders and ethnicities that are preferred equally for recommendations. Demographic parity enables us to ensure that AI systems do not systematically favour one group over the other based on irrelevant characteristics, thus promoting fairness and inclusivity in customer interactions (Xu et al., 2024).
- Equalised Odds: Equalised odds measure how often a prediction is correct in an individual subgroup of the population. This ensures that AI models generate unbiased predictions (e.g., propensity to buy a product) for all customer segments with the same reliability. For example, suppose an AI model renders high accuracy for predicting purchasing behaviour among one demographic but less than the desired level of precision when applied to another. In that case, it indicates a significant gap in the prediction ability that needs targeting (Mogaji and Jain, 2024).
- Disparate Impact: Evaluating whether AI-based decisions inordinately benefit or disadvantage certain demographic groups. It is designed to assess AI outcomes (e.g., approval rates for product financing or eligibility for promotional offers) across customer segments (Cronin, 2024). Disparate impact analysis will easily uncover potential biases in the decision-making process, for example, when more than 50% of one demographic group is implied to obtain favourable outcomes compared to any other due solely or proportionally by determination makers. Algorithms need to be trained, programmed, or, in some way, fine-tuned so that customer demographics that stand intrinsically apart from all others can yield equitable outcomes as well as differences (Wakunuma and Eke, 2024).

5.2 Techniques for Auditing E-Commerce AI Model Outputs

Fairness Evaluations in E-commerce Contexts

Regular audits of AI model outputs are essential to detect and mitigate biases in e-commerce applications. This entails methodically assessing the performance of AI algorithms across different demographic groups and detecting any disparities that might point to biased decision-making. For instance, auditing may show that particular client groups routinely receive less advantageous pricing offers or recommendations, indicating possible biases in AI decision-making.

Testing With Diverse Consumer Data

In order to prevent these biases from going unreported in homogenous datasets, previous studies have addressed effective defences against such bias, such as making sure AI models are evaluated with representative and diverse data samples. By adding data from various demographic groups, geographic locations, and local purchase patterns, developers may evaluate how their AI systems respond to real-world scenarios across a range of client sectors (Chodak, 2024). Early in the e-commerce AI lifecycle, this testing aids in identifying and removing biased behaviour, ensuring that every user has an equal chance of receiving fair treatment.

6. MITIGATING BIAS IN E-COMMERCE AI

- *Data Augmentation and Balancing:* In this strategy, data sets are usually skewed; under-sampling to balance out classes may generate better classifier accuracy. These broadly range from synthetic data creation to text/image augmentation, downsampling majority or upsampling minority classes in the datasets, etc.
- *Re-Sampling Methods for Consumer Data:* In e-commerce datasets, re-sampling, like stratified sampling and algorithms such as SMOTE, ensures that data representation across different demographic groups is balanced. These techniques not only divide the data into internally consistent sets but also generate synthetic examples when necessary in order to avoid seeing biased training and ensure that AI learns from different ranges of realistic input. This indeed creates fairness in the AI results & inclusive practices can be followed even in e-commerce applications using this method.
- *Implementing Fairness Constraints in E-commerce Models:* Fairness constraints are crucial when running e-commerce models. Using fairness con-

straints during model training can help keep the biases under control and thus deliver equitable decisions. Fairness constraints: Ad-hoc rules during model training, like setting up metrics or thresholds that AI models are expected to meet. For instance, such constraints can involve predictions or recommendations being of age - if overbearing information on one demography does not unfairly advantage/disadvantage any other in conditions (fair treatment through explanations across groups). Also, a working example of fairness constraint is demographic parity, in which loan approvals or product tips are similarly dispensed among extraordinary demographics (Sidaoui et al., 2024).

- *Adversarial Training for E-commerce Applications:* Adversarial training represents one approach to eliminating biases from e-commerce AI. It trains two networks at once: an adversarial network that aims to counteract biases introduced by its top peer and a network that predicts outputs given input data. By adopting an adversarial image creation method, it is possible to train the model to disregard sensitive attributes like gender or race, resulting in a choice that is free from bias and produces more equitable outcomes. Through adversarial manipulation of the data being fed into the model, adversarial training creates models that are resilient to biases drawn from it. This will minimise the danger of discrimination while optimising model performance, but it should be complex to generalise this careful balancing act. In short, incorporating fairness constraints and adversarial learning into such AI models used in e-commerce will build trust among users, eliminate discrimination, and ensure fair outcomes. The bleow mentioned figure 1 depicts the mitigating bias in E-commerce AI.

Figure 1. Mitigating bias in E-commerce AI

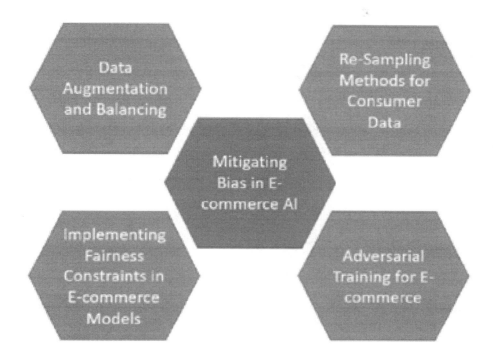

6.1 Post-Processing Methods

Post-processing methods - These methods use re-ranking and fairness-aware reweighting to adjust a model's outputs. Re-ranking reorders the recommended products or services list based on fairness metrics (Demographic parity, equalised odds) to get a fair distribution across different demographics. In the use case of a lending model, fairness-aware reweighting can change how significant or heavy importance on features is placed when deciding (i.e. bias) that could lead to pricing disparities and different product recommendations being made in other machine learning models due to these privileges. This helps e-commerce platforms introduce a higher level of fairness, transparency and credibility within their AI-powered recommendation systems to ensure all consumers are fairly treated according to merit based on preferences & requirements (Rosário, 2024).

Fairness-Aware Generation in Personalized Marketing- In personalised marketing, it is important to take on the role of a fair arbiter and prevent bias in content generation. An overview of the critical elements to consider in creating personalised marketing content using fairness-aware generation techniques is required. This includes pre-emptive bias identification to identify and reduce the biases in marketers' datasets fueling their promotional content. Thus, they are free of stereotypes and non-inclusive language. In addition, inclusion means that the marketing materials themselves are representative of its customer segment - in culture and age. If personalised deals are being suggested, they should be equally handed over on a fair share to all customer segments; the proposal follows equitable norms for a promotional strategy, which would, in turn, strengthen values towards customer engagement strategies and help them increase their loyalty conduct that will eventually assist delight factors while retaining customers, thus uplifting lifetime value of such. Collectively, these nurturing activities enhance moral marketing strategies that support fostering great client relationships and further promote the company as approachable and inclusive in customised communications.

6.2 Best Practices for Reducing Bias in E-Commerce AI

Regular Bias Audits and Monitoring

Regular bias audits are still crucial for identifying the snapped inherent biases in e-commerce AI systems as they continue to be trained and deployed. They are done step-by-step by checking what the datasets consist of, what models predict, and at which points outcomes diverge for different demographic groups. To recognise and correct that, firms ought to analyse how effectively AI systems perform across various segments of their client populations. They even audit for bias, which is especially important given today's landscape of AI decisions carrying more weight over racial or gendered attributes and how fair (or far/right) those rules can go. The last is ongoing monitoring after deployment, which helps to ensure that the deployed model remains fair as it interacts with real-world data and evolves.

Utilising Diverse Training Data and Stakeholder Engagement

Preventing bias in e-commerce: The importance of unbiased, varied training data Diverse datasets - More diversity means spanning a broader spectrum of customer demographics, behaviours and preferences so the AI model trains across as wide inputs. Not that humans are all biased by definition, but eliminating any part of the input (data) upon which they depend to prevent their bias from emerging in response to false or incomplete representations. Furthermore, justice and transparency with respect to human-borne AI also rely on the participation of stakeholders. Customers, data scientists, and ethicists are the different stakeholders who add their perspectives on how each model may be susceptible to biases or ethical implications. Involving stakeholders proactively in varying stages of the AI development lifecycle can support accountability, inclusivity, and openness about AI's ethical standards and dependability.

7. FAIRNESS IN E-COMMERCE AI

Distributive Justice in E-commerce: The equal distribution of advantages and opportunities for various consumer groups while utilizing e-commerce AI solutions is known as distributive justice. The main goal of this principle is to prevent AI systems from unjustly favouring one gender or other demography within the consumer base over another. Artificial intelligence (AI) outputs are quantified and corrected in accordance with a fair distribution using fairness-aware algorithms or methodologies centred around fairness concepts, such as demographic parity or equalized chances. For instance, distributive justice might be used to deliver promotional offers or product recommendations, which guarantees that all client groups have equal possibilities (fairness) to enjoy AI-driven judgments regardless of demographics.

Procedural Fairness in Consumer Interactions: Procedural fairness, or equitable transactions between e-commerce platforms and customers, is at issue. Justice can be attained by a number of methods, such as openness in the decision-making procedures, uniform and easily accessible rule sets, and policies with explicit penalties for non-compliance. Accountability, or something like procedural fairness in AI-customer interactions, is an interpretability guarantee that gives customers an explanation of how decisions made by algorithms were reached. This entails giving clients the ability to challenge or appeal automated judgments, being open and honest about how AI systems use their data, and disclosing the decisions made (price, suggestions). Procedural fairness in AI allows e-commerce companies to create a mechanism that lowers the risk of introducing bias or discrimination in future cases, sets a standard for other tech actors, and increases customer trust through better transparent decision-making.

7.1 Challenges in Achieving Fairness

Trade-Offs in E-commerce Outcomes: One of the most challenging aspects is for e-commerce companies to balance fairness with other essential business objectives (e.g., profit, operational efficiency). Even a perfectly fair AI-driven decision needs to be taken in light of these trade-offs: it may minimise unfairness, reduce revenue generation, hike operational costs, etc. It could mean lowering profits by providing the exact pricing to all customers regardless of purchasing power. Achieving a balance between these trade-offs requires deliberate thought and strategy so that fairness is never compromised and results in maximized business outcomes.

Context-Specific Definitions of Fairness in Online Shopping: To provide a definitive definition of fairness, one must include consumer demographics, market conditions, and cultural norms. The definition of fairness may vary depending on one's perspective and the people it affects. For example, fair pricing techniques will vary depending on the price adjustments specific customer categories might perceive or regional economic differences. Moreover, ethical considerations such as customer privacy and non-discriminatory procedures have contributed to the complexity surrounding the definition and implementation of fairness in online buying. Businesses should continue to communicate with customers while creating a context-specific definition of fairness that complies with legal and moral requirements in order to manage these complications.

8. ETHICAL CONSIDERATIONS IN E-COMMERCE AI

Transparency: The transparency behind e-commerce AI includes proper visibility of operations performed by AI, such as algorithms and decision-making processes, which should be explained in clear terms so that it can be easy to understand for members using those. This enhanced transparency is designed to build trust by helping consumers better understand how AI systems influence their online experiences - say, product recommendations or pricing. It also lets users make smart decisions and punishes businesses for AI-driven actions (Yafei et al., 2024).

Accountability: It refers to being accurate and transparent in keeping records on the AI systems' actions in e-commerce. Part of this entails defining who is responsible for developing, deploying, and overseeing AI within an organisation. Accountability introduces the rules, responsibilities, and structures to ensure that an AI error or unintended consequence is not allowed. Making AI systems accountable will enable organisations to reduce the dangers of biased or unethical usages, promote greater consumer confidence and ensure Compliance with regulatory rules (Mariani and Dwivedi, 2024).

Privacy: Privacy is critical for e-commerce AI; virtual transactions require collecting, recycling, and protecting consumer data. AI-driven changes to customer experiences, such as personalised marketing and dynamic pricing strategies, need an extended depth of data on the customers involved. Data protection: Data must be protected with serious encryption, anonymisation, and storage methods. This includes the requirement of explicit consent before data processing and transparency in how AI uses consumers (Gupta et al., 2023). Protection of privacy rights can eliminate risks to unauthorised data access or misuse, increase customer trust in the security of personal information handling processes and maintain Compliance with such local regulations as GDPR (General Data Protection Regulation) or CCPA (California Consumer Privacy Act).

8.1 Addressing Ethical Dilemmas

Balancing Personalisation and Privacy

The first problem related to AI ethics in e-commerce is the balancing act of personalisation and consumer privacy. Personalisation - Messaging like product recommendations and marketing messages can be highly improved by personalising the user journey to cater to individual needs. However, this practice must be ethically followed by general privacy principles and user consent. For the sake of personalisation, businesses should be transparent when it comes to consumer data collection, use, and sharing. This will include sufficiently explaining data practices, explicit user consent mechanisms to process their private information, and privacy settings for their control. Prioritizing consumer privacy with personalisation will help businesses create trust and improve customer satisfaction by eliminating misuse or unauthorised access risks (Rane, 2023).

Ensuring Fair Access to AI Benefits

Equally crucial ethical consideration is that AI dividends should be democratised, i.e., everyone, irrespective of origin or characteristics. This blog addresses how AI technologies can be leveraged to better the lives of customers and process owners in e-commerce. However, there is a risk that this will deny these advantages to underprivileged or marginalized people, which could lead to a number of problems for them, including economic inequality, algorithm bias, and digital illiteracy. Companies need to take preventative measures by creating AI procedures and systems that promote diversity and provide equal access for all (Chamola et al., 2024).

8.2 Implementing Ethical Guidelines

Industry Standards and Regulations These standards are crucial in assisting the development of responsible AI for e-commerce. Principles and best practices from industry standards put forth by organizations such as IEEE or the Partnership on AI prescribe how to design ethical AI models with transparency, accountability, and fairness. These guides often cover data privacy, bias mitigation, and identification, user consent, algorithmic transparency, Compliance with regulations governing data protection, consumer rights, and algorithmic accountability - such as the GDPR in

Europe or CCPA in California. Complying with industry standards and regulations will help businesses reduce the risk of unethical AI practices, build consumer trust, and show responsible commitment to enforcing their AIs (Linkon et al., 2024).

Corporate Responsibility and Governance: Integrating ethical AI principles into e-commerce operations can be facilitated by ensuring corporate responsibility governance frameworks. This includes developing business policies and defining ethical AI design, deployment, and use. Ethics committees/trusts or advisory boards can be established for oversight and guidance on the ethical aspects of AI in order to guarantee ethical practice in the field. Subject matter experts in consumer protection, data privacy, AI ethics, and other pertinent fields may make up these groups. Integrating justice, accountability, transparency, and consumer primacy norms into AI-powered projects is necessary to do corporate ethics right. This ensures that ethical issues are given top priority at every stage of the artificial intelligence lifecycle, from conception and development to implementation and long-term action monitoring. Companies may foster a culture of ethics that will guarantee that AI's adverse effects are minimized and enhance their standing as ethical users of AI in e-commerce (Krishnan and Mariappan,2024).

9. CONCLUSION

Herein lies the opportunity for generative AI, a formidable technology that has the ability to upend e-commerce. By generating suggestion sections, personalized navigation, and search experiences across websites, as well as improving tailored offers for up-selling and cross-selling, generative AI makes it feasible to create a more engaging buying experience. Yet there are also certain drawbacks to generative AI, chief among them being ethics, justice, and bias issues. In order to enable all consumer segments to benefit from generative AI equitably, solving these issues will need a conscious focus on identifying, reducing, and eliminating biases in AI systems. As we go into a new era of e-commerce, the moral and responsible uses of generative AI in the business sector will be essential to maintaining growth, establishing fair business practices with customers, and fostering trust.

REFERENCES

Afaq, A., & Gaur, L. (2021, November). The rise of robots to help combat covid-19. In *2021 International Conference on Technological Advancements and Innovations (ICTAI)* (pp. 69-74). IEEE. DOI:10.1109/ICTAI53825.2021.9673256

Afaq, A., Gaur, L., & Singh, G. (2022, April). A latent dirichlet allocation technique for opinion mining of online reviews of global chain hotels. In 2022 3rd International Conference on Intelligent Engineering and Management (ICIEM) (pp. 201-206). IEEE. DOI:10.1109/ICIEM54221.2022.9853114

Afaq, A., Gaur, L., & Singh, G. (2023a). A trip down memory lane to travellers' food experiences. *British Food Journal*, 125(4), 1390–1403. DOI:10.1108/BFJ-01-2022-0063

Afaq, A., Gaur, L., & Singh, G. (2023b). Social CRM: Linking the dots of customer service and customer loyalty during COVID-19 in the hotel industry. *International Journal of Contemporary Hospitality Management*, 35(3), 992–1009. DOI:10.1108/IJCHM-04-2022-0428

Afaq, A., Singh, G., Gaur, L., & Kapoor, S. (2023c, November). Aspect-Based Opinion Mining of Customer Reviews in the Hospitality Industry: Leveraging Recursive Neural Tensor Network Algorithm. In 2023 3rd International Conference on Technological Advancements in Computational Sciences (ICTACS) (pp. 1392-1397). IEEE.

Chakraborty, U., Roy, S., & Kumar, S. (2023). *Rise of Generative AI and ChatGPT: Understand how Generative AI and ChatGPT are transforming and reshaping the business world (English Edition)*. BPB Publications.

Chamola, V., Sai, S., Sai, R., Hussain, A., & Sikdar, B. (2024). Generative AI for Consumer Electronics: Enhancing User Experience with Cognitive and Semantic Computing. *IEEE Consumer Electronics Magazine*, 1–9. DOI:10.1109/MCE.2024.3387049

Chaudhary, M., Gaur, L., Singh, G., & Afaq, A. (2024). Introduction to Explainable AI (XAI) in E-Commerce. In *Role of Explainable Artificial Intelligence in E-Commerce* (pp. 1–15). Springer Nature Switzerland. DOI:10.1007/978-3-031-55615-9_1

Chodak, G. (2024). Artificial Intelligence in E-Commerce. In *The Future of E-commerce: Innovations and Developments* (pp. 187–233). Springer Nature Switzerland. DOI:10.1007/978-3-031-55225-0_7

Cronin, I. (2024). The Evolving World of Generative AI. In *Understanding Generative AI Business Applications: A Guide to Technical Principles and Real-World Applications* (pp. 223–242). Apress. DOI:10.1007/979-8-8688-0282-9_15

Gaur, L., & Afaq, A. (2020). Metamorphosis of CRM: incorporation of social media to customer relationship management in the hospitality industry. In *Handbook of Research on Engineering Innovations and Technology Management in Organizations* (pp. 1–23). IGI Global. DOI:10.4018/978-1-7998-2772-6.ch001

Gaur, L., Afaq, A., Arora, G. K., & Khan, N. (2023). Artificial intelligence for carbon emissions using system of systems theory. *Ecological Informatics*, 76, 102165. DOI:10.1016/j.ecoinf.2023.102165

Gaur, L., Afaq, A., Singh, G., & Dwivedi, Y. K. (2021a). Role of artificial intelligence and robotics to foster the touchless travel during a pandemic: A review and research agenda. *International Journal of Contemporary Hospitality Management*, 33(11), 4079–4098. DOI:10.1108/IJCHM-11-2020-1246

Gaur, L., Afaq, A., Solanki, A., Singh, G., Sharma, S., Jhanjhi, N. Z., My, H. T., & Le, D. N. (2021b). Capitalizing on big data and revolutionary 5G technology: Extracting and visualizing ratings and reviews of global chain hotels. *Computers & Electrical Engineering*, 95, 107374. DOI:10.1016/j.compeleceng.2021.107374

Gaur, L., Gaur, D., & Afaq, A. (2024a). Demystifying Metaverse Applications for Intelligent Healthcare. In Metaverse Applications for Intelligent Healthcare (pp. 1-23). IGI Global.

Gaur, L., Gaur, D., & Afaq, A. (2024b). Ethical Considerations in the Use of the Metaverse for Healthcare. In Metaverse Applications for Intelligent Healthcare (pp. 248-273). IGI Global.

Ghaffari, S., Yousefimehr, B., & Ghatee, M. (2024, February). Generative-AI in E-Commerce: Use-Cases and Implementations. In 2024 20th CSI International Symposium on Artificial Intelligence and Signal Processing (AISP) (pp. 1-5). IEEE.

Gupta, M., Akiri, C., Aryal, K., Parker, E., & Praharaj, L. (2023). From chatgpt to threatgpt: Impact of generative ai in cybersecurity and privacy. *IEEE Access : Practical Innovations, Open Solutions*, 11, 80218–80245. DOI:10.1109/ACCESS.2023.3300381

Gupta, R., Nair, K., Mishra, M., Ibrahim, B., & Bhardwaj, S. (2024). Adoption and impacts of generative artificial intelligence: Theoretical underpinnings and research agenda. *International Journal of Information Management Data Insights*, 4(1), 100232. DOI:10.1016/j.jjimei.2024.100232

Harreis, H., Koullias, T., Roberts, R., & Te, K. (2023). *Generative AI: Unlocking the future of fashion*. McKinsey & Company.

Khan, S. (2023). Role of Generative AI for Developing Personalized Content Based Websites. *International Journal of Innovative Science and Research Technology*, 8, 1–5.

Krishnan, C., & Mariappan, J. (2024). The AI Revolution in E-Commerce: Personalization and Predictive Analytics. In *Role of Explainable Artificial Intelligence in E-Commerce* (pp. 53–64). Springer Nature Switzerland. DOI:10.1007/978-3-031-55615-9_4

Linkon, A. A., Shaima, M., Sarker, M. S. U., Nabi, N., Rana, M. N. U., Ghosh, S. K., & Chowdhury, F. R. (2024). Advancements and applications of generative artificial intelligence and large language models on business management: A comprehensive review. *Journal of Computer Science and Technology Studies*, 6(1), 225–232. DOI:10.32996/jcsts.2024.6.1.26

Mariani, M., & Dwivedi, Y. K. (2024). Generative artificial intelligence in innovation management: A preview of future research developments. *Journal of Business Research*, 175, 114542. DOI:10.1016/j.jbusres.2024.114542

Mogaji, E., & Jain, V. (2024). How generative AI is (will) change consumer behaviour: Postulating the potential impact and implications for research, practice, and policy. *Journal of Consumer Behaviour*, cb.2345. DOI:10.1002/cb.2345

Ooi, K. B., Tan, G. W. H., Al-Emran, M., Al-Sharafi, M. A., Capatina, A., Chakraborty, A., Dwivedi, Y. K., Huang, T.-L., Kar, A. K., Lee, V.-H., Loh, X.-M., Micu, A., Mikalef, P., Mogaji, E., Pandey, N., Raman, R., Rana, N. P., Sarker, P., Sharma, A., & Wong, L. W. (2023). The potential of generative artificial intelligence across disciplines: Perspectives and future directions. *Journal of Computer Information Systems*, 1–32. DOI:10.1080/08874417.2023.2261010

Rane, N. (2023). ChatGPT and Similar Generative Artificial Intelligence (AI) for Smart Industry: role, challenges and opportunities for industry 4.0, industry 5.0 and society 5.0. Challenges and Opportunities for Industry, 4.

Rosário, A. T. (2024). Generative ai and generative pre-trained transformer applications: Challenges and opportunities. Making Art With Generative AI Tools, 45-71.

Sharma, S., Singh, G., Gaur, L., & Afaq, A. (2022). Exploring customer adoption of autonomous shopping systems. *Telematics and Informatics*, 73, 101861. DOI:10.1016/j.tele.2022.101861

Sidaoui, K., Mahr, D., & Odekerken-Schröder, G. (2024). Generative AI in Responsible Conversational Agent Integration: Guidelines for Service Managers. *Organizational Dynamics*, 53(2), 101045. DOI:10.1016/j.orgdyn.2024.101045

Sohn, J. J., Guo, N., & Chung, Y. (2024, June). Sustainable E-commerce Marketplace: Reshaping Consumer Purchasing Behavior Through Generative AI (Artificial Intelligence). In International Conference on Human-Computer Interaction (pp. 254-269). Cham: Springer Nature Switzerland. DOI:10.1007/978-3-031-61315-9_18

Wakunuma, K., & Eke, D. (2024). Africa, ChatGPT, and Generative AI Systems: Ethical Benefits, Concerns, and the Need for Governance. *Philosophies*, 9(3), 80. DOI:10.3390/philosophies9030080

Xu, D., Zhang, D., Yang, G., Yang, B., Xu, S., Zheng, L., & Liang, C. (2024). Survey for Landing Generative AI in Social and E-commerce Recsys—the Industry Perspectives. arXiv preprint arXiv:2406.06475.

Yafei, X., Wu, Y., Song, J., Gong, Y., & Lianga, P. (2024). Generative AI in Industrial Revolution: A Comprehensive Research on Transformations, Challenges, and Future Directions. Journal of Knowledge Learning and Science Technology ISSN: 2959-6386 (online), 3(2), 11-20.

Chapter 13
Ethical Use of AI in Criminal Justice System

Aravind Ganesan
https://orcid.org/0009-0009-0936-6678
Hindusthan College of Engineering and Technology, India

ABSTRACT

This chapter explores the utilization of generative AI in the criminal justice system, emphasizing the advantages and disadvantages of its applications in risk assessment, evidence analysis, and predictive policing. The chapter focuses on technical challenges such as data bias, algorithm transparency, and the accuracy of AI predictions, all of which can have a significant impact on the fairness and reliability of legal decisions. The preservation of systemic biases, privacy concerns, and striking a balance between personal freedoms and public safety are among the other ethical problems covered. Proposed remedies encompass the adoption of moral protocols, ongoing evaluation and scrutiny of artificial intelligence systems, heightened clarity, and the encouragement of cooperation across many fields. To guarantee that AI in criminal justice enhances fairness and justice, the chapter promotes ongoing, inclusive discussions on its application. It underlines the technology's ability to improve the system while also highlighting the significance of mitigating its hazards.

1. INTRODUCTION TO AI CRIMINAL JUSTICE

The operations and public-facing interactions of law enforcement, courts, and prisons have seen a substantial transformation with the introduction of Artificial Intelligence (AI) into the criminal justice system. With its ability to handle and analyze vast amounts of data quickly and precisely, artificial intelligence (AI) offers unique prospects to improve public safety, streamline legal procedures, and guarantee

DOI: 10.4018/979-8-3693-9173-0.ch013

more equal justice results. But this integration also brings up important moral, legal, and societal questions, such as those pertaining to accountability, bias, and privacy.

In the field of criminal justice, artificial intelligence (AI) aims to complement human judgment by providing cutting-edge analytical tools and insights. AI technology has the power to completely change the criminal justice system. From automated risk assessment tools that help determine bail, sentencing, and parole decisions, to predictive police algorithms that try to stop crimes before they happen.

1.1 Overview of AI Technologies in Criminal Justice

1. Predictive Policing: AI systems examine crime data to forecast future crime hotspots, including locations and hours. Although these systems can aid in the more efficient use of police resources, there are worries that they could reinforce preexisting prejudices in law enforcement.
2. Facial Recognition: Law enforcement agencies utilize AI-powered facial recognition technology to identify suspects by comparing surveillance images with databases of known individuals. Although it may greatly speed up investigations, privacy rights and the accuracy of matches across different populations have become hot topics of discussion.
3. Automated Risk Assessments: To help judges decide on bail, sentence, and parole, AI systems assess the likelihood of reoffending in courtrooms. These instruments aim to improve the criminal justice system's objectivity and efficiency by making it less subjective and more effective. However, questions regarding the algorithms' openness and the data that was used to train them continue to exist.
4. Digital Forensics: AI methods are employed to look for patterns or evidence relevant to criminal investigations in vast datasets and digital evidence, such as emails, social media posts, and phone records. This can cut down on how long it takes to process the evidence.
5. Correctional Facility Management: AI is employed in prisons to keep an eye on the conduct of the inmates, anticipate any problems, and oversee prison operations. The purpose of these applications is to improve jail operations and safety.
6. Crime Reporting and Analytics: AI-powered technologies help law enforcement organizations make more strategic and informed decisions by helping the public report crimes anonymously and by studying crime trends and patterns in real-time.
7. Legal Document Analysis: Artificial intelligence (AI) systems help attorneys and legal scholars find precedents and pertinent case law more quickly by evaluating and analyzing large volumes of legal documents.

1.2 Historical Context and Evolution of AI Applications

Legal philosophy, technological advancement, and public attitudes toward justice and crime prevention are all reflected in the historical background and development of AI applications in the criminal justice system. Following this development reveals the potential and difficulties that lie ahead while also shedding light on how we got to the current status of AI in criminal justice.

1.2.1 Early Foundations and Theoretical Concepts

The idea of applying computational techniques to support judicial and law enforcement proceedings dates back to the middle of the 20th century, around the time that computer science and information technology were first developed. Rather than concentrating on the intricate analytical tasks that AI systems are utilized for today, early attempts were mainly concerned with automating administrative activities and maintaining data.

1.2.2 The 1970s and 1980s: Pioneering Days

As artificial intelligence started to develop in the 1970s and 1980s, scientists looked into expert systems that were intended to replicate human decision-making in particular fields, such as law. These systems, which were basic by today's standards, were mainly utilized in academic and research contexts to show how artificial intelligence (AI) may be applied to legal reasoning and decision-making.

1.2.3 The 1990s: Expansion of Digital Data

More advanced AI applications were made possible by the internet's and more affordable, potent computer resources' role in the 1990s' digital data explosion. In order to store and retrieve enormous volumes of data, law enforcement organizations began using database management systems, which set the foundation for data-driven police and investigations.

1.2.4 The 2000s: Advances in Machine Learning

The technologies at the core of contemporary AI, machine learning and data analytics, saw tremendous advancements in the 2000s. During this time, risk assessment instruments and predictive police models were developed. These tools used past crime data along with other characteristics to predict future incidents. These changes signaled a shift in the criminal justice system toward more proactive strategies that seek to deter crime before it starts and use data to influence choices about bail and punishment.

1.2.5 The 2010s to the Present: Ethical Concerns and AI Integration

Over the past ten years, there has been a growing integration of artificial intelligence (AI) technologies into the criminal justice system. These technologies range from biometric analysis and face recognition to AI-assisted legal research and autonomous drones for surveillance. Growing knowledge of the ethical ramifications of AI, such as concerns about accountability, justice, and bias, has also occurred throughout this time. High-profile incidents and studies have brought to light the ways in which AI systems can reinforce and magnify preconceived notions, prompting demands for more transparent and equitable use of AI in the criminal justice system.

2. LITERATURE REVIEW

Table 1. Literature review

S.No.	Author(s)	Published Year	Purpose	Results/Findings
1.	Yu et al.	2020	Define artificial intelligence in the context of crime prediction and prevention as the use of technologies applying algorithms to large datasets to support or replace human police work.	Explain the role of AI in enhancing efficiency and providing insights from big data in crime prediction and prevention efforts, emphasizing its ability to find patterns more efficiently.
2.	Rademacher T.	2020	Investigate the potential of AI in crime prediction and prevention, focusing on its efficiency in finding patterns in large datasets and its ability to contribute to police work.	Highlight the efficiency of AI in analyzing large datasets to identify patterns and assist in placing resources more rapidly for crime prevention efforts.
3.	Hajela et al.	2020	Examine the variables involved in crime prediction with AI, including time, weather, location, annual income, and literacy rate, to assess their impact on the probability of crime occurrence.	Explore how AI incorporates various risk factors to predict crime occurrence and contribute to crime prevention efforts by identifying areas at higher risk of criminal activity.
4.	Wiafe et al.	2020	Investigate the potential of AI technologies in cybersecurity and crime prevention, focusing on reducing computational complexity and model training times to enhance effectiveness in criminal investigation.	Highlight the opportunities offered by AI in cybersecurity and crime prevention, emphasizing its ability to reduce computational complexity and training times for more effective crime prevention methods.

continued on following page

Table 1. Continued

S.No.	Author(s)	Published Year	Purpose	Results/Findings
5.	Broadhurst et al.	2019	Explore the application of AI technology in crime prediction and prevention, highlighting its potential to assist human police work or replace it through algorithmic analysis of large datasets.	Discuss the potential of AI to increase efficiency in crime prediction and prevention by analyzing patterns in big data more effectively than human police officers.
6.	Asaro	2019	Discuss the use of AI technologies such as machine learning, deep learning, and data mining in crime prediction and prevention, aiming to improve the effectiveness of these practices.	Explore how AI technologies can enhance crime prediction and prevention by efficiently identifying patterns and accurately predicting future criminal behavior.
7.	Danaher J.	2018	Investigate the role of AI in predicting criminal acts to prevent them, emphasizing the utilization of big data and various AI techniques to identify patterns and predict crime occurrence.	Discuss the process of using big data and AI techniques to predict criminal behavior and prevent crime by identifying risk factors such as time, weather, location, and socio-economic factors.
8.	Hassani et al.	2016	Explore ethical issues and safety risks associated with the use of AI in law enforcement, particularly in predicting future criminal activity and creating profiles of individuals at risk of committing crimes.	Discuss concerns about the ethical implications of creating profiles of individuals at risk of committing crimes and the safety risks of AI technology being used in law enforcement, including potential for criminal exploitation.

3. TECHNICAL FOUNDATIONS OF GENERATIVE AI

Generative AI represents a frontier in artificial intelligence where the focus shifts from understanding and classifying data (as in discriminative models) to actually generating new data that resembles the input data. This feature creates exciting opportunities for creativity and automation in a variety of industries, including music, writing, painting, and more. It is crucial to investigate the technical underpinnings of generative AI, particularly the algorithms and architectures that power it, in order to comprehend the concepts and mechanisms underlying it.

3.1 Understanding Generative AI: Concepts and Mechanisms

3.1.1 Key Concepts of Gen AI

1. Learning from Data: Generative AI models pick up patterns, structures, and subtleties by training on huge datasets. They are able to generate outputs that closely resemble the properties of the input data thanks to this training.
2. Latent Space: The latent space is important to many generative models. A condensed representation of the input data that preserves all of its important characteristics is called latent space. Generative models are able to generate new data instances that are variants on the training data by traveling through this latent space.
3. Generative vs. Discriminative Models: It is fundamental to understand the difference between generative and discriminative models. Generative models learn to create new instances within a category (e.g., producing fresh photographs of cats), whereas discriminative models learn to discriminate between distinct categories or labels (e.g., categorizing images of cats vs. dogs).

3.1.2 Mechanisms of Generative AI

The workings of generative AI are defined by a number of architectures and algorithms, the two most well-known of which are Variational Autoencoders (VAEs) and Generative Adversarial Networks (GANs).

1. Adversarial Networks Generative (GANs): The discriminator and generator neural networks that make up a GAN are trained concurrently in a competitive environment. The discriminator determines whether the data is real—derived from the training set—or false—produced by the generator. The generator generates fake data that mimics the training set. The adversarial process enhances the generator's capacity to generate accurate data.
2. Variational Autoencoders (VAEs): A latent space is used by VAEs to compress data before reconstructing it. Input data is first encoded into a latent representation, which is subsequently decoded back into the original data format. One of the most well-known features of VAEs is their capacity to create new data instances by adjusting values in the latent space.
3. Transformer Models: Although transformer models like as GPT (Generative Pre-trained Transformer) were initially intended for natural language processing tasks, they have demonstrated impressive adaptability in generative applications.

These models can produce content that mimics human writing styles and is coherent and contextually relevant by virtue of their extensive training on text data.

3.2 Data Sources for AI in Criminal Justice: Scope and Limitations

When it comes to creating and utilizing AI in the criminal justice system, data is essential. The quality, variety, and comprehensiveness of the underlying data sources have a significant impact on the efficacy, equity, and transparency of AI applications. These data sources include social media, surveillance footage, public records, and law enforcement databases. Each offers a different perspective and presents a different set of difficulties.

3.2.1 Scope of Data Sources

1. Law Enforcement Records: These comprise incident reports, arrest logs, and investigative conclusions. These kinds of data are essential for teaching AI systems criminal analysis, predictive policing, and pattern identification. They support more strategic resource allocation, the identification of crime hotspots, and a knowledge of criminal behavior.
2. Judicial and Court Records: AI systems may learn about legal results and the judicial process by analyzing data from court sessions, verdicts, and sentences. This data is essential for creating technologies that try to forecast recidivism rates, support parole decisions, or process legal paperwork automatically.
3. Correctional Data: Data on parole histories, rehabilitation outcomes, and inmate populations provide a rich dataset for AI systems aimed at controlling correctional institutions, lowering recidivism, and customizing rehabilitation plans for each individual.
4. Public and Social Media Data: To get information, look into crimes, and comprehend community opinions, law enforcement organizations are increasingly depending on public data sources, including social media. Large volumes of internet material may be analyzed by AI algorithms to spot dangers, find people who have gone missing, or keep an eye on gang activity.
5. Surveillance and Body Camera Footage: Artificial intelligence (AI) may be trained in facial recognition, behavioral analysis, and evidence gathering with the use of video data from police body cameras and public surveillance systems. These apps help speed up the investigation process, increase situational awareness, and increase police accountability.

3.2.2 Limitations and Challenges

1. Bias and Representation Issues: The objectivity of AI systems depends on the quality of the training data. When AI models are trained on historical criminal justice data, there is a chance that systematic biases and inequality may continue. For the development of equitable AI technologies, diversity and equity in data sources are essential.
2. Data Privacy and Consent: Using personal information creates serious privacy issues, particularly when it comes from public and social media sources. Tight data governance and ethical standards are necessary to strike a balance between the advantages of AI applications and the need to safeguard human privacy rights.
3. Data Integrity and Quality: Outdated, inaccurate, or incomplete data can seriously compromise the efficacy of AI systems. Maintaining the high caliber and dependability of data sources is crucial, requiring constant work to update and validate datasets.
4. Data Silos and Interoperability: The fragmented nature of criminal justice data, which is frequently in incompatible forms by several authorities, makes it more difficult to create all-encompassing AI solutions. Standardizing data gathering and sharing procedures is essential to ensuring that AI systems reach their full potential.
5. Ethical and Legal Constraints: There are ethical and legal limitations on the use of AI in criminal justice, particularly with regard to surveillance and the application of predictive algorithms. To overcome these limitations, one must carefully strike a balance between the advancement of technology and the preservation of civil liberties.

4. APPLICATIONS OF AI IN CRIMINAL JUSTICE

4.1 Predictive Policing

The term "predictive policing" describes the application of algorithmic methods and data analysis to foresee and stop possible crimes before they happen. Law enforcement organizations seek to reduce crime rates by more effectively allocating resources through the analysis of trends found in huge databases. The practice's efficacy and moral consequences, however, have generated a great deal of discussion.

4.1.1 Case Studies

1. PredPol: One of the most well-known techniques in this field is Predictive Policing, or PredPol. It predicts where and when future crimes are likely to occur using three data points: crime type, location, and time. Numerous American localities have implemented PredPol, and some have reported decreased crime rates. For example, after putting the system in place, the Los Angeles Police Department saw a significant drop in burglaries.
2. HunchLab: This other predictive police tool uses a broader range of information to identify crime hotspots, such as socioeconomic data and meteorological conditions. The goal of HunchLab's methodology is to not only forecast crimes but also recommend actions that can lessen the likelihood of crime in particular locations.

4.1.2 Effectiveness

Research and discussion on the efficacy of predictive policing algorithms are continuing. Certain jurisdictions claim favorable results, such as declines in particular categories of criminal activity, such auto theft and break-ins. Predictive policing proponents contend that by concentrating efforts where they are most needed, police resources may be allocated more wisely.

However, because there are so many variables that affect crime rates, such as the state of the economy, community policing initiatives, and modifications to law enforcement procedures unrelated to predictive policing, it is difficult to quantify the precise effect of predictive policing.

4.1.3 Controversies

1. Bias and Discrimination: The possibility of algorithmic bias is a major point of contention in predictive policing. Predictive models, according to its detractors, have the potential to institutionalize and reinforce preexisting prejudices in law enforcement since they are trained on past crime data. For instance, a model may unfairly target minority populations in its projections if it was trained on data from an area where police have traditionally been more aggressive against such communities.
2. Privacy Concerns: There are serious privacy issues raised by the collection and analysis of enormous volumes of data, particularly personal data. Predictive policing's use of surveillance technologies raises the possibility that they may be abused to spy people or groups without reason.

3. Accountability and Transparency: The methods by which predictive policing algorithms operate and formulate their forecasts are frequently not made clear. External parties find it challenging to assess the fairness and accuracy of these systems due to their opacity, which has prompted requests for more monitoring and responsibility.
4. Community Relations: Predictive policing may cause a rift in the law enforcement community's relationships with the communities they serve. Tensions can rise and public confidence in the police might be damaged by the belief that particular communities are being unfairly singled out for attention.

4.2 Evidence Analysis

The integration of Artificial Intelligence (AI) into digital forensics represents a significant leap forward in the ability to process and analyze the vast amounts of data involved in criminal investigations. The recovery and examination of information discovered on digital devices, frequently in connection with computer crime, is the focus of digital forensics. AI can speed up and improve the accuracy of investigations by automating the processing of data from a variety of sources, such as emails, social media, and mobile devices. But there are difficulties in using AI in this field, particularly when working with large amounts of data.

4.2.1 AI in Digital Forensics

1. Pattern Recognition: Artificial intelligence (AI) can spot patterns and irregularities in data that might point to criminal activity like fraud, hacking, or the dissemination of illicit content.
2. Image and Video Analysis: AI systems are able to automatically analyze photos and videos by using computer vision to detect important items, faces, or actions. This is particularly useful in situations involving child exploitation or security breaches.
3. Data Clustering: AI is capable of grouping enormous volumes of data into clusters, which facilitates the identification of connections between disparate bits of information, such as connecting several occurrences to a single offender.

4.2.2 Challenges in Handling Big Data

Although AI offers digital forensics some amazing prospects, processing large amounts of data brings with it a number of difficulties.

1. Volume: Digital gadgets produce an astounding volume of data. It is difficult to process and analyze this data quickly, and it takes a lot of computer power.
2. Variety: Digital data may be found in many different forms, ranging from intricate databases to text and graphics. It is extremely challenging to develop AI models that can handle this range of data kinds efficiently.
3. Veracity: Forensic analysis depends on the precision and consistency of digital data. Artificial intelligence (AI) systems need to be able to recognize and lessen the impact of tampered or corrupted data.
4. Velocity: There might be an overwhelming amount of data being created at once that has to be processed. For AI systems to be useful in real-time analysis or almost real-time circumstances, they need to be able to work at this speed.
5. Privacy and Ethical Considerations: There are serious privacy issues when using AI in digital forensics. Analyzing personal data via digital devices carries a particular danger of violating people's right to privacy. It is crucial to make sure AI systems follow moral and legal requirements.
6. Bias and Fairness: AI algorithms can inadvertently perpetuate or amplify biases present in the training data. This may result in biased conclusions from forensic analysis, such as falsely designating some people as suspects because of faulty data.

4.3 Risk Assessment Tools

The criminal justice system uses risk assessment methods to determine how likely it is that a defendant would commit another crime or miss court. These technologies provide a risk score by using algorithms to evaluate data on a variety of variables, such as age, criminal history, and occasionally socioeconomic information. Then, based on this score, choices are made about parole, bail, and punishment. There is a lot of discussion and interest in the development, application, and effects of these tools; these subjects include accuracy, fairness, and potential systemic bias.

4.3.1 Design and Implementation

A key component of developing risk assessment tools is choosing factors that have a statistical relationship to the probability of reoffending or missing court dates. These may consist of:

(i) Static Factors: Unalterable traits of the accused, such age upon initial arrest and prior criminal activity.

(ii) Dynamic Factors: Things like present employment position or drug misuse concerns are examples of variables that might vary over time.

The use of these technologies varies greatly among jurisdictions, with some regions implementing proprietary systems produced by commercial corporations and others employing open-source or custom-made solutions. When implementing risk assessment techniques, the following factors should be taken into account:

Transparency: In order to guarantee equity and for public examination and supervision, the standards and algorithms employed should be visible.

Validation: To guarantee that tools' predictions continue to be impartial and accurate, they must be routinely verified on the population to which they are applied.

Training: In order for judges, probation officers, and other tool users to comprehend the capabilities and limits of the tool, they must get the appropriate training.

4.3.2 Positive Impact on Sentencing and Bail Decisions

1. Decision-Making Standardization: The goal of risk assessment tools is to bring some consistency and impartiality to judgments about bail and punishment, which have traditionally been made with a great deal of judicial discretion. These techniques aim to lessen the possibility of personal bias by ensuring that instances that are similar are handled similarly through the use of data and algorithms.
2. Public Safety and Resource Allocation: Courts may be able to reduce the number of people in pretrial custody without endangering public safety if they are able to precisely identify those who represent a minimal risk to the public. Additionally, by concentrating on high-risk persons and redirecting low-risk ones to less restrictive measures or support programs, this may result in a more effective use of resources.
3. Increased Attention to Rehab: Using risk assessment tools at sentencing might assist in identifying criminals who would benefit more from rehabilitation than from jail. Through addressing the underlying reasons of criminal conduct, such as substance addiction or mental health difficulties, this strategy helps achieve the larger objective of reducing recidivism.

4.3.3 Challenges and Controversies

1. Potential for Bias: The possibility that risk assessment instruments could reinforce preexisting prejudices in the criminal justice system raises serious concerns. These tools may unjustly classify members of marginalized populations as higher

risk if the historical data that was used to create them reflects systematic biases, such as racial differences in arrest or conviction rates.

2. Accountability and Transparency: Since many risk assessment systems are private, it can be challenging for outside parties to carefully examine their data sources and algorithms. This lack of transparency calls into question who is responsible, particularly since these instruments are central to choices that impact people's futures and freedoms.

3. Overreliance on Algorithmic Decisions: There's a chance that decision-makers, such as judges, would depend too much on the results of risk assessment instruments, viewing them as final determinations of a person's risk rather than as one aspect to take into account among many.

4. Impact on Pretrial Detention: Although designed to support the release of low-risk individuals, in practice, risk assessment tools may contribute to higher rates of pretrial detention for those deemed high risk, potentially exacerbating issues of overcrowding and the negative impacts of detention on individuals' employment, family, and community ties. Such an overreliance on these tools can undermine the role of judicial discretion and the consideration of the unique circumstances of each case.

5. TECHNICAL CHALLENGES IN AI DEPLOYMENT

The deployment of Artificial Intelligence (AI) systems across multiple industries promises transformational prospects, but it also highlights considerable technical obstacles, notably in terms of data quality and bias. These difficulties must be addressed, as AI systems are only as good as the data on which they are taught. Poor data quality and inherent biases might result in erroneous, unfair, or even discriminatory consequences. Identifying and reducing data bias is thus essential for the responsible deployment of AI technology.

5.1 Data Quality and Bias

1. Data Quality: AI models rely significantly on vast datasets for training. This data must be of the highest quality—complete, accurate, and relevant. Poor data quality can result from a variety of causes, including mistakes during data collection, obsolete information, and missing numbers. These difficulties can impair the functioning of AI systems, resulting in incorrect predictions or choices.

2. Data bias: Data bias is a widespread issue that presents a substantial hurdle in AI adoption. It happens when the training datasets aren't indicative of the real-world settings in which the AI system is supposed to work, or when the data reflects historical bias. Biases can emerge in different forms, including selection bias, measurement bias, and prejudice bias, all of which influence the AI's decision-making process and possibly perpetuating or amplifying existing inequalities.

5.1.1 Identifying Bias

1. Data Auditing: Conducting regular dataset audits can assist detect biases. These audits should determine the data's representativeness, looking for overrepresented or underrepresented groups.
2. Bias Metrics: Measuring bias in datasets and AI model outputs is critical. These indicators might be useful for evaluating and tracking performance across different demographic groupings.
3. Expert Involvement: Including domain experts and sociologists in the AI development process might help identify potential sources of bias that solely technical teams may miss.

5.1.2 Mitigating Bias

1. Diverse and Inclusive Data: Ensuring that training datasets are diverse and representative of all relevant groups is fundamental. This might include gathering more data from underrepresented groups or using strategies to rebalance databases.
2. Algorithmic Fairness Methods: There are several technical approaches to reducing bias, including pre-processing techniques that aim to remove bias from data before it is used to train the AI model, in-processing methods that incorporate fairness constraints directly into the training process, and post-processing techniques that adjust the model's outputs for fairness.
3. Transparency and Explainability: Creating transparent and explainable AI models may assist stakeholders understand how choices are made, which is critical for detecting potential biases in AI decision-making processes.
4. Continuous Monitoring and Evaluation: AI systems should be constantly monitored and analyzed after deployment to guarantee that they stay fair and accurate over time. This entails periodically updating the model with fresh data and recalibrating it as needed to accommodate changing real-world situations.

5.2 Case Studies of Bias in Criminal Justice AI

The use of artificial intelligence (AI) in criminal justice systems has been praised for its potential to improve efficiency while also being chastised for maintaining and worsening existing prejudices. The case studies below describe situations in which AI tools have displayed prejudice in the criminal justice system, emphasizing the significance of careful evaluation and mitigating techniques before deploying these technologies.

5.2.1 COMPAS (Correctional Offender Management Profiling for Alternative Sanctions)

Background: COMPAS is a risk assessment methodology used by US courts to determine a defendant's chance of reoffending. It has been a source of contention regarding algorithmic bias.

Issue: A 2016 ProPublica study revealed that the COMPAS tool was biased against black defendants. The analysis found that the algorithm was nearly twice as likely to mistakenly categorize black defendants as having a higher risk of reoffending, whereas white defendants were more likely to be incorrectly labeled as low risk.

Impact: The findings generated a national discussion about the use of algorithmic decision-making in the criminal justice system, particularly in terms of openness and fairness.

5.2.2 PredPol

Background: PredPol, a predictive policing system, analyzes historical crime data to forecast future crime locations.

Issue: Critics say that predictive policing systems such as PredPol might create a feedback cycle of over policing in historically oppressed groups. If police are assigned to these regions more frequently based on prior crime statistics, the chance of uncovering criminality increases, reinforcing the evidence indicating that these places are at high risk.

Impact: This loop can worsen racial inequalities in enforcement, resulting in greater rates of arrests and convictions in minority neighborhoods and reinforcing the bias in predictive policing statistics.

5.2.3 Face Recognition Technology in Law Enforcement

Background: Law enforcement organizations have become more reliant on face recognition technology to identify suspects. This system analyzes photos from surveillance footage to a database of recognized faces.

Issue: Many facial recognition systems exhibit severe racial, ethnic, and gender biases, according to studies, including a critical one conducted by the National Institute of Standards and Technology. These algorithms have been found to have greater rates of false positives for individuals of color, notably black and Asian faces, as well as women.

Impact: The deployment of biased face recognition technology can result in erroneous arrests and a greater burden of monitoring on minority populations. Notably, cases have emerged of individuals being falsely arrested based on incorrect matches from facial recognition systems, highlighting the severe consequences of relying on biased AI technologies.

5.3 Transparency and Explainability

Transparency and explainability in Artificial Intelligence (AI) are top priorities, especially as AI systems grow more common in crucial industries such as healthcare, banking, and criminal justice. The difficulty of making AI systems visible and their judgments explainable is sometimes referred to as the "black box" problem. This phrase refers to a situation in which AI systems' decision-making processes are opaque, making it difficult or impossible for humans to grasp how the AI reached its conclusions or predicted outcomes.

5.3.1 The Black Box Problem

The "black box" problem in artificial intelligence (AI) captures the issues and concerns associated with the opacity of AI decision-making processes, particularly in complicated models such as those employed in deep learning. This topic is fundamental to talks on AI ethics and governance because it raises important concerns about accountability, transparency, and justice. The term "black box" alludes to the difficulty of comprehending how an AI system converts inputs into outputs, making it practically impossible to follow the decision-making route.

This opacity is especially concerning in high-stakes fields like healthcare, criminal justice, and autonomous driving, where AI judgments can have a significant impact on human lives. The inability to comprehend AI judgments challenges efforts to create responsibility when mistakes occur, to assure justice by discovering and correcting biases in AI systems, and to foster trust among users and the general public.

Trust in AI is critical for its adoption and acceptance, particularly in industries that directly affect well-being and societal order.

Efforts to minimize the black box problem include the development of Explainable AI (XAI) approaches, which aim to make AI systems more transparent and their judgments easier to explain for humans. This includes both designing intrinsically interpretable models and developing tools to explain the decisions made by existing complicated models. Furthermore, regulatory regimes are beginning to mandate that AI systems, particularly those deployed in sensitive or important areas, fulfill certain levels of openness and explainability.

Addressing the black box dilemma is critical to the proper development and deployment of AI systems. By making AI systems more intelligible and decision-making procedures more transparent, stakeholders want to improve accountability, assure fairness, and encourage confidence in AI across several industries. This endeavor is critical not only for ethical grounds, but also for the practical viability and public acceptability of AI technology.

The process typically includes the following steps:

1. Sophisticated algorithms examine extensive data sets to find patterns. To achieve this, a large number of data examples are fed to an algorithm, enabling it to experiment and learn on its own through trial and error. The model learns to change its internal parameters until it can predict the exact output for new inputs using a large sample of inputs and expected outputs.
2. As a result of this training, the machine learning model is finally ready to make predictions using real-world data. Fraud detection using a risk score is an example of a use case for this mechanism.
3. The model scales its method, approaches and body of knowledge as additional data is gathered over time.

5.3.2 Advances in Explainable AI (XAI)

Explainable AI (XAI) has advanced significantly in recent years, reflecting the rising relevance of openness, trust, and understandability in artificial intelligence systems. These advancements are being pushed by both technology improvements and growing regulatory and social demands for AI that can be examined and understood by humans. The following are significant areas of advancement in XAI.

1. Developing Interpretable Models: One of the key goals of XAI has been the creation of models that are intrinsically interpretable. This entails creating algorithms that give explicit insights into how they analyze inputs and make

judgments or predictions. For example, decision trees and linear models, which give simple approaches to comprehend the link between input factors and outputs, have been upgraded to accommodate more complicated data, while maintaining their interpretability.

2. Post-hoc Explanation Techniques: Post-hoc explanation approaches have advanced significantly for AI models that are not intrinsically interpretable, such as deep neural networks. These approaches seek to provide insight on the decision-making process of complicated models after a forecast. Techniques such as LIME (Local Interpretable Model-agnostic Explanations) and SHAP (SHapley Additive Explanations) reveal how specific characteristics contribute to the model's output, making AI judgments more visible.

3. Visualization Tools: Advances in visualization tools have also helped to explain AI more clearly. These tools assist in viewing the inner workings of complicated models, such as how data flows through neural networks, how different layers of the network activate in response to specific inputs, and how changes in input data impact the output. Visualization tools like TensorBoard and saliency maps help comprehend and interpret AI algorithms.

4. Interactive and User-Centered explanations: Recognizing that various stakeholders may require varying levels of explanation, there has been a push to create interactive and user-centered explanation systems. These technologies allow users to question AI models and obtain explanations tailored to their degree of knowledge and individual needs. This strategy recognizes the vast spectrum of AI users, from domain specialists to laypeople, and seeks to make AI insights available to everyone.

5. Regulatory and Ethical Frameworks for XAI: Along with technical advancements, regulatory and ethical frameworks have evolved to require specific levels of explainability. The European Union's General Data Protection Regulation (GDPR), for example, has clauses that might be read as requiring explanations for AI-driven choices in some circumstances. Such rules encourage widespread use of XAI approaches while also guaranteeing that AI systems deployed in the market fulfill basic transparency and accountability requirements.

6. Benchmarking and Metrics for Explainability: Efforts to standardize how explainability is assessed and evaluated have led in the creation of benchmarking and assessment measures. These measures aid in evaluating various XAI strategies and determining the quality of explanations supplied by AI systems. Establishing consistent standards and measurements is critical for moving the industry forward and ensuring that explainability gains are real and verifiable.

In a nutshell, progress in explainable AI is complex, encompassing the development of new models and methodologies, the production of tools and platforms for interpretation and visualization, and the adoption of regulatory and ethical norms. These advancements are important to ensuring that AI systems are not just strong and effective, but also transparent, understandable, and reliable.

5.4 Accuracy and Reliability

Accuracy and reliability are critical to the successful deployment of artificial intelligence (AI) systems in a variety of fields, including healthcare and finance, autonomous cars, and criminal justice. Maintaining public trust and safety requires AI systems to have high levels of accuracy and dependability. This part digs into approaches for assessing and improving AI performance, emphasizing the significance of these procedures in the development and implementation of AI technology.

5.4.1 Measuring AI Performance

1. Accuracy Metrics: The selection of accuracy measures is critical in evaluating AI systems. Precision, recall, F1 score, and accuracy rate are common metrics for classification tasks, whereas mean squared error (MSE) and mean absolute error (MAE) are used for regression tasks. The selection of an acceptable measure is determined by the individual application and the relative relevance of various sorts of mistakes.
2. Validation Techniques: Cross-validation and split-test procedures are used to evaluate an AI model's generalizability. These strategies include splitting the data into training and testing sets to ensure that the model works well on previously unknown data, hence reducing overfitting.
3. Benchmarking: Benchmarking against standard datasets and comparing performance to current state-of-the-art models can offer insight into an AI model's relative performance. This approach is critical in research and development for determining progress in the sector.

5.4.2 Improving Performance

1. Data Quality and Quantity: AI model performance is highly dependent on the quality and amount of training data. It is critical to ensure that data is clean, labeled correctly, varied, and reflective of real-world events. Data augmentation techniques can improve model resilience by artificially extending the training dataset.

2. Model Complexity: Increasing the complexity of the model can boost performance. While more complicated models may detect nuanced patterns in data, they also risk overfitting. Regularization approaches, model trimming, and dropout are the methodologies employed.

3. Optimizing the Algorithm Choosing the proper algorithm and adjusting its parameters is crucial to AI performance. Hyperparameter tweaking, whether done by grid search, random search, or automated approaches such as Bayesian optimization, may dramatically improve model accuracy.

4. Ensemble Procedures: Ensemble approaches, which use several models to produce predictions, can enhance accuracy and dependability. Bagging, boosting, and stacking techniques are used to minimize variance and bias, respectively, hence improving model performance.

5. Continuous learning: Implementing continuous learning methods enables AI systems to adapt to new data over time while maintaining or increasing accuracy and dependability. This strategy is especially relevant in dynamic contexts where data distribution may alter.

6. Ethical and Bias Consideration: Addressing ethical concerns and biases in AI models is critical for their dependability and fairness. Techniques for detecting and mitigating bias guarantee that AI systems function consistently across diverse groups and contexts.

5.4.3 Limitations of Predictive Models

Because of its potential to guide decision-making and foresee future occurrences, predictive modeling has become a staple in a variety of sectors, including banking, healthcare, marketing, and environmental research. However, despite its many uses and benefits, predictive modeling has limits. Understanding these constraints is critical for correctly interpreting model predictions and making sound judgments.

1. Data Quality and Availability:
   - Incomplete or Biased Data: Predictive models rely significantly on past data. If the data is inadequate, faulty, or biased, it might result in incorrect forecasts. Data collected from biased samples or without representation from all subgroups might lead to models that underperform for underrepresented groups.
   - Data overfitting: Models that are excessively complicated may overfit the training data, collecting noise rather than the underlying pattern. This overfitting reduces the model's capacity to generalize to new, previously unknown data.

2. Dynamic Environments and Concept Drift
   - Changing Environments: In dynamic contexts where fundamental patterns and connections change over time, models based on historical data can soon become out of date. This process, known as idea drift, presents a considerable difficulty for sustaining model fidelity over time.
   - Adaptability: Predictive models may not adapt well to quick changes in the environment or context, necessitating frequent updates and retraining to remain relevant.
3. Complexity and Interpretability:
   - Black Box Models Some of the most effective prediction models, such as deep neural networks are sometimes referred to as "black boxes" due to their intricate underlying architecture that are difficult to understand. This lack of openness can be a major disadvantage in situations where understanding the decision-making process is critical.
   - Simplified Reality: Predictive models always reduce the intricacies of the actual world. This simplification may result in the omission of key factors and interactions that are difficult to quantify or were not included in the dataset.
4. Ethical and Privacy problems:
   - Privacy: Using personal data in predictive modeling poses privacy problems, particularly with sensitive information. Ensuring data privacy and following standards such as GDPR are critical, but they can limit the extent of data acquisition and utilization.
   - Bias and Fairness: Predictive algorithms may unwittingly perpetuate or even amplify existing biases in training data, resulting in unjust outcomes or discrimination. Identifying and correcting these biases is necessary but difficult.
5. Reliance on Models for Decision-Making:
   - Decision Making: Overreliance on predictive models in decision-making can be dangerous, particularly if the model's limitations are not fully recognized or taken into account. It is critical to supplement model predictions with human judgment and domain knowledge.
   - Risk and Uncertainty Predictive models frequently create a sense of confidence about the future, which can be deceptive. It is critical to account for the inherent uncertainty in projections, which might arise from model inaccuracy, unknown future conditions, or both.

5. ETHICAL AND SOCIAL CONSIDERATIONS

The use of new technology into public safety measures, notably monitoring and data analysis, highlights the delicate relationship between increasing security and protecting personal rights. This duality raises ethical and societal concerns that require careful examination to guarantee that attempts to increase public safety do not jeopardize individual liberties and privacy. To successfully traverse this delicate balance of opposing interests, a sophisticated strategy is required.

5.1 Balancing Public Safety and Personal Rights

The need to safeguard civilians and maintain public order frequently justifies the use of surveillance technologies and predictive policing techniques. These technologies have the potential to dramatically improve law enforcement organizations' abilities to prevent crime and respond to emergencies quickly. However, the use of such measures must be balanced against the risk of intruding on personal rights and freedoms. The ethical difficulty is defining how much monitoring and data collection on people is appropriate in the sake of public safety. Striking a balance entails ensuring that monitoring is not only efficient in reaching the desired safety objectives, but also minimally invasive, respecting persons' privacy and rights.

5.1.1 Surveillance and Privacy Concerns

In an increasingly technologically advanced society, the deployment of surveillance devices in public places has spawned a complicated discussion about how to strike a balance between guaranteeing public safety and respecting individual privacy rights. This argument is critical for understanding the ethical and societal consequences of employing advanced techniques such as artificial intelligence (AI), face recognition, and big data analytics for surveillance.

Surveillance technologies, previously considered science fiction, have become commonplace. Cities all around the world are installing more CCTV cameras, while governments and companies collect massive quantities of data from internet activity, mobile devices, and IoT (Internet of Things) apps. These tools have the potential to dramatically boost public safety by assisting with crime prevention, emergency response, and national security efforts.

The main worry is the concept of ongoing monitoring, which creates an environment in which people may believe they are constantly being watched. This widespread monitoring has the potential to stifle free speech and assembly by causing people to change their conduct out of fear of being observed or judged. Furthermore, widespread gathering and analysis of personal data without strict protections might result in abuse, such as unlawful commercial usage, political targeting, or discriminatory profiling.

The problem is to strike a careful balance between maximizing the benefits of monitoring for public safety and firmly preserving individuals' privacy and fundamental freedoms. This balance needs strong legislative frameworks and ethical principles to guarantee that surveillance technologies are utilized responsibly, transparently, and with regard for human rights.

One way to address these issues is to impose rigorous laws on data collection, usage, and storage, ensuring that surveillance activities are appropriate, required, and subject to monitoring. Transparency regarding the use of surveillance technology, along with clear, understandable information about data protection measures, can assist to foster public confidence. Furthermore, engaging communities in surveillance debates and asking consent when possible, helps foster a feeling of shared responsibility for public safety.

Furthermore, the development of privacy-enhancing technologies (PETs) is a viable route for reducing privacy breaches. PETs can allow for the efficient use of data for public safety while maintaining individual privacy using techniques such as anonymization, encryption, and data reduction.

5.1.2 Ethical Implications on Mass Data Collection

The ethical implications of large data collecting are varied and profound, addressing questions of privacy, consent, security, and the balance of power among individuals, businesses, and governments. As digital technologies pervade every part of contemporary life, massive volumes of personal data are gathered, processed, and stored at unprecedented rates. This data-driven paradigm brings several advantages, including tailored services, economic development, and advances in healthcare and public safety. However, it creates serious ethical considerations that need careful analysis and accountable administration.

Privacy Concerns: The right to privacy is central to the ethical issue around bulk data collecting. In an age where people's actions, conversations, and tastes are continually observed and evaluated, the line between public and private life blurs. This surveillance may violate individuals' rights to keep personal information private and make autonomous decisions without undue influence or observation. The ethical dilemma is to determine what constitutes an appropriate amount of privacy in the digital era and how to maintain it without impeding innovation or societal benefits.

Consent and Autonomy: The idea of permission is essential for ethical data processing. However, the complexity and opacity of data gathering methods can jeopardize informed consent. Terms of service and privacy policies are usually lengthy, convoluted, and inaccessible to the ordinary user, making legitimate consent difficult. This circumstance raises ethical concerns regarding autonomy and people' power to control and choose how their personal information is utilized.

Data Security and Misuse: The buildup of massive data repositories raises the potential of breaches and misuse. Cyberattacks can compromise sensitive personal information, while illegal access by other parties can result in identity theft, financial loss, and bodily injury. Furthermore, the potential for data misuse includes discriminatory profiling, spying, and public opinion manipulation, as seen by a number of incidents involving social media platforms and political consultancy businesses. Ethical data stewardship necessitates strong security measures and open standards to avoid misuse and protect persons from damage.

Equity and Justice: Large-scale data collecting can increase socioeconomic imbalances and injustices. Biased data sets can provide biased results in fields including law enforcement, employment, credit scoring, and healthcare. For example, predictive policing systems may disproportionately target minority populations, reinforcing cycles of monitoring and punishment. Ethically, biases in data usage must be identified and remedied to provide fair and equal outcomes for all persons, regardless of race, gender, or socioeconomic background.

Governance and Accountability: The governance of vast data collecting raises ethical concerns about accountability, supervision, and balance of power. Data is a source of power, and its accumulation in the hands of a few firms or government agencies may jeopardize individual liberties and democratic processes. Ethical governance necessitates accountability systems such as regulatory scrutiny, transparency reports, and ways for individuals to own their data and seek remedies for violations.

5.2 Consent and Data Governance

Consent and data governance are critical components in the ethical use of artificial intelligence (AI) and data analytics. In an era of ubiquitous data collecting and analysis, obtaining individuals' consent to the use of their data and developing strong governance structures are critical for preserving privacy rights and instilling confidence in AI systems.

5.2.1 Consent for Data Use

Obtaining informed consent from individuals before collecting, processing, or disclosing their data is an essential ethical concept. In the context of artificial intelligence, where computers rely on massive volumes of data to make predictions and judgments, having express consent is even more important. Individuals should be aware of how their data will be used, who will have access to it, and for what purposes. However, the complexity of AI systems and the opacity of their decision-making processes might make it difficult to get meaningful consent. Efforts to increase AI transparency and explainability can help users make more educated data sharing decisions.

5.2.3 Data Governance Frameworks

Effective data governance frameworks are critical for guaranteeing the appropriate and ethical usage of data in AI systems. These frameworks define policies, methods, and techniques for managing data over its entire lifespan, from collection and storage to processing and sharing. Data stewardship, data quality management, data security measures, and legal and regulatory compliance are all key components of data governance. Organizations that employ strong data governance processes can reduce the risks of data abuse, breaches, and privacy violations while also fostering accountability and transparency.

5.3 Privacy Laws and AI

Privacy laws vary greatly over the world, reflecting various cultural values, legal traditions, and policy agendas. Despite these disparities, the majority of privacy laws seek to safeguard individuals' personal data from unlawful use, maintain data security, and grant individual data rights. To demonstrate the various approaches to data privacy, we will look at three significant privacy legislation from different nations.

5.3.1 General Data Protection Regulation (GDPR) – European Union

The GDPR, which took effect in May 2018, is one of the world's most strict privacy rules. It applies to all firms that process personal data of persons resident in the EU, regardless of location. The key provisions include:

- Consent: Individuals must give explicit consent for their data to be processed, and they can withdraw it at any time.
- Right to Access: Individuals can access their personal data and obtain information about how it is processed.
- Data Portability: Individuals can receive their personal data in a structured, commonly used format and transfer it to another controller.
- Right to view: Individuals have the right to view their personal data and learn about how it is handled.

- Data Portability: Individuals have the right to obtain their personal information in a structured, frequently used format and transfer it to another controller.
- Right to Be Forgotten: Individuals have the right to have their personal information wiped under certain circumstances.
- Data Protection Officers (DPOs): Organizations must appoint a DPO if they handle substantial volumes of sensitive data or conduct wide-scale surveillance of individuals.
- Data Breach Notifications: Organizations must inform the competent supervisory body of data breaches within 72 hours, if possible, where the breach threatens people' rights and freedoms.

5.3.2 California Consumer Privacy Act (CCPA) – United States

The California Consumer Privacy Act, which went into effect in January 2020, gives California citizens extensive control over their personal information. Key features include:

- Right to Know: Consumers have the right to obtain information about how their personal data is collected, used, and shared.
- Right to Delete: Subject to certain conditions, consumers may request that their personal information be removed from a company's database.
- Opt-Out Right: Consumers have the right not to allow firms to sell their personal information.
- Non-Discrimination: Businesses cannot discriminate against consumers who use their CCPA rights.

5.3.3 Personal Information Protection and Electronic Documents Act (PIPEDA) – Canada

PIPEDA restricts how private-sector companies acquire, use, and disclose personal information in the course of commercial operations. The key concepts include:

- Consent: Organizations must get individuals' consent before collecting, using, or disclosing their personal information.
- Limiting Collection: Personal information shall be collected only to the extent necessary for the organization's specified purposes.
- Accountability: Organizations are liable for personal information under their control and must appoint a PIPEDA compliance officer.

5.3.4 General Data Protection Law (LGPD) of Brazil

The LGPD, which went into effect in September 2020, is modeled after the GDPR and provides a legal framework for the use of personal data in Brazil. The key features include:

- Legal Basis for Processing: The LGPD, like the GDPR, includes many legal basis for data processing, such as permission, compliance with a legal duty, and the data controller's legitimate interests.
- Data Subject Rights: Individuals have rights comparable to those under the GDPR, including access to data, correction, erasure, and data transfer.
- Data Protection Officer: Organizations are required to employ a Data Protection Officer to guarantee legal compliance.

5.3.5 Data Protection Act 2018, United Kingdom

The UK has implemented GDPR through the Data Protection Act 2018 (DPA 2018). It enhances the GDPR with particular rules for the handling of personal data, including:

- Law Enforcement Processing: The DPA 2018 establishes particular rules for processing personal data for law enforcement purposes.
- National Security: It provides exclusions and changes to GDPR principles for national security and intelligence services.

5.4.6 Personal Data Protection Bill (PDP) - India

India has been striving to enact comprehensive data protection laws known as the Personal Data Protection Bill (PDP Bill). The PDP Bill, modeled after the GDPR, seeks to provide a legislative framework for digital privacy and data protection in India. While the measure has gone through multiple revisions and encountered delays in enactment, its primary elements include:

- Data Principal's Rights: Individuals whose data is processed are referred to as "Data Principals" under the bill, and they are granted a variety of rights, including the ability to view, rectify, and erase their personal data.
- Consent: Similar to the GDPR, the PDP Bill mandates that permission be free, informed, explicit, and unambiguous. Data fiduciaries must treat personal data for purposes that are explicit, lawful, and consented to by the individual.

- Data Localization: One of the bill's most contentious provisions is the need for some types of vital personal data to be maintained in India. It also requires that a copy of some types of personal data be kept within the nation.
- The Data Protection Authority: The bill proposes the formation of an Indian Data Protection Authority (DPA), which would be in charge of implementing and administering the legislation, as well as resolving data protection-related issues.
- Data Fiduciaries and Processors: Organizations that determine the purpose and method of processing personal data are classed as data fiduciaries, and they must comply with a variety of data processing responsibilities. These include implementing privacy-by-design techniques, keeping records of data processing operations, and completing data protection impact assessments.
- Penalties: The law provides provisions for sanctions and compensation in the event of noncompliance or data breaches, which can be considerable depending on the type and severity of the violation.

6. CONCLUSIONS

The ethical application of artificial intelligence in criminal justice is a complicated and diverse topic that highlights the need to strike a balance between leveraging technology breakthroughs for public safety and respecting individual rights.

First and foremost, AI has the ability to significantly improve the efficiency and efficacy of the criminal justice system. AI can help handle large volumes of data for evidence analysis, identify crime hotspots, and make educated bail and sentence choices. These skills, if used effectively, might result in more fast and equitable judicial systems.

Furthermore, comprehending the ethical environment of AI in criminal justice has required a focus on privacy and surveillance issues. The balance between using AI for public safety and protecting individual private rights is still a difficult subject, demanding clear legislative frameworks and strong privacy protections.

Finally, the ethical application of AI in criminal justice necessitates a joint effort among engineers, legal experts, ethicists, and community stakeholders. It demands the creation of transparent, responsible AI systems that have been carefully vetted for bias and fairness. Only through such joint efforts can the benefits of AI be fully realized while upholding justice and protecting individuals' rights and dignity.

REFERENCES

Barabas, C. (2020). Beyond Bias: Re-Imagining the Terms of" Ethical AI" in Criminal Law. *Georgetown Journal of Law & Modern Critical Race Perspectives*, 12, 83.

Cortez, E. K. (Ed.). (2020). *Data Protection Around the World: An Introduction* (Vol. 33). Springer Nature.

Feuerriegel, S., Hartmann, J., Janiesch, C., & Zschech, P. (2023, September 12). Generative AI. *Business & Information Systems Engineering*. Advance online publication. DOI:10.1007/s12599-023-00834-7

Gawali, P., & Sony, R. (2020, December 17). *The Role of Artificial Intelligence in Improving Criminal Justice System: Indian Perspective*. Legal Issues in the Digital Age. DOI:10.17323/2713-2749.2020.3.78.96

Hamilton, M. (n.d.). *A 'black box' AI system has been influencing criminal justice decisions for over two decades – it's time to open it up*. The Conversation. https://theconversation.com/a-black-box-ai-system-has-been-influencing-criminal-justice-decisions-for-over-two-decades-its-time-to-open-it-up-200594

Kabol, A. F. (2022, November 5). *The Use Of Artificial Intelligence In The Criminal Justice System (A Comparative Study)*. ResearchGate. https://www.researchgate.net/publication/365027297_The_Use_Of_Artificial_Intelligence_In_The_Criminal_Justice_System_A_Comparative_Study

Santosh, K. C., & Wall, C. (2022, January 1). *AI and Ethical Issues*. SpringerBriefs in Applied Sciences and Technology. DOI:10.1007/978-981-19-3935-8_1

Stahl, B. C. (2021, January 1). *Ethical Issues of AI*. SpringerBriefs in Research and Innovation Governance. DOI:10.1007/978-3-030-69978-9_4

Stahl, B. C. (2021, January 1). *Addressing Ethical Issues in AI*. SpringerBriefs in Research and Innovation Governance. DOI:10.1007/978-3-030-69978-9_5

Sushina, T., & Sobenin, A. (2020, May). Artificial intelligence in the criminal justice system: leading trends and possibilities. In 6th International Conference on Social, economic, and academic leadership (ICSEAL-6-2019) (pp. 432-437). Atlantis Press. DOI:10.2991/assehr.k.200526.062

Chapter 14
Responsible AI Implementation in the Hospitality Sector:
Ethical Challenges and Solutions for VR and AR Applications

Milan Sharma
https://orcid.org/0000-0002-3320-8129
Lovely Professional University, India

Amrik Singh
https://orcid.org/0000-0003-3598-8787
Lovely Professional University, India

Rohit
Faculty of Hotel and Tourism Management, SGT University, Gurgaon, India

ABSTRACT

The hotel sector is leading the charge to harness the power of artificial intelligence (AI), virtual reality (VR), and augmented reality to improve client experiences and streamline operations. The moral weight of artificial intelligence (AI) in virtual reality (VR) and augmented reality (AR) for the hotel industry is the subject of this study. The effects of artificial intelligence and automation on hotel employees, as well as the significance of honest and ethical data collecting, are discussed in the study. This study adds to the continuing conversation on ethical AI deployment in the hospitality industry by doing a thorough examination of the technological obstacles presented by AI in virtual reality and augmented reality.

DOI: 10.4018/979-8-3693-9173-0.ch014

INTRODUCTION

The hospitality sector has consistently been at the forefront of embracing cutting-edge technologies to improve customer experiences and optimize operational efficiency. Artificial intelligence (AI), virtual reality (VR), and augmented reality (AR) have become more common in recent years, offering the potential to transform how hotels engage with customers and handle their internal operations. Nevertheless, as the use of these technologies expands, the ethical considerations about their deployment also increase.

Artificial Intelligence (AI) refers to a collection of technologies capable of identifying, examining, responding, acquiring knowledge, and exhibiting sophisticated aspects of human intelligence while solving problems (McCartney & McCartney, 2020). The latest advancements in generative conversational AI highlight the potential, difficulties, and consequences of AI in all areas of life (Dwivedi et al., 2023). Artificial intelligence (AI) is revolutionizing the operational and marketing aspects of tourist locations and companies (Inanc-Demir & Kozak, 2019). AI systems enhance personalization and recommender systems, robotics, conversational systems (such as chatbots and voice assistants), forecasting systems, smart travel aides, language translation applications, and smart tourism and smart destination systems.

Modern technology like augmented and virtual reality (AR/VR) offers exciting new possibilities to teach visitors about sustainability through interactive and immersive experiences. Visitors can gain a deeper understanding and appreciation for sites by using augmented reality programs to overlay information on their ecological and cultural significance. Virtual reality, on the other hand, allows for virtual excursions to environmentally delicate regions, mitigating both the environmental impact and the need for actual footprints. Virtual reality has the potential to educate the public about conservation initiatives, foster empathy, and promote responsible behavior by allowing visitors to fully immerse themselves in virtual environments. This chapter explores the ethical consequences of artificial intelligence (AI) in virtual reality (VR) and augmented reality (AR) for the hotel industry. It analyses the effects on hotel staff and emphasizes the significance of truthful and ethical data gathering.

The incorporation of artificial intelligence (AI), virtual reality (VR), and augmented reality (AR) in the hotel sector is motivated by the aim to enhance client contentment and operational effectiveness. Artificial intelligence (AI) chatbots can provide support for consumer concerns, while virtual reality (VR) and augmented reality (AR) can be used to generate captivating and immersive experiences for guests. Nevertheless, the implementation of these technologies also gives rise to noteworthy ethical considerations. For example, the implementation of automation in tasks could result in unemployment and alterations in the character of work for hotel staff, thereby causing disruptions to their means of living and overall welfare. In addition, the gathering and utilization of data in virtual reality (VR) and augmented reality (AR) applications might give rise to concerns regarding privacy and security. This is because hotels may be acquiring sensitive information about their guests without obtaining their explicit consent.

The ethical dilemmas presented by artificial intelligence (AI) in virtual reality (VR) and augmented reality (AR) for the hotel industry are complex and extensive. chapter seeks to conduct a thorough analysis of these difficulties and put up solutions for the responsible application of AI in the hotel business. This study examines the ethical implications of AI, VR, and AR and adds to the ongoing discussion about the ethical deployment of AI in the hospitality industry. It highlights the technological challenges posed by these technologies and provides practical recommendations for hotels to effectively address these challenges.

LITERATURE REVIEW

In order to better serve their customers and save costs, businesses in the hotel industry have been quick to embrace new technology. There has been a recent uptick in the application of VR, AR, and AI to enhance customer service, grow operations, and save expenses. However, there are major ethical questions that arise with the implementation of these technologies, such as the potential loss of jobs, invasions of privacy, and security breaches. This chapter discusses the ethical concerns with artificial intelligence, virtual reality, and augmented reality in the hotel industry and offers suggestions for how to responsibly use these technologies. Aims to boost customer happiness and operational efficiency are propelling the hotel industry towards AI, VR, and AR integration. Chatbots powered by AI can answer customer questions, and virtual reality and augmented reality can immerse visitors. On the other hand, there are moral questions that arise from using these technology. For example, hotel workers may see alterations to their jobs or maybe lose their jobs altogether as a result of automation, which might have a devastating impact on their financial stability and quality of life. In addition, hotels may be gathering personal information about their customers without their knowledge or consent when using virtual reality and augmented reality apps, which can lead to privacy and security issues.

Gonzalez and Costa (2024) set out to investigate how the hospitality sector might benefit from immersive augmentation with AI service robots in order to increase social inclusion. Perceived ethics and the moderating effect of consumers' AI knowledge are both uncovered by the studies. Immersive augmentation with inclusive-AI service robots increases the likelihood that customers will pay a premium price, improves their purchasing intentions, and increases levels of supporting tipping behaviour, according to the data. Perceived ethics mediates these consequences.

Many businesses are implementing ML/DL algorithms for various tasks like as BPA, fraud protection, malware detection, spam filtering, and predictive maintenance of recommender systems (Engel et al., 2022; Romao et al., 2019). Because these technologies can analyze email content and notify business practitioners to reply to the most essential communications, they are also beneficial for customer relationship management (CRM) systems. These changes necessitate AI regulation and corporate accountability (Li et al., 2021) to guarantee that people can safely and easily use AI systems for their benefit. Organizations should connect their responsible AI initiatives to their CSR objectives, according to Renieris et al. (2022). They made it seem as though the basic concepts of CSR are already in line with the essential notions of responsible AI, which include preventing bias, being transparent, and being fair.

AI and Impact on Marketing

A collection of computer programs designed to mimic human intellect and solve problems in a similar fashion is known as artificial intelligence (AI). Artificial intelligence (AI) is comparable to humans in that it can learn and adapt to new situations by applying rules and gaining experience (Russell & Norvig, 2016). The original definition of artificial intelligence (AI) was based on a system's ability to do simple tasks (Buhalis et al., 2019). But standards have evolved, and modern intelligent systems must exhibit certain autonomous behaviors (Sterne, 2017). According to Kaplan and Haenlein (2020), AI should possess the following abilities: autonomy, self-awareness, creativity, and social skills. Therefore, for the sake of this study, artificial intelligence systems are defined as those that can reason like humans, make judgments, and execute complex operations on massive datasets without human intervention. According to Huang and Rust (2022), AI will initially supplement human intelligence (HI) before completely replacing it at a certain level of intelligence, and the two will work best together as a team. It is believed that AI will possess transformative and substitutive powers comparable to those of machines or IT, and it is also one of the factors propelling the fourth industrial revolution (Schwab, 2017; Dwivedi et al., 2021). AI is especially useful in marketing because the field often relies on analyzing massive volumes of data to spot trends, behaviors, and possibilities (Mustak et al., 2021). (Davenport et al., 2020). In their three-stage framework for strategic marketing planning, Huang and Rust (2021) show how AI affects each stage: mechanical, which handles repetitive activities; thinking, which processes data to arrive at decisions; and feeling, which analyzes interactions and human emotions. Among the first uses for AI have been recommendation and personalization systems, conversational AI like chatbots and voice assistants, AI prediction and forecasting, AI smart travel assistants, and AI language translation apps (Buhalis & Moldavska, 2022; Bulchand-Gidumal, 2020; Cheng & Jiang, 2022). The publication from 2021 by Vlačić et al.Online review sentiment analysis (Antonio et al., 2018; Ma et al., 2018), marketing management enhancement (Claveria et al., 2015; Stylos et al., 2021), and forecasting are all areas that have benefited from the integration of AI and data science.Automated AI marketing is attracting clients, finding them more efficiently, and creating solutions that meet their specific needs (Grossberg, 2016).Dwivedi et al. (2021) and Gross- berg (2016) both predict that AI will greatly improve the efficiency of marketing campaigns.As consumers get real-time messaging tailored to their unique wants and needs, the conventional customer journey may even become obsolete (Buhalis & Sinarta, 2019).AI acts as a go-between in interactions with service providers.e. Lastly, in encounters aided by AI, both employees and AI collaborate to meet customer service needs.r. As a result, once thought-safe service industry occupations are now in jeopardy due to AI's capacity to augment, complement, or even replace people (Huang & Rust, 2020).Important questions that need answering include how marketing staff will adjust to the new system and how to decide which jobs to automate and which to retain in the hands of humans (Mustak et al., 2021).Furthermore, there are a number of significant drawbacks to AI marketing that need to be considered (Kozinets & Gretzel, 2021), the most notable of which are incomprehensibility, disconnection, and vulnerability.

Figure 1. AI trends for stakeholders

Referring to Figure 1. The primary effect of current AI trends on stakeholders is the ROI of AI investments. Both focus groups spent a lot of time talking about ROI in relation to AI. Members of major hotel chains spearheaded the conversation. Some have stated that essential organizational indicators in the tourist industry center on revenue since shareholders place a premium on long-term profitability. In order for investments in AI to take place, there must be a proven correlation between its use and higher profits and revenue. People also thought that partnerships were really important for staying competitive. As for the second consideration, both focus groups spent considerable time talking about sustainability and AI as a tool to make things better. As a result of AI, participants said, businesses will be able to provide consumers with more accurate information about their organizational footprint, which may lead to consumption reduction or compensation efforts. They may also charge customers based on the amount of damage they do to the environment. Incorporating sustainability into product ideation, participants envisioned a system in which the service might mitigate negative externalities. Data ethics and legal considerations are two more important considerations. Legal and ethical concerns complicate various phases of adopting and implementing AI, according to participants in both focus groups. These stages range from data collecting and processing to the transformation of jobs. Some say that the lack of a workable legal structure that can accommodate the AI-advanced industry is due to legacy legislation. The participants said that the media and regulators were making users too worried. Even more concerning was the suggestion that the privacy and data protection regulations in Europe and Spain, such as the General Data Protection Regulation (GDPR), were overly stringent.

Ethical Challenges

In an effort to better serve customers and increase efficiency, businesses in the hotel industry have been quick to embrace AI, VR, and AR. There are serious ethical questions that need answering, despite the fact that these technologies have many positive uses. An exhaustive summary of the ethical issues surrounding the use of artificial intelligence, virtual reality, and augmented reality in the hotel industry is the goal of this literature review.

Numerous rules and regulations are in place to safeguard customers' rights, guarantee honest dealings, and uphold moral principles in the tourist sector. Data protection standards are a source of concern when integrating technology like AI, as these judgments frequently involve sensitive data. While blockchain technology does increase openness, it also raises concerns about meeting regulatory requirements in the banking sector. Industry stakeholders and lawmakers must take the initiative to navigate the intricate web of legislation and ensure the ethical use of new technologies.

Layoffs and Job Migration

The potential loss of jobs is a major ethical worry when it comes to artificial intelligence, virtual reality, and augmented reality in the hospitality industry. Workers in the hotel industry may see a decline in their standard of living and other negative consequences as a result of job cuts and other changes brought about by automation. Chatbots and virtual assistants powered by artificial intelligence (AI) pose a unique threat since they can take over jobs that humans used to do.

Security and Privacy of Data

There may be security and privacy issues with the data collecting and usage in AR, VR, and AI apps. Data breaches and other security risks might occur if hotels collect sensitive information about their guests without their explicit authorization. Protecting guests' personal information is essential for keeping their faith and staying in line with data protection laws.

Equity and Prejudice

When taught on biased data, AI algorithms may unwittingly reinforce those biases. Some groups, such as those with impairments or varied histories, may experience prejudice as a result of this. Consistently checking AI systems for biases and implementing measures to reduce them is crucial.

Clear Communication and Easy Understanding

Questions of openness and explainability may arise from decision-making procedures that incorporate VR, AR, and AI. Uncertainty and discontent among guests could result from their lack of knowledge about the usage of their data and the decision-making processes employed by AI systems. Keeping guests' trust requires transparent AI decision-making procedures and straightforward explanations.

Accessibility and Inclusivity

All guests, regardless of age, language, or ability, should be able to easily access and use AI technologies. We must be cautious that AI systems do not unintentionally leave out specific demographics.

Standards of Conduct and Preparedness

Understanding and addressing ethical AI rules is a big barrier for small and medium-sized organisations (SMEs) in the hospitality sector. More sector-specific rules are required, as is assistance for SMEs to implement AI in a responsible and ethical manner.

Educational Barriers

A competent workforce that can comprehend and handle the intricacies of these technologies is essential for their effective deployment. But experts in artificial intelligence, blockchain, the internet of things, augmented reality, and virtual reality are in limited supply right now. To ensure that enterprises in the tourism sector can properly exploit new technologies, it is necessary to bridge the talent gap through education and training programs. In order to overcome this obstacle, partnerships between educational institutions and businesses are crucial.

Solution - Artificial Intelligence Governance

A combination of the words "artificial intelligence" and "corporate governance," the phrase "artificial intelligence governance" encompasses both concepts. Legal mandates and enforceable regulations constitute the basis of artificial intelligence (AI) governance, which also relies on voluntary principles to direct researchers, developers, and maintainers of AI systems (Butcher & Beridze, 2019; Gonzalez et al., 2020). It is basically a legislative framework that can help AI professionals with both the big picture and the details of their day-to-day job. Responsible AI governance aims to protect the interests of all stakeholders while ensuring that automated systems, including ML/DL technologies, help individuals and organizations achieve their long-term objectives (Corea et al., 2022; Hickok, 2022). AI governance necessitates adherence to applicable laws, regulations, and statutes by organizational executives (Mäntymäki et al., 2022).Also, according to Koniakou (2023), they should act in accordance with established ethical principles. When dealing with information technology, money, and employees, practitioners should be trustworthy, diligent, and accountable so that they can overcome obstacles, reduce risks, and mitigate uncertainties (such as less human oversight in decision making, among other things).

Figure 2. An ethical structure for the governance of AI

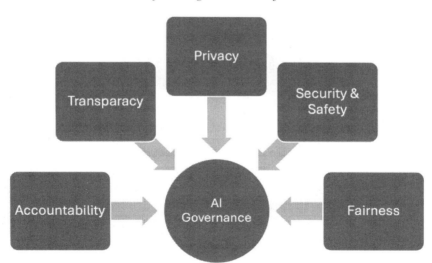

Accountability

It is imperative that hotels establish transparent channels of responsibility for the creation, implementation, and utilization of AI systems. Part of this process is figuring out who makes choices regarding AI and how to keep tabs on how well those systems are doing. The appropriate and ethical usage of AI systems, as well as the fast resolution of any problems or concerns that may arise, are both aided by accountability.

Transparency

Establishing confidence in AI systems requires openness. Hotels should be forthright about their AI practices, including the data they gather and its intended purpose. Offering guests easily understandable information regarding the usage of their data and the decisions taken from that data is an important part of this. Guests should be able to make educated choices regarding their engagement with AI systems, and transparency aids in this process.

Privacy

When it comes to artificial intelligence (AI) in the hotel industry, one of the biggest concerns is guest privacy. A hotel's data collection and use practices should be in line with applicable privacy rules and regulations, such as the EU's General Data Protection Regulation (GDPR). Acquiring guests' informed consent, being transparent about data usage, and storing and using data securely are all part of this.

Security and Safety

The design and deployment of AI systems should prioritize security and safety. It is the hotel's responsibility to safeguard its artificial intelligence systems from cybercriminals and make sure they won't hurt anybody. This involves keeping an eye on AI systems to fix any weaknesses or new risks, as well as testing and validating them thoroughly before deployment.

Fairness

Fairness and non-discrimination should guide the design and implementation of AI systems. This necessitates checking that AI systems are fair and equitable in their treatment of customers and do not reinforce preexisting biases. On a regular basis, hotels should check their AI systems for bias and fix any problems they find.

Using AI technologies in a fair way helps to make sure that everyone is welcome and treated fairly.

CONCLUSION

This review shows that more and more people in the academic and professional communities are focusing on artificial intelligence (AI), discussing it in depth, and drawing conclusions about its advantages and disadvantages. It said that those who support it are trying to get the word out about how great AI systems are for both people and businesses. Meanwhile, it implies that several academics and other interested parties, including policymakers, are voicing worries about its potential dangers (e.g., Berente et al., 2021; Gonzalez et al., 2020; Zhang & Lu, 2021).

Several studies have pointed to potential dangers associated with AI (Li et al., 2021). They frequently voiced fears that AI would cause the spread of false information, encourage bigotry and discrimination, compromise personal privacy, and even cause people to lose their jobs (Butcher & Beridze, 2019).

In order to improve customer experiences and streamline operations, the hospitality sector has been at the forefront of implementing cutting-edge technology, such as augmented reality (AR), virtual reality (VR), and artificial intelligence (AI). The incorporation of these technologies does, however, also bring up important moral and legal issues, which demand attention. This study has offered a thorough analysis of the moral dilemmas and legal ramifications related to the application of AI, VR, and AR in the hotel industry. The main ethical issues noted are loss of employment and displacement, security and privacy of data, bias and equity, openness and explainability, accessibility and inclusion, and the requirement for ethical standards and preparedness. Concerns about data privacy and security, justice and bias, employment loss and displacement, intellectual property and copyright, accountability and responsibility, and regulatory compliance are some of the legal issues that arise from the application of AI in the hospitality sector. Hotels need to make sure they have proper insurance coverage and risk management procedures in place, as well as that they are abiding by applicable data privacy laws like the GDPR.

The research paper has suggested a paradigm for responsible AI application in the hospitality industry in order to overcome these ethical and legal challenges. In developing and implementing AI, VR, and AR systems, this paradigm highlights the significance of accountability, transparency, privacy, security and safety, and justice. Hotels can guarantee the appropriate and ethical integration of AI technology while fostering transparency and trust with their guests by giving priority to three essential AI governance elements. In addition to assisting in reducing the risks and difficulties that could arise from using AI, VR, and AR, this will also allow hotels to fully utilize the advantages of these technologies in order to improve both the visitor experience and operational effectiveness.

Hotels must be cautious when it comes to handling the moral and legal ramifications of AI, VR, and AR as the hospitality sector continues to harness their potential. By doing this, they can make sure that these technologies are integrated in a creative and responsible way, which will eventually help the industry succeed and remain sustainable over the long run. The expectation is that AI developers employed by other companies will establish connections with external stakeholders, such as legislators, and with individuals and groups with comparable interests in AI. Better AI systems may come from creative clusters and network developments, which may also lessen the likelihood of potential hazards. As Camilleri et al. (2023) point out, practitioners may find themselves in a better position if they collaborate with stakeholders to develop trustworthy AI and if they foster human capacity building to enhance intellectual properties. They can improve their possibilities for growth and competitiveness in this way. In order to foster responsible, transparent, interpretable, repeatable, equitable, inclusive, and secure AI solutions, it is arguably in their best interest to actively involve internal stakeholders (including employees) and educate them about AI governance aspects. As a result, they might reduce their risks and related expenses while maximizing the benefits of AI.

REFERENCES

Berente, N., Gu, B., Recker, J., & Santhanam, R. (2021). Managing Artificial Intelligence. *Management Information Systems Quarterly*, 45(3), 1433–1450.

Bertino, E., Kantarcioglu, M., Akcora, C. G., Samtani, S., Mittal, S., & Gupta, M. (2021). AI for security and security for AI. In proceedings of the eleventh ACM conference on data and application security and privacy (pp. 333–334). ACM Digital Library. DOI:10.1145/3422337.3450357

Buhalis, D., & Sinarta, Y. (2019). Real-time co-creation and nowness service: Lessons from tourism and hospitality. *Journal of Travel & Tourism Marketing*, 36(5), 563–582. DOI:10.1080/10548408.2019.1592059

Butcher, J., & Beridze, I. (2019). What is the state of artificial intelligence governance globally? *RUSI Journal*, 164(5–6), 88–96. DOI:10.1080/03071847.2019.1694260

Camilleri, M. A., Troise, C., Strazzullo, S., & Bresciani, S. (2023). Creating shared value through open innovation approaches: Opportunities and challenges for corporate sustainability. *Business Strategy and the Environment*, 32(7), 4485–4502. Advance online publication. DOI:10.1002/bse.3377

Corea, F., Fossa, F., Loreggia, A., Quintarelli, S., & Sapienza, S. (2022). A principle-based approach to AI: The case for European Union and Italy. *AI & Society*, 38(2), 521–535. DOI:10.1007/s00146-022-01453-8

Davenport, T., Guha, A., Grewal, D., & Bressgott, T. (2020). How artificial intelligence will change the future of marketing. *Journal of the Academy of Marketing Science*, 48(1), 24–42. DOI:10.1007/s11747-019-00696-0

Dwivedi, Y. K., Hughes, L., Ismagilova, E., Aarts, G., Coombs, C., Crick, T., Duan, Y., Dwivedi, R., Edwards, J., Eirug, A., Galanos, V., Ilavarasan, P. V., Janssen, M., Jones, P., Kar, A. K., Kizgin, H., Kronemann, B., Lal, B., Lucini, B., & Williams, M. D. (2021). Artificial Intelligence (AI): Multidisciplinary perspectives on emerging challenges, opportunities, and agenda for research, practice and policy. *International Journal of Information Management*, 57, 101994. DOI:10.1016/j.ijinfomgt.2019.08.002

Dwivedi, Y. K., Kshetri, N., Hughes, L., Slade, E. L., Jeyaraj, A., Kar, A. K., Baab-dullah, A. M., Koohang, A., Raghavan, V., Ahuja, M., Albanna, H., Albashrawi, M. A., Al-Busaidi, A. S., Balakrishnan, J., Barlette, Y., Basu, S., Bose, I., Brooks, L., Buhalis, D., & Wright, R. (2023). Opinion Paper: "So what if ChatGPT wrote it?" Multidisciplinary perspectives on opportunities, challenges and implications of generative conversational AI for research, practice and policy. *International Journal of Information Management*, 71, 102642. DOI:10.1016/j.ijinfomgt.2023.102642

Engel, C., Ebel, P., & Leimeister, J. M. (2022). Cognitive automation. *Electronic Markets*, 32(1), 339–350. DOI:10.1007/s12525-021-00519-7

Gonzalez, R. A., Ferro, R. E., & Liberona, D. (2020). Government and governance in intelligent cities, smart transportation study case in Bogotá Colombia. *Ain Shams Engineering Journal*, 11(1), 25–34. DOI:10.1016/j.asej.2019.05.002

Gonzalez-Jimenez, H., & Costa Pinto, D. (2024). Can AI robots foster social inclusion? Exploring the role of immersive augmentation in hospitality. *International Journal of Contemporary Hospitality Management*. Advance online publication. DOI:10.1108/IJCHM-09-2023-1459

Hickok, M. (2022). Public procurement of artificial intelligence systems: New risks and future proofing. *AI & Society*, 1–15. DOI:10.1007/s00146-022-01572-2 PMID:36212228

Inanc-Demir, M., & Kozak, M. (2019). Big data and its supporting elements: Implications for tourism and hospitality marketing. In Sigala, M., Rahimi, R., & Thelwall, M. (Eds.), *Big data and innovation in tourism, travel, and hospitality* (pp. 213–223). Springer. DOI:10.1007/978-981-13-6339-9_13

Kaplan, A., & Haenlein, M. (2020). Rulers of the world, unite! The challenges and opportunities of artificial intelligence. *Business Horizons*, 63(1), 37–50. DOI:10.1016/j.bushor.2019.09.003

Koniakou, V. (2023). From the "rush to ethics" to the "race for governance" in artificial intelligence. *Information Systems Frontiers*, 25(1), 71–102. DOI:10.1007/s10796-022-10300-6

Kozinets, R. V., & Gretzel, U. (2021). Commentary: Artificial intelligence: The marketer's dilemma. *Journal of Marketing*, 85(1), 156–159. DOI:10.1177/0022242920972933

Li, G., Li, N., & Sethi, S. P. (2021). Does CSR reduce idiosyncratic risk? Roles of operational efficiency and AI innovation. *Production and Operations Management*, 30(7), 2027–2045. DOI:10.1111/poms.13483

Mäntymäki, M., Minkkinen, M., Birkstedt, T., & Viljanen, M. (2022). Defining organizational AI governance. *AI and Ethics*, 2(4), 603–609. DOI:10.1007/s43681-022-00143-x

McCartney, G., & McCartney, A. (2020). Rise of the machines: Towards a conceptual service-robot research framework for the hospitality and tourism industry. *International Journal of Contemporary Hospitality Management*, 32(12), 3835–3851. DOI:10.1108/IJCHM-05-2020-0450

Mustak, M., Salminen, J., Plé, L., & Wirtz, J. (2021). Artificial intelligence in marketing: Topic modeling, scientometric analysis, and research agenda. *Journal of Business Research*, 124, 389–404. DOI:10.1016/j.jbusres.2020.10.044

Renieris, E. M., Kiron, D., & Mills, S. (2022). Should organizations link responsible AI and corporate social responsibility?It's Complicated. MIT Sloan https://sloanreview.mit.edu/article/should-organizations-link-responsible-ai-and-corporate-social-responsibility-its-complicated/

Romao, M., Costa, J., & Costa, C. J. (2019). Robotic process automation: A case study in the banking industry. In 2019 14th Iberian conference on information systems and technologies (CISTI) (pp. 1–6). IEEE. DOI:10.23919/CISTI.2019.8760733

Russell, S. J., & Norvig, P. (2016). *Artificial intelligence: A modern approach.* Pearson Education Limited.

Schwab, K. (2017). *The fourth industrial revolution.* Crown Business.

Sethi, K., Sharma, M., & Gusain, A. (2023, June). The Acceptance of Machine module in the Hospitality Industry with Prospects and Challenges: A Review. In 2023 2nd International Conference on Computational Modelling, Simulation and Optimization (ICCMSO) (pp. 283-288). IEEE.

Sharma, M., Bathla, G., Kaushik, A., & Rana, S. (2023, June). A Study on Impact of Adaptation of AI-Artificial intelligence services on Business Performance of Hotels. In 2023 2nd International Conference on Computational Modelling, Simulation and Optimization (ICCMSO) (pp. 28-32). IEEE.

Sharma, M., & Singh, A. (2024). Enhancing Competitive Advantages Through Virtual Reality Technology in the Hotels of India. In Utilizing Smart Technology and AI in Hybrid Tourism and Hospitality (pp. 243-256). IGI Global. DOI:10.4018/979-8-3693-1978-9.ch011

Sterne, J. (2017). *Artificial intelligence for marketing: Practical applications.* John Wiley and Sons. DOI:10.1002/9781119406341

Zhang, C., & Lu, Y. (2021). Study on artificial intelligence: The state of the art and future prospects. *Journal of Industrial Information Integration*, 23, 100224. DOI:10.1016/j.jii.2021.100224

Compilation of References

Abasaheb, S. A., & Subashini, R. (2023). Maneuvering of Digital Transformation: Role of Artificial Intelligence in Empowering Leadership-An Empirical Overview. *International Journal of Professional Business Review: Int.J. Prof. Bus. Rev.*, 8(5), 20.

Abina, A., Batkovic, T., Cestnik, B., Kikaj, A., Kovacic, L. R., Kurbus, M., & Zidansek, A. (2022). Decision support concept for improvement of sustainability-related competences. *Sustainability (Basel)*, 14(14), 8539. DOI:10.3390/su14148539

Abrokwah-Larbi, K. (2023). The role of generative artificial intelligence (GAI) in customer personalisation (CP) development in SMEs: A theoretical framework and research propositions. *Industrial Artificial Intelligence*, 1(1), 11. DOI:10.1007/s44244-023-00012-4

Afaq, A., Singh, G., Gaur, L., & Kapoor, S. (2023, November). Aspect-Based Opinion Mining of Customer Reviews in the Hospitality Industry: Leveraging Recursive Neural Tensor Network Algorithm. In 2023 3rd International Conference on Technological Advancements in Computational Sciences (ICTACS) (pp. 1392-1397). IEEE.

Afaq, A., Singh, G., Gaur, L., & Kapoor, S. (2023c, November). Aspect-Based Opinion Mining of Customer Reviews in the Hospitality Industry: Leveraging Recursive Neural Tensor Network Algorithm. In 2023 3rd International Conference on Technological Advancements in Computational Sciences (ICTACS) (pp. 1392-1397). IEEE.

Afaq, A., & Gaur, L. (2021). The Rise of Robots to Help Combat Covid-19. *International Conference on Technological Advancements and Innovations (ICTAI)*. IEEE. DOI:10.1109/ICTAI53825.2021.9673256

Afaq, A., Gaur, L., & Singh, G. (2022). A latent dirichlet allocation technique for opinion mining of online reviews of global chain hotels. *3rd International Conference on Intelligent Engineering and Management (ICIEM)* (pp. 201-206). IEEE. DOI:10.1109/ICIEM54221.2022.9853114

Afaq, A., Gaur, L., & Singh, G. (2023). A trip down memory lane to travellers' food experiences. *British Food Journal*, 125(4), 1390–1403. DOI:10.1108/BFJ-01-2022-0063

Afaq, A., Gaur, L., & Singh, G. (2023). Social CRM: Linking the dots of customer service and customer loyalty during COVID-19 in the hotel industry. *International Journal of Contemporary Hospitality Management*, 35(3), 992–1009. DOI:10.1108/IJCHM-04-2022-0428

Afaq, A., Singh, G., Gaur, L., & Kapoor, S. (2023). Aspect-Based Opinion Mining of Customer Reviews in the Hospitality Industry: Leveraging Recursive Neural Tensor Network Algorithm. *3rd International Conference on Technological Advancements in Computational Sciences (ICTACS)* (pp. 1392-1397). IEEE. DOI:10.1109/ICTACS59847.2023.10390384

Agbese, M., Alanen, H. K., Antikainen, J., Erika, H., Isomaki, H., Jantunen, M., Kemell, K.-K., Rousi, R., Vainio-Pekka, H., & Vakkuri, V. (2023). Governance in ethical and trustworthy AI systems: Extension of the ECCOLA method for AI ethics governance using GARP. *E-Informatica Software Engineering Journal*, 17(1), 230101. DOI:10.37190/e-Inf230101

Ahmad, M., Naeem, M. K. H., Mobo, F. D., Tahir, M. W., & Akram, M. (2024). Navigating the Journey: How Chatbots Are Transforming Tourism and Hospitality. In Darwish, D. (Ed.), *Design and Development of Emerging Chatbot Technology* (pp. 236–255). IGI Global., DOI:10.4018/979-8-3693-1830-0.ch014

Ahn, J., Kim, J., & Sung, Y. (2021). AI-powered recommendations: The roles of perceived similarity and psychological distance on persuasion. *International Journal of Advertising*, 40(8), 1366–1384. Advance online publication. DOI:10.1080/02650487.2021.1982529

AI Now Institute. (2018, December). *AI Now Report 2018*. Retrieved from https://ainowinstitute.org/AI_Now_2018_Report.pdf

Ai, D., Jiang, G., Lam, S. K., & Li, C. (2023). Computer vision framework for crack detection of civil infrastructure—A review. *Engineering Applications of Artificial Intelligence*, 117, 105478. DOI:10.1016/j.engappai.2022.105478

Aitken, M., Ng, M., Horsfall, D., Coopamootoo, K. P. L., van Moorsel, A., & Elliott, K. (2021). In pursuit of socially-minded data-intensive innovation in banking: A focus group study of public expectations of digital innovation in banking. *Technology in Society*, 66, 101666. DOI:10.1016/j.techsoc.2021.101666

Ajder, H., Patrini, G., Cavalli, F., & Cullen, L. (2019, September). *The State of Deepfakes: Landscape, Threats, and Impact.* Retrieved from https://regmedia.co .uk/2019/10/08/deepfake_report.pdf

Alajaji, S. A., Khoury, Z. H., Elgharib, M., Saeed, M., Ahmed, A. R., Khan, M. B., & Sultan, A. S. (2023). Generative Adversarial Networks In Digital Histopathology: Current Applications, Limitations, Ethical Considerations, and Future Directions. *Modern Pathology*, ●●●, 100369. PMID:37890670

Albreem, M. A., Sheikh, A. M., Alsharif, M. H., Jusoh, M., & Mohd Yasin, M. N. (2021). Green Internet of Things (GIoT): Applications, Practices, Awareness, and Challenges. *IEEE Access : Practical Innovations, Open Solutions*, 9, 38833–38858. Advance online publication. DOI:10.1109/ACCESS.2021.3061697

Alexsoft. (2021). *Comparing Machine Learning as a Service: Amazon, Microsoft Azure, Google Cloud AI, IBM Watson.* Alexsoft.

Ali, S. I., Kale, G. P., Shaikh, M. S., Ponnusamy, S., & Chouhan, P. S. (2024). AI Applications and Digital Twin Technology Have the Ability to Completely Transform the Future. In S. Ponnusamy, M. Assaf, J. Antari, S. Singh, & S. Kalyanaraman (Eds.), Harnessing AI and Digital Twin Technologies in Businesses (pp. 26-39). IGI Global. https://doi.org/. J. Denning et al., "Computing as a Discipline," Computer, Vol. 22, No. 2, 1989, pp. 63–70, DOI:10.4018/979-8-3693-3234-4.ch003

Ali, S. I.. (2023). Marketing policy in service enterprises using deep learning model. *International Journal of Intelligent Systems and Applications in Engineering*, 12, 239–243. Retrieved January 4, 2024, from https://ijisae.org/index.php/IJISAE/article/view/4066

Alnamrouti, A., Rjoub, H., & Ozgit, H. (2022). Do strategic human resources and artificial intelligence help to make organisations more sustainable? Evidence from non-governmental organisations. *Sustainability (Basel)*, 14(12), 7327. DOI:10.3390/su14127327

Alpaydin, E. (2016). *Machine Learning: The New AI.* MIT Press.

Al-Sharafi, M. A., Al-Emran, M., Arpaci, I., Iahad, N. A., AlQudah, A. A., Iranmanesh, M., & Al-Qaysi, N. (2023). Generation Z use of artificial intelligence products and its impact on environmental sustainability: A cross-cultural comparison. *Computers in Human Behavior*, 143, 107708. Advance online publication. DOI:10.1016/j.chb.2023.107708

Amato, G., Behrmann, M., Bimbot, F., Caramiaux, B., Falchi, F., Garcia, A., . . . Vincent, E. (2019). AI in the media and creative industries. arXiv preprint arXiv:1905.04175.

Anantrasirichai, N., & Bull, D. (2022). Artificial intelligence in the creative industries: A review. *Artificial Intelligence Review*, 55(1), 1–68. DOI:10.1007/s10462-021-10039-7

Anderson, M. (2011). *Machine ethics*. Cambridge University Press. DOI:10.1017/CBO9780511978036

Andrada, G., Clowes, R. W., & Smart, P. R. (2022). Varieties of transparency: Exploring agency within AI systems. *AI & Society*, ●●●, 1–11. DOI:10.1007/s00146-021-01326-6 PMID:35035112

Andriulli, F., Chen, P. Y., Erricolo, D., & Jin, J. M. (2022). Guest editorial machine learning in antenna design, modeling, and measurements. *IEEE Transactions on Antennas and Propagation*, 70(7), 4948–4952. DOI:10.1109/TAP.2022.3189963

Angwin, J., Larson, J., Mattu, S., & Kirchner, L. (2016, May 23). *ProPublica*. Retrieved from Machine Bias: https://www.propublica.org/article/machine-bias-risk-assessments-in-criminal-sentencing

Antwarg, L., Miller, R. M., Shapira, B., & Rokach, L. (2021). Explaining anomalies detected by autoencoders using Shapley Additive Explanations. *Expert Systems with Applications*, 186, 115736. DOI:10.1016/j.eswa.2021.115736

Arenander, M. (2023). Technology Acceptance for AI implementations: A case study in the Defense Industry about 3D Generative Models.

Arya, V., Bellamy, R. K., Chen, P. Y., Dhurandhar, A., Hind, M., Hoffman, S. C., Houde, S., Liao, Q. V., Luss, R., Mojsilovic, A., Mourad, S., Pedemonte, P., Raghavendra, R., Richards, J., Sattigeri, P., Shanmugam, K., Singh, M., Varshney, K. R., Wei, D., & Zhang, Y. (2019). One explanation does not fit all: A toolkit and taxonomy of ai explainability techniques. Cornell University. https://doi.org//arXiv.1909.03012.DOI:10.48550

Asimov, I. (1976). *Robot Visions*. Penguin Books.

Assayed, S. K., Woods, D., Alkahtib, M., & Shaalan, K. (2024). Effective Human-Chatbot Interaction: A High School Student Perspective. *International Journal of Computing and Digital Systems*, 15(1).

Astobiza, A. M. (2022). Ethical Governance of AI in the Global South: A Human Rights Approach to Responsible Use of AI. *MDPI Proceedings*, 1-5.

Augusto, L. M. (2021). From symbols to knowledge systems: A. Newell and HA Simon's contribution to symbolic. *AI*, 2(1), 29–62.

Austin, D. (2023, May 16). *AI music could revolutionize the industry — and this artist is leading the way*. Retrieved April 30, 2024, from Business Insider India: https://www.businessinsider.in/cryptocurrency/news/ai-music-could-revolutionize-the-industry-and-this-artist-is-leading-the-way/articleshow/100285845.cms

Autor, D. H. (2015). Why Are There Still So Many Jobs? The History and Future of Workplace Automation. *The Journal of Economic Perspectives*, 29(3), 3–30. DOI:10.1257/jep.29.3.3

Baía Reis, A., & Ashmore, M. (2022). From video streaming to virtual reality worlds: An academic, reflective, and creative study on live theatre and performance in the metaverse. *International Journal of Performance Arts and Digital Media*, 18(1), 7–28. DOI:10.1080/14794713.2021.2024398

Baidoo-Anu, D., & Owusu Ansah, L. Education in the era of generative artificial intelligence (AI): Understanding the potential benefits of ChatGPT in promoting teaching and learning. *SSRN* 2023. [CrossRef] DOI:10.2139/ssrn.4337484

Bandi, A., Adapa, P. V. S. R., & Kuchi, Y. E. V. P. K. (2023). The Power of Generative Ai: A Review of Requirements, Models, Input–Output Formats, Evaluation Metrics, And Challenges. *Future Internet*, 15(8), 260. DOI:10.3390/fi15080260

Bankins, S., & Formosa, P. (2023). The ethical implications of artificial intelligence (AI) for meaningful work. *Journal of Business Ethics*, 185(4), 725–740. DOI:10.1007/s10551-023-05339-7

Barabas, C. (2020). Beyond Bias: Re-Imagining the Terms of" Ethical AI" in Criminal Law. *Georgetown Journal of Law & Modern Critical Race Perspectives*, 12, 83.

Baraniuk, C. (2018, November 8). *China's Xinhua agency unveils AI news presenter*. Retrieved April 30, 2024, from BBC: https://www.bbc.com/news/technology-46136504

Barros, A., Prasad, A., & Śliwa, M. (2023). Generative artificial intelligence and academia: Implication for research, teaching and service. *Management Learning*, 54(5), 597–604. DOI:10.1177/13505076231201445

Barrot, J. S. (2023). Using ChatGPT for second language writing: Pitfalls and potentials. *Assessing Writing*, 57, 100745. DOI:10.1016/j.asw.2023.100745

Bassett, C. (2019). The computational therapeutic: Exploring Weizenbaum's ELIZA as a history of the present. *AI & Society*, 34(4), 803–812. Advance online publication. DOI:10.1007/s00146-018-0825-9

Bauer, J. M. (2022). Toward new guardrails for the information society. *Telecommunications Policy*, 46(5), 102350. DOI:10.1016/j.telpol.2022.102350

Bayamlıoğlu, I. E., Baraliuc, I., Janssens, L., & Hildebrandt, M. (2018). *Being Profiled: Cogitas Ergo Sum: 10 years of "Profiling the European Citizen"*. Amsterdam University Press. DOI:10.5117/9789463722124

Bellamy, R. K., Dey, K., Hind, M., Hoffman, S. C., Houde, S., Kannan, K., Lohia, P., Martino, J., Mehta, S., Mojsilovic, A., Nagar, S., Ramamurthy, K. N., Richards, J., Saha, D., Sattigeri, P., Singh, M., Varshney, K. R., & Zhang, Y. (2019). AI fairness 360: An extensible toolkit for detecting and mitigating algorithmic bias. *IBM Journal of Research and Development*, 63(4/5), 1–4. DOI:10.1147/JRD.2019.2942287

Bello, A., Ng, S. C., & Leung, M. F. (2023). A BERT Framework to Sentiment Analysis of Tweets. *Sensors (Basel)*, 23(1), 506. Advance online publication. DOI:10.3390/s23010506 PMID:36617101

Bender, E. M., Gebru, T., McMillan-Major, A., & Shmitchell, S. (2021). On the Dangers of Stochastic Parrots: Can Language Models Be Too Big? *Proceedings of the 2021 ACM Conference on Fairness, Accountability, and Transparency* (pp. 610-623.). New York City: Association for Computing Machinery. DOI:10.1145/3442188.3445922

Bengesi, S., El-Sayed, H., Sarker, M. K., Houkpati, Y., Irungu, J., & Oladunni, T. (2023). Advancements in Generative AI: A Comprehensive Review of GANs, GPT, Autoencoders, Diffusion Model, and Transformers. *arXiv preprint arXiv:2311.10242*.

Berente, N., Gu, B., Recker, J., & Santhanam, R. (2021). Managing Artificial Intelligence. *Management Information Systems Quarterly*, 45(3), 1433–1450.

Berthelot, A., Caron, E., Jay, M., & Lefèvre, L. (2024). Estimating The Environmental Impact of Generative-AI Services Using An LCA-Based Methodology. [Energy consumption]. *Procedia CIRP*, 122, 707–712. DOI:10.1016/j.procir.2024.01.098

Bertino, E., Kantarcioglu, M., Akcora, C. G., Samtani, S., Mittal, S., & Gupta, M. (2021). AI for security and security for AI. In proceedings of the eleventh ACM conference on data and application security and privacy (pp. 333–334). ACM Digital Library. DOI:10.1145/3422337.3450357

Bishop, J. M. (2021). Artificial intelligence is stupid and causal reasoning will not fix it. *Frontiers in Psychology*, 11, 513474. DOI:10.3389/fpsyg.2020.513474 PMID:33584394

Blauth, T. F., Gstrein, O. J., & Zwitter, A. (2022). Artificial Intelligence Crime: An Overview of Malicious Use and Abuse of AI. *IEEE Access : Practical Innovations, Open Solutions*, 10, 77110–77122. Advance online publication. DOI:10.1109/ACCESS.2022.3191790

Bockting, C. L., van Dis, E. A. M., van Rooij, R., Zuidema, W., & Bollen, J. (2023, October 19). Living guidelines for generative AI—Why scientists must oversee its use. *Nature*, 622(7984), 03266–1. https://www.nature.com/articles/d41586-023-. DOI:10.1038/d41586-023-03266-1 PMID:37857895

Boddington, P. (2017). *Towards a Code of Ethics for Artificial Intelligence*. Springer. DOI:10.1007/978-3-319-60648-4

Boden, M. A. (1998). Creativity and artificial intelligence. *Artificial Intelligence*, 103(1-2), 347–356. DOI:10.1016/S0004-3702(98)00055-1

Bogdan, R., Tatu, A., Crisan-Vida, M. M., Popa, M., & Stoicu-Tivadar, L. (2021). A practical experience on the amazon alexa integration in smart offices. *Sensors (Basel)*, 21(3), 734. Advance online publication. DOI:10.3390/s21030734 PMID:33499092

Bolton, T., Dargahi, T., Belguith, S., Al-Rakhami, M. S., & Sodhro, A. H. (2021). On the security and privacy challenges of virtual assistants. *Sensors (Basel)*, 21(7), 2312. Advance online publication. DOI:10.3390/s21072312 PMID:33810212

Bolukbasi, T., Chang, K. W., Zou, J. Y., Saligrama, V., & Kalai, A. T. (2016). Man is to Computer Programmer as Woman is to Homemaker? Debiasing Word Embeddings. *Advances in Neural Information Processing Systems*, ●●●, 4349–4357.

Bors, L., Samajdwer, A., & van Oosterhout, M. (2020). Introduction to Oracle Digital Assistant. In *Oracle Digital Assistant*. DOI:10.1007/978-1-4842-5422-6_1

Bostrom, N. (2014). *Superintelligence- Paths, Dangers and Strategies*. Oxford University Press.

Boucher, E. M., Harake, N. R., Ward, H. E., Stoeckl, S. E., Vargas, J., Minkel, J., Parks, A. C., & Zilca, R. (2021). Artificially intelligent chatbots in digital mental health interventions: a review. *Expert Review of Medical Devices, 18*(sup1). DOI: 10.1080/17434440.2021.2013200

Brandom, R. (2018, July 3). *Self-driving cars are headed toward an AI roadblock*. Retrieved from The Verge: https://www.theverge.com/2018/7/3/17530232/self-driving-ai-winter-full-autonomy-waymo-tesla-uber

Broadbent, E., Stafford, R., & MacDonald, B. (2018). Acceptance of Healthcare Robots for the Older Population: Review and Future Directions. *International Journal of Social Robotics*, ●●●, 257–271.

Broer, T. (2022). The Googlization of health: Invasiveness and corporate responsibility in media discourses on Facebook's algorithmic programme for suicide prevention. *Social Science & Medicine*, 306, 115131. DOI:10.1016/j.socscimed.2022.115131 PMID:35714428

Brown, M. (2016, April 5). *'New Rembrandt' to be unveiled in Amsterdam*. Retrieved April 30, 2024, from The Guardian: https://www.theguardian.com/artanddesign/2016/apr/05/new-rembrandt-to-be-unveiled-in-amsterdam

Brown, C. V. (1999). Horizontal mechanisms under differing IS organization contexts. *Management Information Systems Quarterly*, 23(3), 421–454. DOI:10.2307/249470

Brown, T. e. (2020, July 22). *Language Models are Few-Shot Learners*. Retrieved from ARXIV: https://arxiv.org/abs/2005.14165

Brundage, M. e. (2018). *The Malicious Use of Artificial Intelligence: Forecasting, Prevention, and Mitigation*. Oxford: Future of Humanity Institute.

Brynjolfsson, E., & McAfee, A. (2014). *The Second Machine Age: Work, Progress, and Prosperity in a Time of Brilliant Technologies*. W.W. Norton & Company.

Bryson, J. (2019). *The Past Decade and Future of AI's Impact on Society*. Retrieved from Openmind BBVA: https://www.bbvaopenmind.com/en/articles/the-past-decade-and-future-of-ais-impact-on-society/

Buchanan, B., & Shortliffe, E. (1984). *Rule-based expert systems: The MYCIN experiments of the Stanford heuristic programming project*. Addison-Wesley.

Buhalis, D., & Sinarta, Y. (2019). Real-time co-creation and nowness service: Lessons from tourism and hospitality. *Journal of Travel & Tourism Marketing*, 36(5), 563–582. DOI:10.1080/10548408.2019.1592059

Buhmann, A., & Fieseler, C. (2023). Deep learning meets deep democracy: Deliberative governance and responsible innovation in artificial intelligence. *Business Ethics Quarterly*, 33(1), 146–179. DOI:10.1017/beq.2021.42

Buolamwini, J., & Gebru, T. (2018). Gender Shades: Intersectional Accuracy Disparities in Commercial Gender Classification. *Proceedings of the 1st Conference on Fairness, Accountability and Transparency*, (pp. 77-91.).

Butcher, J., & Beridze, I. (2019). What is the state of artificial intelligence governance globally? *RUSI Journal*, 164(5–6), 88–96. DOI:10.1080/03071847.2019.1694260

Cadwalladr, C., & Graham-Harrison, E. (2018, March 17). *Revealed: 50 million Facebook profiles harvested for Cambridge Analytica in major data breach.* Retrieved from The Guardian: https://www.theguardian.com/news/2018/mar/17/cambridge-analytica-facebook-influence-us-election

Cadwalladr, C., & Graham-Harrison, E. (2018, March 7). *Revealed: 50 million Facebook profiles harvested for Cambridge Analytica in major data breach.* Retrieved from The Guardian: https://www.theguardian.com/news/2018/mar/17/cambridge-analytica-facebook-influence-us-election

Cahlan, S. (2020, February 13). *How misinformation helped spark an attempted coup in Gabon.* Retrieved April 30, 2024, from The Washington Post: https://www.washingtonpost.com/politics/2020/02/13/how-sick-president-suspect-video-helped-sparked-an-attempted-coup-gabon/

Cai, Q., Wang, H., Li, Z., & Liu, X. (2019). A survey on multi-modal data-driven smart healthcare systems: Approaches and applications. *IEEE Access : Practical Innovations, Open Solutions, 7,* 133583–133599. DOI:10.1109/ACCESS.2019.2941419

California Legislative Assembly. (2018, July 1). *California Civil Code, Section 17940 (2018). Bot Disclosure Law.* Retrieved from Casetext-Thomson Reuters: https://casetext.com/statute/california-codes/california-business-and-professions-code/division-7-general-business-regulations/part-3-representations-to-the-public/chapter-6-bots/section-17941-unlawful-use-of-bots

Cameron, K. (2012). Responsible leadership as virtuous leadership. *Responsible leadership,* 25-35.

Camilleri, M. A. (2017). *Corporate sustainability, social responsibility and environmental management.* Springer Nature. DOI:10.1007/978-3-319-46849-5

Camilleri, M. A. (2019). Measuring the corporate managers' attitudes towards ISO's social responsibility standard. *Total Quality Management & Business Excellence,* 30(13–14), 1549–1561. DOI:10.1080/14783363.2017.1413344

Camilleri, M. A. (2023). Artificial intelligence governance: Ethical considerations and implications for social responsibility. *Expert Systems: International Journal of Knowledge Engineering and Neural Networks,* 13406. Advance online publication. DOI:10.1111/exsy.13406

Camilleri, M. A., & Troise, C. (2023). Live support by chatbots with artificial intelligence: A future research agenda. *Service Business,* 17(1), 61–80. DOI:10.1007/s11628-022-00513-9

Camilleri, M. A., Troise, C., Strazzullo, S., & Bresciani, S. (2023). Creating shared value through open innovation approaches: Opportunities and challenges for corporate sustainability. *Business Strategy and the Environment*, 32(7), 4485–4502. Advance online publication. DOI:10.1002/bse.3377

Campbell, M., Hoane, A. J.Jr, & Hsu, F. H. (2002). Deep Blue. *Artificial Intelligence*, 134(1-2), 57–83. DOI:10.1016/S0004-3702(01)00129-1

Cao, Y., Li, S., Liu, Y., Yan, Z., Dai, Y., Yu, P. S., & Sun, L. (2023). A Comprehensive Survey of Ai-Generated Content (Aigc): A History of Generative AI From GAN To Chatgpt. *arXiv preprint arXiv:2303.04226*.

Cao, L. (2022). Ai in finance: Challenges, techniques, and opportunities. *ACM Computing Surveys*, 55(3), 1–38. DOI:10.1145/3502289

Cappuccio, M. L. (2016). The Seminal Speculation of a Precursor: Elements of Embodied Cognition and Situated AI in Alan Turing. *Fundamental Issues of Artificial Intelligence*, 479-496.

Card, D., & Smith, N. A. (2020). On consequentialism and fairness. *Frontiers in Artificial Intelligence*, 3, 34. DOI:10.3389/frai.2020.00034 PMID:33733152

Carmody, J., Shringarpure, S., & Van de Venter, G. (2021). AI and privacy concerns: A smart meter case study. *Journal of Information. Communication and Ethics in Society*, 19(4), 492–505. DOI:10.1108/JICES-04-2021-0042

Carvalho, A., Levitt, A., Levitt, S., Khaddam, E., & Benamati, J. (2019). Off-the-shelf artificial intelligence technologies for sentiment and emotion analysis: A tutorial on using IBM natural language processing. *Communications of the Association for Information Systems*, 44, 918–943. DOI:10.17705/1CAIS.04443

Case, M. (2022). Google, Big Data, & Antitrust. *Delaware Journal of Corporate Law, 46*(2).

Casini, L., & Roccetti, M. (2018). The impact of AI on the musical world: will musicians be obsolete?. Studi di estetica, (12).

Castelli, M., & Manzoni, L. (2022). Generative models in artificial intelligence and their applications. *Applied Sciences (Basel, Switzerland)*, 12(9), 4127. DOI:10.3390/app12094127

Cave, S., & Dihal, K. (2019). Hopes and fears for intelligent machines in fiction and reality. *Nature Machine Intelligence*, 1(2), 74–78. DOI:10.1038/s42256-019-0020-9

Cetinic, E., & She, J. (2022). Understanding and creating art with AI: Review and outlook. [TOMM]. *ACM Transactions on Multimedia Computing Communications and Applications*, 18(2), 1–22. DOI:10.1145/3475799

Chakraborty, U., Roy, S., & Kumar, S. (2023). *Rise of Generative AI and ChatGPT: Understand how Generative AI and ChatGPT are transforming and reshaping the business world (English Edition)*. BPB Publications.

Chamola, V., Sai, S., Sai, R., Hussain, A., & Sikdar, B. (2024). Generative AI for Consumer Electronics: Enhancing User Experience with Cognitive and Semantic Computing. *IEEE Consumer Electronics Magazine*, 1–9. DOI:10.1109/MCE.2024.3387049

Chan, C. D. (2022, October 4). *Are Virtual Influencers the Real Deal*. Retrieved from The Hollywood Reporter: https://www.hollywoodreporter.com/business/digital/virtual-influencers-digital-world-1235228125/

Chan, C. K. Y., & Hu, W. (2023). Students' voices on generative AI: Perceptions, benefits, and challenges in higher education. *International Journal of Educational Technology in Higher Education*, 20(1), 43. DOI:10.1186/s41239-023-00411-8

Chang, C. H., & Kidman, G. (2023). The rise of generative artificial intelligence (AI) language models-challenges and opportunities for geographical and environmental education. *International Research in Geographical and Environmental Education*, 32(2), 85–89. DOI:10.1080/10382046.2023.2194036

Chang, W. (2022). *ISO/IEC JTC 1/SC 42 (AI)/WG 2 (data) data quality for analytics and machine learning (ML)*. Information Technology Laboratory.

Chatila, R., & Havens, J. C. (2019). The IEEE global initiative on ethics of autonomous and intelligent systems. Robotics and well-being, 11-16.

Chaudhary, M., & Gaur, L. (2021). COVID-19 a "BIG RESET"—Role of GHRM in Achieving Organisational Sustainability in Context to Asian Market. In Proceedings of Second International Conference on Computing, Communications, and Cyber-Security: IC4S 2020 (pp. 607-625). Springer Singapore.

Chaudhary, M., Gaur, L., & Chakrabarti, A. (2022, April). Comparative Analysis of Entropy Weight Method and C5 Classifier for Predicting Employee Churn. In 2022 3rd International Conference on Intelligent Engineering and Management (ICIEM) (pp. 232-236). IEEE. DOI:10.1109/ICIEM54221.2022.9853181

Chaudhary, M., Gaur, L., & Chakrabarti, A. (2022, November). Detecting the Employee Satisfaction in Retail: A Latent Dirichlet Allocation and Machine Learning approach. In 2022 3rd International Conference on Computation, Automation and Knowledge Management (ICCAKM) (pp. 1-6). IEEE. DOI:10.1109/ICCAKM54721.2022.9990186

Chaudhary, M., Singh, G., Gaur, L., Mathur, N., & Kapoor, S. (2023, November). Leveraging Unity 3D and Vuforia Engine for Augmented Reality Application Development. In 2023 3rd International Conference on Technological Advancements in Computational Sciences (ICTACS) (pp. 1139-1144). IEEE.

Chaudhary, M., Gaur, L., Jhanjhi, N. Z., Masud, M., & Aljahdali, S. (2022). Envisaging Employee Churn Using MCDM and Machine Learning. *Intelligent Automation & Soft Computing*, 33(2), 1009–1024. DOI:10.32604/iasc.2022.023417

Chaudhary, M., Gaur, L., Singh, G., & Afaq, A. (2024). Introduction to Explainable AI (XAI) in E-Commerce. In *Role of Explainable Artificial Intelligence in E-Commerce* (pp. 1–15). Springer Nature Switzerland. DOI:10.1007/978-3-031-55615-9_1

Chen, P., Wu, L., & Wang, L. (2023). AI Fairness in Data Management and Analytics: A Review on Challenges, Methodologies and Applications. In *Applied Sciences (Switzerland)* (Vol. 13, Issue 18). DOI:10.3390/app131810258

Cherukuri, A. K., Jonnalagadda, A., & Murugesan, S. (2021, May). AI in education: Applications and impact. *Amplify*, 34(5), 26–33.

Chesney, B., & Citron, D. (2019). Deep Fakes: A Looming Challenge for Privacy, Democracy, and National Security. *California Law Review*, ●●●, 1753–1820.

Chesterman, S. (2023, May). *From Ethics to Law: Why, When, and How to Regulate AI*. Retrieved from NUS Law Working Paper: https://law.nus.edu.sg/wp-content/uploads/2023/05/014_SimonChesterman.pdf

Chin, H., Song, H., Baek, G., Shin, M., Jung, C., Cha, M., Choi, J., & Cha, C. (2023). The Potential of Chatbots for Emotional Support and Promoting Mental Well-Being in Different Cultures: Mixed Methods Study. *Journal of Medical Internet Research*, 25, e51712. Advance online publication. DOI:10.2196/51712 PMID:37862063

Chodak, G. (2024). Artificial Intelligence in E-Commerce. In *The Future of E-commerce: Innovations and Developments* (pp. 187–233). Springer Nature Switzerland. DOI:10.1007/978-3-031-55225-0_7

Chopkar, P., Wanjari, M., Jumle, P., Chandankhede, P., Mungale, S., & Shaikh, M. S. (2024), A Comprehensive Review on Cotton Leaf Disease Detection using Machine Learning Method, Grenze International Journal of Engineering and Technology, June Issue, Grenze ID: 01.GIJET.10.2.537, Grenze Scientific Society, 2024

Chouldechova, A. (2017). Fair prediction with disparate impact: A study of bias in recidivism prediction instruments. *Big Data*, 5(2), 153–163. DOI:10.1089/big.2016.0047 PMID:28632438

Chow, J. C. L., Wong, V., Sanders, L., & Li, K. (2023). Developing an AI-Assisted Educational Chatbot for Radiotherapy Using the IBM Watson Assistant Platform. *Healthcare (Basel)*, 11(17), 2417. Advance online publication. DOI:10.3390/healthcare11172417 PMID:37685452

Choy, L. T. (2014). The strengths and weaknesses of research methodology: Comparison and complimentary between qualitative and quantitative approaches. *IOSR journal of humanities and social science, 19*(4), 99-104.

Ciesla, R. (2024). The Book of Chatbots. In *The Book of Chatbots*. DOI:10.1007/978-3-031-51004-5

CNBC. (2023). OpenAI CEO Sam Altman says he's a 'little bit scared' of A.I. https://www.cnbc.com/2023/03/20/openai-ceo-sam-altman-says-hes-a-littlebit-scared-of-ai.html

Coeckelbergh, M. (2020). *AI Ethics*. The MIT Press. DOI:10.7551/mitpress/12549.001.0001

Cohn, G. (2018, October 25). *AI Art at Christie's Sells for $432,500*. Retrieved April 30, 2024, from The New York Times: https://www.nytimes.com/2018/10/25/arts/design/ai-art-sold-christies.html

Cooper, G. (2023). Examining Science Education in ChatGPT: An Exploratory Study of Generative Artificial Intelligence. *Journal of Science Education and Technology*, 32(3), 444–452. DOI:10.1007/s10956-023-10039-y

Copeland, J. B. (2004). *The Essential Turing-Seminal Writings in Computing, Logic, Philosophy, Artificial Intelligence, and Artificial Life plus The Secrets of Enigma*. Oxford University Press.

Corea, F. (2017). *Artificial intelligence and exponential technologies: Business models evolution and new investment opportunities*. Springer.

Corea, F., Fossa, F., Loreggia, A., Quintarelli, S., & Sapienza, S. (2022). A principle-based approach to AI: The case for European Union and Italy. *AI & Society*, 38(2), 521–535. DOI:10.1007/s00146-022-01453-8

Cortes, C., & Vapnik, V. (1995). Support-vector networks. *Machine Learning*, 20(3), 273–297. DOI:10.1007/BF00994018

Cortez, E. K. (Ed.). (2020). *Data Protection Around the World: An Introduction* (Vol. 33). Springer Nature.

Crawford, K., & Paglen, T. (2021). Excavating AI: The politics of images in machine learning training sets. *AI & Society*, 36(4), 1105–1116. DOI:10.1007/s00146-021-01301-1

Cronin, I. (2024). The Evolving World of Generative AI. In *Understanding Generative AI Business Applications: A Guide to Technical Principles and Real-World Applications* (pp. 223–242). Apress. DOI:10.1007/979-8-8688-0282-9_15

Crowder, J. (2023). *AI chatbots: The good, the bad, and the ugly*. Springer Nature.

Damiani, J. (2019, September 3). *A Voice Deepfake Was Used To Scam A CEO Out Of $243,000*. Retrieved April 30, 2024, from Forbes: https://www.forbes.com/sites/jessedamiani/2019/09/03/a-voice-deepfake-was-used-to-scam-a-ceo-out-of-243000/?sh=2354d82f2241

Damoah, I. S., Ayakwah, A., & Tingbani, I. (2021). Artificial intelligence (AI)-enhanced medical drones in the healthcare supply chain (HSC) for sustainability development: A case study. *Journal of Cleaner Production*, 328, 129598. DOI:10.1016/j.jclepro.2021.129598

Data & Society. (2018, April 18). *Algorithmic Accountability: A Primer*. Retrieved from Data & Society: https://datasociety.net/library/algorithmic-accountability-a-primer/

DataEthics. (2021). *Addressing Ethical Dilemmas in AI: Listening to Engineers*. Association of Nordic Engineers.

Dauvergne, P. (2020). *AI in the Wild: Sustainability in the Age of Artificial Intelligence*. MIT Press. DOI:10.7551/mitpress/12350.001.0001

Dauvergne, P. (2022). Is artificial intelligence greening global supply chains? Exposing the political economy of environmental costs. *Review of International Political Economy*, 29(3), 696–718. DOI:10.1080/09692290.2020.1814381

Davenport, T., Guha, A., Grewal, D., & Bressgott, T. (2020). How artificial intelligence will change the future of marketing. *Journal of the Academy of Marketing Science*, 48(1), 24–42. DOI:10.1007/s11747-019-00696-0

De Cremer, D. (2022). With AI entering organizations, responsible leadership may slip! *AI and Ethics*, 2(1), 49–51. DOI:10.1007/s43681-021-00094-9

De Cremer, D., & Narayanan, D. (2023). How AI tools can—and cannot—help organizations become more ethical. *Frontiers in Artificial Intelligence*, 6, 1093712. DOI:10.3389/frai.2023.1093712 PMID:37426304

Deckers, N., Fröbe, M., Kiesel, J., Pandolfo, G., Schröder, C., Stein, B., & Potthast, M. (2023, March). The Infinite Index: Information Retrieval on Generative Text-To-Image Models. In *Proceedings of the 2023 Conference on Human Information Interaction and Retrieval* (pp. 172-186). DOI:10.1145/3576840.3578327

Deep, G., & Verma, J. (2024). Textual Alchemy: Unleashing the Power of Generative Models for Advanced Text Generation. In *Advanced Applications of Generative AI and Natural Language Processing Models* (pp. 124-143). IGI Global.

Deng, L. (2018). Artificial intelligence in the rising wave of deep learning: The historical path and future outlook [perspectives]. *IEEE Signal Processing Magazine*, 35(1), 180–177. DOI:10.1109/MSP.2017.2762725

Devillers, L., Fogelman-Soulié, F., & Baeza-Yates, R. (2021). AI & Human Values: Inequalities, Biases, Fairness, Nudge, and Feedback Loops. In *Lecture Notes in Computer Science (including subseries Lecture Notes in Artificial Intelligence and Lecture Notes in Bioinformatics): Vol. 12600 LNCS*. DOI:10.1007/978-3-030-69128-8_6

Dhimolea, T. K., Kaplan-Rakowski, R., & Lin, L. (2022). *Supporting Social and Emotional Well-Being with Artificial Intelligence*. DOI:10.1007/978-3-030-84729-6_8

Diakopoulos, N. (2019). *Automating the news-how algorithms are rewriting the media*. Harvard University Press. DOI:10.4159/9780674239302

Doshi, R., Amin, K., Khosla, P., Bajaj, S., Chheang, S., & Forman, H. P. (2023). Utilizing Large Language Models to Simplify Radiology Reports: a comparative analysis of ChatGPT3.5, ChatGPT4.0, Google Bard, and Microsoft Bing. *MedRxiv*. DOI:10.1101/2023.06.04.23290786

Doshi-Velez, F., & Kim, B. (2017, March 2). *Towards a Rigorous Science of Interpretable Machine Learning*. Retrieved from ARXIV: https://arxiv.org/abs/1702.08608

Dourish, P., & Bell, G. (2011). *Divining a Digital Future: Mess and Mythology in Ubiquitous Computing.* MIT Press. DOI:10.7551/mitpress/9780262015554.001.0001

Dr Ali, S. I.. (2023) An Innovation of Algebraic Mathematical based statistical Modelfor complex number Theory. In: *ICDT 2023 IEEE Explorer Conference Proceedings* DOI: DOI:10.1109/ICDT57929.2023.10151169

Dr Ali, S. I.. (2023). *Causal convolution employing Almeida–pineda recurrent backpropagation for mobile network design.* ICTACT Journals., DOI:10.21917/ijct.2023.0460

Drafke, M. (2009). *Human side of organizations* (10th ed.). PHI Learning Private Ltd.

Durgadevi, M. (2021, July). Generative adversarial network (gan): a general review on different variants of gan and applications. In 2021 6th International Conference on Communication and Electronics Systems (ICCES) (pp. 1-8). IEEE.

Du, S., El Akremi, A., & Jia, M. (2022). Quantitative research on corporate social responsibility: A quest for relevance and rigor in a quickly evolving, turbulent world. *Journal of Business Ethics*, 187(1), 1–15. DOI:10.1007/s10551-022-05297-6 PMID:36465988

Du, S., & Xie, C. (2021). Paradoxes of artificial intelligence in consumer markets: Ethical challenges and opportunities. *Journal of Business Research*, 129, 961–974. DOI:10.1016/j.jbusres.2020.08.024

Du, W., & Han, Q. (2021). Research on application of artificial intelligence in movie industry. *Proc. SPIE 12076, 2021 International Conference on Image, Video Processing, and Artificial Intelligence,* DOI:10.1117/12.2619500

Dwivedi, Y. K., Hughes, L., Ismagilova, E., Aarts, G., Coombs, C., Crick, T., Duan, Y., Dwivedi, R., Edwards, J., Eirug, A., Galanos, V., Ilavarasan, P. V., Janssen, M., Jones, P., Kar, A. K., Kizgin, H., Kronemann, B., Lal, B., Lucini, B., & Williams, M. D. (2021). Artificial intelligence (AI): Multidisciplinary perspectives on emerging challenges, opportunities, and agenda for research, practice and policy. *International Journal of Information Management*, 57, 101994. DOI:10.1016/j.ijinfomgt.2019.08.002

Dwivedi, Y. K., Kshetri, N., Hughes, L., Slade, E. L., Jeyaraj, A., Kar, A. K., Baab-dullah, A. M., Koohang, A., Raghavan, V., Ahuja, M., Albanna, H., Albashrawi, M. A., Al-Busaidi, A. S., Balakrishnan, J., Barlette, Y., Basu, S., Bose, I., Brooks, L., Buhalis, D., & Wright, R. (2023). Opinion Paper: "So what if ChatGPT wrote it?" Multidisciplinary perspectives on opportunities, challenges and implications of generative conversational AI for research, practice and policy. *International Journal of Information Management*, 71, 102642. DOI:10.1016/j.ijinfomgt.2023.102642

Dwivedi, Y. K., Kshetri, N., Hughes, L., Slade, E. L., Jeyaraj, A., Kar, A. K., & Wright, R. (2023). Opinion Paper:"So what if ChatGPT wrote it?" Multidisciplinary perspectives on opportunities, challenges and implications of generative conversational AI for research, practice and policy. *International Journal of Information Management*, 71, 102642.

Dyer, G. (2023). Yet Another Article on AI (p. 7). Winnipeg Free Press.

Ebers, M. (2019). *Algorithms and Law (Regulating AI and Robotics: Ethical and Legal Challenges)*. Cambridge University Press.

Ebert, C., & Lourida, P. (2023). Generative AI for Software Practitioners, (2023). *IEEE Software*, 40(4), 30–38. DOI:10.1109/MS.2023.3265877

Edu, J. S., Such, J. M., & Suarez-Tangil, G. (2021). Smart Home Personal Assistants: A Security and Privacy Review. In *ACM Computing Surveys* (Vol. 53, Issue 6). DOI:10.1145/3412383

Elahi, H., Wang, G., Peng, T., & Chen, J. (2019). On transparency and accountability of smart assistants in smart cities. *Applied Sciences (Basel, Switzerland)*, 9(24), 5344. Advance online publication. DOI:10.3390/app9245344

Elkhatat, A. M., Elsaid, K., & Almeer, S. (2023). Evaluating the efficacy of AI content detection tools in differentiating between human and AI-generated text. *International Journal for Educational Integrity*, 19(1), 17. DOI:10.1007/s40979-023-00140-5

Elliott, A. (2022). *The Routledge Social Science Handbook of AI*. Routledge.

Ellison, H., & Asimov, I. (1978). *I, Robot: The Illustrated Screenplay*. Warner Books.

Engel, C., Ebel, P., & Leimeister, J. M. (2022). Cognitive automation. *Electronic Markets*, 32(1), 339–350. DOI:10.1007/s12525-021-00519-7

Erdelyi, O. J., & Goldsmith, J. (2022). Regulating Artificial Intelligence: Proposal for a Global Solution. *Government Information Quarterly*, 39(4), 1–13. DOI:10.1016/j.giq.2022.101748

EU. (2021). Regulation of the European Parliament and the council laying down harmonized rules on artificial intelligence (artificial intelligence act) and amending certain union legislative acts. European Commission https://eur-lex.europa.eu/legal -content/EN/TXT/?uri=celex%3A52021PC0206

Eubanks, V. (2018). *Automating Inequality: How High-Tech Tools Profile, Police, and Punish the Poor*. St. Martin's Press.

European Commission. (2021, April 21). *Proposal for a Regulation laying down harmonised rules on artificial intelligence*. Retrieved from European Commission: https://digital-strategy.ec.europa.eu/en/library/proposal-regulation-laying-down -harmonised-rules-artificial-intelligence

European Commission. (2021, April 4). *Proposal for a Regulation laying down harmonised rules on artificial intelligence (Artificial Intelligence Act)*. Retrieved from EUR-LEX: https://eur-lex.europa.eu/legal-content/EN/TXT/?uri=CELEX %3A52021PC0206

EY. (October 24, 2023). CEO perspectives on the current state of artificial intelligence (AI) at their organization worldwide in 2023 [Graph]. In *Statista*. Retrieved June 16, 2024, from https://www.statista.com/statistics/1445865/ceo-perspectives -on-ai-usage-at-their-organization/

Eze, M. O. (2010). *Intellectual History in Contemporary South Africa*. Palgrave Macmillan. DOI:10.1057/9780230109698

Farina, M., Yu, X., & Lavazza, A. (2024). Ethical considerations and policy interventions concerning the impact of generative AI tools in the economy and in society. *AI and Ethics*, ●●●, 1–9. DOI:10.1007/s43681-023-00405-2

Farina, M., Zhdanov, P., Karimov, A., & Lavazza, A. (2022). AI and society: A virtue ethics approach. *AI & Society*, 1–14.

Farrelly, T., & Baker, N. (2023, November 04). Generative Artificial Intelligence: Implications and Considerations for Higher Education Practice. *Education Sciences*, 13(11), 1109. DOI:10.3390/educsci13111109

Faruqi, F., Katary, A., Hasic, T., Abdel-Rahman, A., Rahman, N., Tejedor, L., & Mueller, S. (2023, October). Style2Fab: Functionality-Aware Segmentation for Fabricating Personalized 3D Models with Generative AI. In *Proceedings of the 36th Annual ACM Symposium on User Interface Software and Technology* (pp. 1-13). DOI:10.1145/3586183.3606723

Felzmann, H., Fosch-Villaronga, E., Lutz, C., & Tamò-Larrieux, A. (2020). Towards Transparency by Design for Artificial Intelligence. *Science and Engineering Ethics*, 26(6), 3333–3361. Advance online publication. DOI:10.1007/s11948-020-00276-4 PMID:33196975

Ferrara, E. (2024). Fairness and Bias in Artificial Intelligence: A Brief Survey of Sources, Impacts, and Mitigation Strategies. In *Sci* (Vol. 6, Issue 1). DOI:10.3390/sci6010003

Ferrucci, D. e., Brown, E., Chu-Carroll, J., Fan, J., Gondek, D., Kalyanpur, A. A., Lally, A., Murdock, J. W., Nyberg, E., Prager, J., Schlaefer, N., & Welty, C. (2010). Building Watson: An Overview of the DeepQA Project. *AI Magazine*, 31(3), 59–79. DOI:10.1609/aimag.v31i3.2303

Feuerriegel, S., Hartmann, J., Janiesch, C., & Zschech, P. (2024). Generative AI. *Business & Information Systems Engineering*, 66(1), 111–126. DOI:10.1007/s12599-023-00834-7

Filgueiras, F. (2022). New Pythias of public administration: Ambiguity and choice in AI systems as challenges for governance. *AI & Society*, 37(4), 1473–1486. DOI:10.1007/s00146-021-01201-4

Floridi, L., & Cowls, J. (2022). A unified framework of five principles for AI in society. Machine learning and the city: Applications in architecture and urban design, 535-545.

Floridi, L. (2013). *The Ethics of Information*. Oxford University Press. DOI:10.1093/acprof:oso/9780199641321.001.0001

Floridi, L., & Chiriatti, M. (2020). GPT-3: Its nature, scope, limits, and consequences. *Minds and Machines*, 30(4), 681–694. DOI:10.1007/s11023-020-09548-1

Floridi, L., Cowls, J., Beltrametti, M., Chatila, R., Chazerand, P., Dignum, V., Luetge, C., Madelin, R., Pagallo, U., Rossi, F., Schafer, B., Valcke, P., & Vayena, E. (2018). AI4People—An Ethical Framework for a Good AI Society: Opportunities, Risks, Principles, and Recommendations. *Minds and Machines*, 28(4), 689–707. DOI:10.1007/s11023-018-9482-5 PMID:30930541

Fosch-Villaronga, E., Drukarch, H., Khanna, P., Verhoef, T., & Custers, B. (2022). Accounting for diversity in AI for medicine. *Computer Law & Security Report*, 47, 105735. Advance online publication. DOI:10.1016/j.clsr.2022.105735

Franco, S. F., Graña, J. M., Flacher, D., & Rikap, C. (2023). Producing and using artificial intelligence: What can Europe learn from Siemens's experience? *Competition & Change*, 27(2), 302–331. DOI:10.1177/10245294221097066

Frank, B. (2021). Artificial intelligence-enabled environmental sustainability of products: Marketing benefits and their variation by consumer, location, and product types. *Journal of Cleaner Production*, 285, 125242. DOI:10.1016/j.jclepro.2020.125242

Frankish, K., & Ramsey, W. M. (2014). *The Cambridge Handbook of Artificial Intelligence*. Cambridge University Press. DOI:10.1017/CBO9781139046855

Frey, C. B., & Osborne, M. (2023). Generative AI and the Future of Work: A Reappraisal. *The Brown Journal of World Affairs*, ●●●, 1–12.

Fui-Hoon Nah, F., Zheng, R., Cai, J., Siau, K., & Chen, L. (2023). Generative AI and ChatGPT: Applications, challenges, and AI-human collaboration. *Journal of Information Technology Case and Application Research*, 25(3), 277–304. DOI:10.1080/15228053.2023.2233814

Fulmer, R. M. (2004). The challenge of ethical leadership. *Organizational Dynamics*, 33(3), 307–317. DOI:10.1016/j.orgdyn.2004.06.007

Furman, J., & Seamans, R. (2019). AI and the Economy. *Innovation Policy and the Economy*, 19(1), 161–191. DOI:10.1086/699936

Future of Life Institute. (2017, August 11). *Asilomar AI Principles*. Retrieved from Future of Life Institute: https://futureoflife.org/open-letter/ai-principles/

Gaafar, A. A. (2021). Metaverse in architectural heritage documentation & education. *Adv. Ecol. and Environ. Res.*, 6(10), 66–86.

Galaz, V., Centeno, M. A., Callahan, P. W., Causevic, A., Patterson, T., Brass, I., Baum, S., Farber, D., Fischer, J., Garcia, D., McPhearson, T., Jimenez, D., King, B., Larcey, P., & Levy, K. (2021). Artificial intelligence, systemic risks, and sustainability. *Technology in Society*, 67, 101741. DOI:10.1016/j.techsoc.2021.101741

Gao, F., Wang, C., Li, L., & Zhang, D. (2020). Altitude information acquisition of uav based on monocular vision and mems. *Journal of Intelligent & Robotic Systems*, 98(3-4), 807–818. DOI:10.1007/s10846-019-01018-w

Garcia Valencia, O. A., Suppadungsuk, S., Thongprayoon, C., Miao, J., Tangpanithandee, S., Craici, I. M., & Cheungpasitporn, W. (2023). Ethical Implications of Chatbot Utilization in Nephrology. In *Journal of Personalized Medicine* (Vol. 13, Issue 9). DOI:10.3390/jpm13091363

Gasser, U., & Almeida, V. A. (2017). A Layered Model for AI Governance. *IEEE Internet Computing*, 21(6), 58–62. DOI:10.1109/MIC.2017.4180835

Gaur, L., Gaur, D., & Afaq, A. (2024). Demystifying Metaverse Applications for Intelligent Healthcare. In Metaverse Applications for Intelligent Healthcare (pp. 1-23). IGI Global.

Gaur, L., Gaur, D., & Afaq, A. (2024). Ethical Considerations in the Use of the Metaverse for Healthcare. In Metaverse Applications for Intelligent Healthcare (pp. 248-273). IGI Global.

Gaur, L., Gaur, D., & Afaq, A. (2024a). Demystifying Metaverse Applications for Intelligent Healthcare. In Metaverse Applications for Intelligent Healthcare (pp. 1-23). IGI Global.

Gaur, L., Gaur, D., & Afaq, A. (2024b). Ethical Considerations in the Use of the Metaverse for Healthcare. In Metaverse Applications for Intelligent Healthcare (pp. 248-273). IGI Global.

Gaur, L., & Afaq, A. (2020). Metamorphosis of CRM: incorporation of social media to customer relationship management in the hospitality industry. In *Handbook of Research on Engineering Innovations and Technology Management in Organizations* (pp. 1–23). IGI. DOI:10.4018/978-1-7998-2772-6.ch001

Gaur, L., Afaq, A., Arora, G. K., & Khan, N. (2023). Artificial intelligence for carbon emissions using system of systems theory. *Ecological Informatics*, 76, 102165. DOI:10.1016/j.ecoinf.2023.102165

Gaur, L., Afaq, A., Singh, G., & Dwivedi, Y. K. (2021). Role of artificial intelligence and robotics to foster the touchless travel during a pandemic: A review and research agenda. *International Journal of Contemporary Hospitality Management*, 33(11), 4079–4098. DOI:10.1108/IJCHM-11-2020-1246

Gaur, L., Afaq, A., Solanki, A., Singh, G., Sharma, S., Jhanjhi, N. Z., My, H. T., & Le, D. N. (2021b). Capitalizing on big data and revolutionary 5G technology: Extracting and visualizing ratings and reviews of global chain hotels. *Computers & Electrical Engineering*, 95, 107374. DOI:10.1016/j.compeleceng.2021.107374

Gaur, L., Gaur, D., & Afaq, A. (2024). Ethical Considerations in the Use of the Metaverse for Healthcare. In *Metaverse Applications for Intelligent Healthcare* (pp. 248–273). IGI Global.

Gawali, P., & Sony, R. (2020, December 17). *The Role of Artificial Intelligence in Improving Criminal Justice System: Indian Perspective*. Legal Issues in the Digital Age. DOI:10.17323/2713-2749.2020.3.78.96

Gehr, T., Mirman, M., Drachsler-Cohen, D., Tsankov, P., Chaudhuri, S., & Vechev, M. (2018). Ai2: Safety and robustness certification of neural networks with abstract interpretation. In In 2018 IEEE symposium on security and privacy (SP) (pp. 3–18). IEEE.

Generative AI Market Size to Hit around USD 118.06 Bn by 2032. 2023. Available online: https://www.globenewswire.com/en/news- release/2023/05/15/2668369/0/en/Generative-AI-Market-Size-to-Hit-Around-USD-118-06-Bn-By-2032.html/ (accessed on 29 June 2023).

Georgieff, A., & Hyee, R. (2022). Artificial Intelligence and Employment: New Cross-Country Evidence. *Frontiers in Artificial Intelligence*, 5, 832736. Advance online publication. DOI:10.3389/frai.2022.832736 PMID:35620279

Ghaffari, S., Yousefimehr, B., & Ghatee, M. (2024, February). Generative-AI in E-Commerce: Use-Cases and Implementations. In 2024 20th CSI International Symposium on Artificial Intelligence and Signal Processing (AISP) (pp. 1-5). IEEE.

Global Information & Communications Team. (2023). *How Generative AI is Transforming Digital Content Services (Report No K88D-70)*. Frost & Sullivan.

Godwin-Jones, R. (2023). Emerging spaces for language learning: AI bots, ambient intelligence, and the metaverse. *Language Learning & Technology*, 27(2), 6–27.

Goel, A. K. (2022). Looking back, looking ahead: Humans, ethics, and AI. *AI Magazine*, 43(2), 267–269. DOI:10.1002/aaai.12052

Goksel Canbek, N., & Mutlu, M. E. (2016). On the track of Artificial Intelligence: Learning with Intelligent Personal Assistants. *Uluslararas Insan Bilimleri Dergisi*, 13(1), 592. Advance online publication. DOI:10.14687/ijhs.v13i1.3549

Gonzalez-Jimenez, H., & Costa Pinto, D. (2024). Can AI robots foster social inclusion? Exploring the role of immersive augmentation in hospitality. *International Journal of Contemporary Hospitality Management*. Advance online publication. DOI:10.1108/IJCHM-09-2023-1459

Gonzalez, R. A., Ferro, R. E., & Liberona, D. (2020). Government and governance in intelligent cities, smart transportation study case in Bogota Colombia. *Ain Shams Engineering Journal*, 11(1), 25–34. DOI:10.1016/j.asej.2019.05.002

Goodfellow, I. J. (2014). *Generative Adversarial Nets. Proceedings of Advances in Neural Information Processing Systems 2014*. Curran Associates.

Goodfellow, I., Bengio, Y., & Courville, A. (2016). *Deep Learning*. MIT Press.

Goodfellow, I., Pouget-Abadie, J., Mirza, M., Xu, B., Warde-Farley, D., Ozair, S., Courville, A., & Bengio, Y. (2014). Generative Adversarial Nets (PDF). *Proceedings of the International Conference on Neural Information Processing Systems (NIPS 2014).* pp. 2672–2680.

Google. (2019). Google Assistant | Your own personal Google. In *Google.*

Google. (2023). *RESPONSIBILITY: Our Principles.* Retrieved from Google AI: https://ai.google/responsibility/principles/

Google. "Responsible AI practices": https://ai.google/responsibility/responsible-ai -practices/, accessed November 5, 2023.

Gordon, C. (2023). ChatGPT Is the Fastest Growing App in the History of Web Applications.

Gordon, S. (2022). Virtual navigation and geospatial mapping tools, customer data analytics, and computer vision and simulation optimization algorithms in the blockchain-based metaverse. *Rev. Contemp. Philos.*, (21), 89–104.

Gosain, M. T., Bhatia, M. K., Sharma, M. R., Bhanvra, M. S., Shaw, A., Singh, T., & Kashyap, B. H. (2024). Artificial Intelligence and the Privacy Paradox: Challenges and Opportunities in Legal Adaptations. *Educational Administration: Theory and Practice*, 30(5), 10384–10394.

Gozalo-Brizuela, R., & Garrido-Merchán, E. C. (2023, June 14). *A survey of Generative AI Applications.* Retrieved from ARXIV: https://arxiv.org/abs/2306.02781

Graves, L. (2016). *Deciding what's true: the rise of political fact-checking in American journalism.* Columbia University Press. DOI:10.7312/grav17506

Greenberg, J. (2011). *Behaviour in organizations* (10th ed.). PHI Learning Private Ltd.

Greenleaf, R. K. (2002). *Servant leadership: A journey into the nature of legitimate power and greatness.* Paulist Press.

Guo, X., Wang, Z., Zhu, W., He, G., Deng, H. B., Lv, C. X., & Zhang, Z. H. (2022). Research on DSO vision positioning technology based on binocular stereo panoramic vision system. *Defence Technol.*, 18(4), 593–603. DOI:10.1016/j.dt.2021.12.010

Guo, Y., Yu, T., Wu, J., Wang, Y., Wan, S., Zheng, J., & Dai, Q. (2022). Artificial Intelligence for Metaverse: A Framework, CAAI. *Artificial Intelligence Research*, 1(1), 54–67.

Gupta, A., & Heath, V. (2020, September 14). AI ethics groups are repeating one of society's classic mistakes. *MIT Technology Review.* Retrieved from .

Gupta, N. (2021, June 7). Retrieved April 30, 2024, from World Association of News Publishers: https://wan-ifra.org/2021/06/how-bbc-news-labs-uses-ai-powered -content-automation-to-engage-young-audiences/

Gupta, M., Akiri, C., Aryal, K., Parker, E., & Praharaj, L. (2023). From chatgpt to threatgpt: Impact of generative ai in cybersecurity and privacy. *IEEE Access : Practical Innovations, Open Solutions*, 11, 80218–80245. DOI:10.1109/AC-CESS.2023.3300381

Gupta, R., Nair, K., Mishra, M., Ibrahim, B., & Bhardwaj, S. (2024). Adoption and impacts of generative artificial intelligence: Theoretical underpinnings and research agenda. *International Journal of Information Management Data Insights*, 4(1), 100232. DOI:10.1016/j.jjimei.2024.100232

H.R. 6216 - 116th congress (2019-2020): National Artificial Intelligence Initiative Act of 2020 | congress.gov | library of Congress. Available at: https://www.congress .gov/bill/116th-congress/house-bill/6216 (Accessed: 25 May 2024).

Hacker, P., Engel, A., & Mauer, M. (2023, June). Regulating ChatGPT and other large generative AI models. In *Proceedings of the 2023 ACM Conference on Fairness, Accountability, and Transparency* (pp. 1112-1123). DOI:10.1145/3593013.3594067

Hagendorf, T. (2020). The Ethics of AI Ethics: An Evaluation of Guidelines. *Minds and Machines*, 30(1), 99–120. DOI:10.1007/s11023-020-09517-8

Hagerty, A., & Rubinov, I. (2019). Global AI ethics: a review of the social impacts and ethical implications of artificial intelligence. arXiv preprint arXiv:1907.07892.

Haidt, J. (2008). Morality. *Perspectives on Psychological Science*, 3(1), 65–72. DOI:10.1111/j.1745-6916.2008.00063.x PMID:26158671

Hale, J. R., & Fields, D. L. (2007). Exploring servant leadership across cultures: A study of followers in Ghana and the USA. *Leadership*, 3(4), 397–417. DOI:10.1177/1742715007082964

Hamilton, M. (n.d.). *A 'black box' AI system has been influencing criminal justice decisions for over two decades – it's time to open it up*. The Conversation. https:// theconversation.com/a-black-box-ai-system-has-been-influencing-criminal-justice -decisions-for-over-two-decades-its-time-to-open-it-up-200594

Hamon, R., Junklewitz, H., & Sanchez, I. (2020). *Robustness and explainability of artificial intelligence*. Publications Office of the European Union.

Han, S., Mao, H., & Dally, W. J. (2016, February 15). *Deep Compression: Compressing Deep Neural Networks with Pruning, Trained Quantization and Huffman Coding*. Retrieved from ARXIV: https://arxiv.org/abs/1510.00149

Hao, K. (2019, May-June). The biggest threat of deepfakes isn't the deepfakes themselves. *MIT Technology Review*.

Harreis, H., Koullias, T., Roberts, R., & Te, K. (2023). *Generative AI: Unlocking the future of fashion*. McKinsey & Company.

Hartman, T., Kennedy, H., Steedman, R., & Jones, R. (2020). Public perceptions of good data management: Findings from a UK-based survey. *Big Data & Society*, 7(1), 2053951720935616. DOI:10.1177/2053951720935616

Hasal, M., Nowaková, J., Ahmed Saghair, K., Abdulla, H., Snášel, V., & Ogiela, L. (2021). Chatbots: Security, privacy, data protection, and social aspects. *Concurrency and Computation*, 33(19), e6426. Advance online publication. DOI:10.1002/cpe.6426

Hawkins, M. (2022). Virtual employee training and skill development, workplace technologies, and deep learning computer vision algorithms in the immersive metaverse environment. *Psychosociological Issues in Human Resource Management*, 10(1), 106–120. DOI:10.22381/pihrm10120228

Hawkins, M. (2023). Metaverse live shopping analytics: Retail data measurement tools, computer vision and deep learning algorithms, and decision intelligence and modeling, J. Self-Governance Manage. *Econ.*, 10(2), 22–36.

Heaven, W. D. (2023) Predictive policing is still racist-whatever data it uses, MIT Technology Review. Available at: https://www.technologyreview.com/2021/02/05/1017560/predictive-policing-racist-algorithmic-bias-data-crime-predpol/ (Accessed: 25 May 2024).

Hecht-Nielsen, R. (1992). Theory of the Backpropagation Neural Network. In Wechsler, H. (Ed.), *Neural Networks for Perception* (Vol. 2, pp. 65–93). Academic Press. DOI:10.1016/B978-0-12-741252-8.50010-8

Heidrick & Struggles. (October 17, 2023). American and European companies' responsible person for the artificial intelligence (AI) strategy in 2023, by industry [Graph]. In *Statista*. Retrieved June 16, 2024, from https://www.statista.com/statistics/1455283/ai-strategy-leaders-by-industry/

Hepburn, G. (2009). Alternatives to traditional regulation. Organization for Economic Cooperation and Development https://www.oecd.org/gov/regulatorypolicy/42245468.pdf

Hickok, M. (2022). Public procurement of artificial intelligence systems: New risks and future proofing. *AI & Society*, ●●●, 1–15. DOI:10.1007/s00146-022-01572-2 PMID:36212228

High-Level Expert Group on AI. (2019, April 8). *Ethics guidelines for trustworthy AI*. Retrieved from European Commission: https://digital-strategy.ec.europa.eu/en/library/ethics-guidelines-trustworthy-ai

Hollander, E. P. (1995). Ethical challenges in the leader-follower relationship. *Business Ethics Quarterly*, 5(1), 55–65. DOI:10.2307/3857272

Hollanek, T. (2020). AI transparency: A matter of reconciling design with critique. *AI & Society*. Advance online publication. DOI:10.1007/s00146-020-01110-y

Holmes, A. F. (2007). *Ethics: Approaching moral decisions*. InterVarsity Press.

Hong, G., Folcarelli, A., Less, J., Wang, C., Erbasi, N., & Lin, S. (2021). Voice assistants and cancer screening: A comparison of alexa, siri, google assistant, and cortana. *Annals of Family Medicine*, 19(5), 447–449. Advance online publication. DOI:10.1370/afm.2713 PMID:34546951

Hornigold, T. (2018, October 25). *The First Novel Written by AI Is Here—and It's as Weird as You'd Expect It to Be*. Retrieved April 30, 2024, from Singularityhub: https://singularityhub.com/2018/10/25/ai-wrote-a-road-trip-novel-is-it-a-good-read/

Hornung, O., & Smolnik, S. (2022). AI invading the workplace: Negative emotions towards the organizational use of personal virtual assistants. *Electronic Markets*, 32(1), 123–138. Advance online publication. DOI:10.1007/s12525-021-00493-0

Howell, J. M., & Avolio, B. J. (1992). The ethics of charismatic leadership: Submission or liberation? *The Academy of Management Perspectives*, 6(2), 43–54. DOI:10.5465/ame.1992.4274395

Huang, C., Zhang, Z., Mao, B., & Yao, X. "An Overview of Artificial Intelligence Ethics," IEEE Transactions on Artificial Intelligence (2024), Vol. 4, Issue 4, pp. 799-819, 2022.

Huang, M. H., & Rust, R. T. (2022). AI as customer. *Journal of Service Management*, 33(2), 210–220. DOI:10.1108/JOSM-11-2021-0425

Huh, J., Whang, C., & Kim, H. Y. (2023). Building trust with voice assistants for apparel shopping: The effects of social role and user autonomy. *Journal of Global Fashion Marketing*, 14(1), 5–19. Advance online publication. DOI:10.1080/20932685.2022.2085603

Hu, Q., Lu, Y., Pan, Z., Gong, Y., & Yang, Z. (2021). Can AI artifacts influence human cognition? The effects of artificial autonomy in intelligent personal assistants. *International Journal of Information Management*, 56, 102250. Advance online publication. DOI:10.1016/j.ijinfomgt.2020.102250

Hutson, J., & Harper-Nichols, M. "Generative AI and Algorithmic Art: Disrupting the Framing of Meaning and Rethinking the Subject- Object Dilemma," Global Journal of Computer Science and Technology (2023): https://digitalcommons.lindenwood.edu/faculty-researchpapers/461

IBM. (2003). AI Ethics Governance Framework. https://www.ibm.com/blog/a-look-into-ibms-ai-ethics-governance-framework/

IBM. (2022). Introducing IBM AI governance: IBM AI governance is a new, one-stop solution built on IBM cloud Pak® for data. Armonk https://www.ibm.com/cloud/blog/announcements/introducing-ibm-ai-governanceIEEE. (2023). IEEE introduces new program for free access to AI ethics and governance standards. Institute of Electrical and Electronics Engineers https://standards.ieee.org/news/get-program-ai-ethics/

IEEE Global Initiative on Ethics of Autonomous and Intelligent Systems. (2019, March 25). *Ethically Aligned Design: A Vision for Prioritizing Human Well-being with Autonomous and Intelligent Systems (A/IS)*. Retrieved from IEEE Standards Association: https://standards.ieee.org/news/ieee-ead1e/

Ienca, M. (2023). On Artificial Intelligence and Manipulation. *Topoi*, 42(3), 833–842. Advance online publication. DOI:10.1007/s11245-023-09940-3

Inanc-Demir, M., & Kozak, M. (2019). Big data and its supporting elements: Implications for tourism and hospitality marketing. In Sigala, M., Rahimi, R., & Thelwall, M. (Eds.), *Big data and innovation in tourism, travel, and hospitality* (pp. 213–223). Springer. DOI:10.1007/978-981-13-6339-9_13

Irani, S. (2023, March 19). *This is not the Luke Skywalker you're looking for*. Retrieved April 30, 2024, from The Michigan Daily: https://www.michigandaily.com/arts/b-side/this-is-not-the-luke-skywalker-youre-looking-for/

Ischen, C., Araujo, T. B., Voorveld, H. A. M., Van Noort, G., & Smit, E. G. (2022). Is voice really persuasive? The influence of modality in virtual assistant interactions and two alternative explanations. *Internet Research*, 32(7), 402–425. Advance online publication. DOI:10.1108/INTR-03-2022-0160

Jain, S., & Bhargav, S. (2010). *Human resource management*. Knowledge Book Distributors.

Janiesch, C., Zschech, P., & Heinrich, K. (2021). Machine learning and deep learning. *Electronic Markets*, 31(3), 685–695. DOI:10.1007/s12525-021-00475-2

Javaid, M., Haleem, A., Singh, R. P., & Suman, R. (2021). Substantial capabilities of robotics in enhancing industry 4.0 implementation. *Cognitive Robotics*, 1, 58–75. DOI:10.1016/j.cogr.2021.06.001

Jennings, P., & Greenberg, M. (2011). The Prosocial Classroom: Teacher Social and Emotional Competence about Student and Classroom Outcomes. Review of Educational Research -. *Review of Educational Research*, 79. Advance online publication. DOI:10.3102/0034654308325693

Jiang, F., Jiang, Y., Zhi, H., Dong, Y., Li, H., Ma, S., Wang, Y., Dong, Q., Shen, H., & Wang, Y. (2017). Artificial intelligence in healthcare: Past, present and future. *Stroke and Vascular Neurology*, 2(4), 230–243. DOI:10.1136/svn-2017-000101 PMID:29507784

Jigsaw (Google). (2021, February 8). *Google's Jigsaw Announces Toxicity-Reducing API, Perspective, is Processing 500M Requests Daily*. Retrieved April 30, 2024, from PR Newswire: https://www.prnewswire.com/news-releases/googles-jigsaw -announces-toxicity-reducing-api-perspective-is-processing-500m-requests-daily -301223600.html

Jobin, A., Ienca, M., & Vayena, E. (2019). The global landscape of AI ethics guidelines. *Nature Machine Intelligence*, 1(9), 389–399. DOI:10.1038/s42256-019-0088-2

John-Mathews, J. M. (2022). Some critical and ethical perspectives on the empirical turn of AI interpretability. *Technological Forecasting and Social Change*, 174, 121209. DOI:10.1016/j.techfore.2021.121209

John-Mathews, J. M., Cardon, D., & Balagué, C. (2022). From reality to world. A critical perspective on AI fairness. *Journal of Business Ethics*, 178(4), 945–959. DOI:10.1007/s10551-022-05055-8

Jones, M. L., Kaufman, E., & Edenberg, E. (2018). AI and the ethics of automating consent. *IEEE Security and Privacy*, 16(3), 64–72. DOI:10.1109/MSP.2018.2701155

Jovanovic, M., & Campbell, M. (2022). Generative artificial intelligence: Trends and prospects. *Computer (Long Beach Calif)*, 55(10), 107–112.

Jumper, J. e., Evans, R., Pritzel, A., Green, T., Figurnov, M., Ronneberger, O., Tunyasuvunakool, K., Bates, R., Žídek, A., Potapenko, A., Bridgland, A., Meyer, C., Kohl, S. A. A., Ballard, A. J., Cowie, A., Romera-Paredes, B., Nikolov, S., Jain, R., Adler, J., & Hassabis, D. (2021). Highly accurate protein structure prediction with AlphaFold. *Nature*, 596(7873), 583–589. DOI:10.1038/s41586-021-03819-2 PMID:34265844

Kabol, A. F. (2022, November 5). *The Use Of Artificial Intelligence In The Criminal Justice System (A Comparative Study)*. ResearchGate. https://www.researchgate.net/ publication/365027297_The_Use_Of_Artificial_Intelligence_In_The_Criminal _Justice_System_A_Comparative_Study

Kamoonpuri, S. Z., & Sengar, A. (2023). Hi, May AI help you? An analysis of the barriers impeding the implementation and use of artificial intelligence-enabled virtual assistants in retail. *Journal of Retailing and Consumer Services*, 72, 103258. Advance online publication. DOI:10.1016/j.jretconser.2023.103258

Kang, W., & Shao, B. (2023). The impact of voice assistants' intelligent attributes on consumer well-being: Findings from PLS-SEM and fsQCA. *Journal of Retailing and Consumer Services*, 70, 103130. Advance online publication. DOI:10.1016/j. jretconser.2022.103130

Kaplan, A., & Haenlein, M. (2020). Rulers of the world, unite! The challenges and opportunities of artificial intelligence. *Business Horizons*, 63(1), 37–50. DOI:10.1016/j.bushor.2019.09.003

Kazim, E., & Koshiyama, A. S. (2021). A high-level overview of AI ethics. *Patterns (New York, N.Y.)*, ●●●, 1–12. PMID:34553166

Keller, P., & Drake, A. (2021). Exclusivity and paternalism in the public governance of explainable AI. *Computer Law & Security Report*, 40, 105490. DOI:10.1016/j. clsr.2020.105490

Kenthapadi, K., Lakkaraju, H., & Rajani, N. (2023, August). Generative AI Meets Responsible Ai: Practical Challenges And Opportunities. In *Proceedings of the 29th ACM SIGKDD Conference on Knowledge Discovery and Data Mining* (pp. 5805-5806). DOI:10.1145/3580305.3599557

Kertysova, K. (2018). Artificial Intelligence and Disinformation: How AI Changes the Way Disinformation is Produced, Disseminated, and Can Be Countered. *Security and Human Rights, 29*(1–4).

Kettle, L., & Lee, Y. C. (2023). User Experiences of Well-Being Chatbots. *Human Factors*. Advance online publication. DOI:10.1177/00187208231162453 PMID:36916743

Khamparia, A., Gupta, D., Rodrigues, J. J., & de Albuquerque, V. H. C. (2021). Dcavn: Cervical Cancer Prediction and Classification Using Deep Convolutional And Variational Autoencoder Network. *Multimedia Tools and Applications*, 80(20), 30399–30415. DOI:10.1007/s11042-020-09607-w

Khan, S. (2023). Role of Generative AI for Developing Personalized Content Based Websites. *International Journal of Innovative Science and Research Technology*, 8, 1–5.

Khoo, B., Phan, R. C. W., & Lim, C. H. (2022). Deepfake attribution: On the source identification of artificially generated images. *Wiley Interdisciplinary Reviews. Data Mining and Knowledge Discovery*, 12(3), e1438. DOI:10.1002/widm.1438

Kietzmann, J., Lee, L., McCarthy, I. P., & Kietzmann, T. C. (2020). Deepfakes: Trick or treat? *Business Horizons*, 63(2), 135–146. DOI:10.1016/j.bushor.2019.11.006

King, D. (2024). *Legal & Humble AI: Addressing the Legal, Ethical, and Societal Dilemmas of Generative AI*. Ingene Publications.

Kingma, D. P., & Welling, M. (2013b). Auto-encoding variational bayes. arXiv preprint arXiv:1312.6114.

Kingma, D. P., & Welling, M. (2013, December 20). *Auto-Encoding Variational Bayes*. Retrieved from ARXIV: https://arxiv.org/abs/1312.6114

Kingma, D. P., & Welling, M. (2019). An introduction to variational auto-encoders. *Foundations and Trends in Machine Learning*, 12(4), 307–392. DOI:10.1561/2200000056

Knight, S., Shibani, A., & Vincent, N. (2024). Ethical AI governance: Mapping a research ecosystem. *AI and Ethics*. Advance online publication. DOI:10.1007/s43681-023-00416-z

Kodish, S. (2020). The Age of Artificial Intelligence and Robotics: Challenges and Issues. *Journal of Leadership, Accountability and Ethics*, 17(5), 42–52. https://articlearchives.co/index.php/JLAE/article/view/3954

Kolade, O., Owoseni, A., & Egbetokun, A. (2024). Is AI changing learning and assessment as we know it? Evidence from a ChatGPT experiment and a conceptual framework. *Heliyon*, 10(4), e25953. Advance online publication. DOI:10.1016/j.heliyon.2024.e25953 PMID:38379960

Koniakou, V. (2023). From the "rush to ethics" to the "race for governance" in artificial intelligence. *Information Systems Frontiers*, 25(1), 71–102. DOI:10.1007/s10796-022-10300-6

Konya, A., & Nematzadeh, P. (2024). Recent applications of AI to environmental disciplines: A review. *The Science of the Total Environment*, 906, 167705. DOI:10.1016/j.scitotenv.2023.167705 PMID:37820816

Kooli, C. (2023). Chatbots in Education and Research: A Critical Examination of Ethical Implications and Solutions. *Sustainability (Basel)*, 15(7), 5614. Advance online publication. DOI:10.3390/su15075614

Kozinets, R. V., & Gretzel, U. (2021). Commentary: Artificial intelligence: The marketer's dilemma. *Journal of Marketing*, 85(1), 156–159. DOI:10.1177/0022242920972933

KPMG. (January 1, 2024). CEO perspectives on generative artificial intelligence (AI) in their workplace worldwide in 2023 [Graph]. In *Statista*. Retrieved June 16, 2024, from https://www.statista.com/statistics/1445762/ceo-perspectives-on-generative-ai-worldwide/

Krizhevsky, A., Sutskever, I., & Hinton, G. E. (2012). *ImageNet Classification with Deep Convolutional Neural Networks*. Advances in Neural Information Processing Systems. Curran Associates.

Krkac, K. (2019). Corporate social irresponsibility: Humans vs artificial intelligence. *Social Responsibility Journal*, 15(6), 786–802. DOI:10.1108/SRJ-09-2018-0219

Kumaraguru Diderot, P., Sakthidasan Sankaran, K., Jawarneh, M., Pallathadka, H., Arias-Gonzáles, J. L., & Sanchez, D. T. (2024). Evaluation of Chabot Text Classification Using Machine Learning. In *Conversational Artificial Intelligence*. DOI:10.1002/9781394200801.ch13

Kumar, L., & Singh, D. K. (2023). A Comprehensive Survey on Generative Adversarial Networks Used For Synthesizing Multimedia Content. *Multimedia Tools and Applications*, 82(26), 40585–40624. DOI:10.1007/s11042-023-15138-x

Kumar, S., & Yadav, M. (2024). Occupation Stress and Leadership in the Hospitality Industry. In Valeri, M., & Sousa, B. (Eds.), *Human Relations Management in Tourism* (pp. 163–187). IGI Global., DOI:10.4018/979-8-3693-1322-0.ch008

Kumar, S., Yadav, M., & Kumar, D. (2024). Viability of Man and Machine as Co-Workers in the Hotel Industry. In Nozari, H. (Ed.), *Building Smart and Sustainable Businesses With Transformative Technologies* (pp. 189–204). IGI Global., DOI:10.4018/979-8-3693-0210-1.ch011

Kuziemski, M., & Misuraca, G. (2020). AI governance in the public sector: Three tales from the frontiers of automated decision-making in democratic settings. *Telecommunications Policy*, 44(6), 101976. DOI:10.1016/j.telpol.2020.101976 PMID:32313360

Labazanova, S. K., Aygumov, T. G., & Mursaliev, M. K. (2024). Issues with generative artificial intelligence tools. In ITM Web of Conferences (Vol. 59, p. 04007). EDP Sciences. DOI:10.1051/itmconf/20245904007

Lai, K. L. (2008). *An Introduction to Chinese Philosophy*. Cambridge University Press. DOI:10.1017/CBO9780511800832

Laitinen, A., & Sahlgren, O. (2021). AI Systems and Respect for Human Autonomy. *Frontiers in Artificial Intelligence*, 4, 705164. Advance online publication. DOI:10.3389/frai.2021.705164 PMID:34765969

Laux, J. (2024). Three pathways for standardisation and ethical disclosure by default under the European union artificial intelligence act. *Computer Law & Security Review: The International Journal of Technology Law and Practice*.

Laux, J., Wachter, S., & Mittelstadt, B. (2024). Three pathways for standardisation and ethical disclosure by default under the European Union Artificial Intelligence Act. *Computer Law & Security Report*, 53, 105957. DOI:10.1016/j.clsr.2024.105957

LeCun, Y., Bengio, Y., & Hinton, G. (2015). Deep Learning. *Nature*, 521(7553), 436–444. DOI:10.1038/nature14539 PMID:26017442

Lee, R. S. T. (2020). Artificial Intelligence in Daily Life. In *Artificial Intelligence in Daily Life*. DOI:10.1007/978-981-15-7695-9

Lee, S., Lee, M., & Lee, S. (2023). What If Artificial Intelligence Become Completely Ambient in Our Daily Lives? Exploring Future Human-AI Interaction through High Fidelity Illustrations. *International Journal of Human-Computer Interaction*, 39(7), 1371–1389. Advance online publication. DOI:10.1080/10447318.2022.2080155

Leiker, D., Gyllen, A. R., Eldesouky, I., & Cukurova, M. (2023, June). Generative AI for Learning: Investigating the Potential of Learning Videos with Synthetic Virtual Instructors. In *International conference on artificial intelligence in education* (pp. 523-529). Cham: Springer Nature Switzerland. DOI:10.1007/978-3-031-36336-8_81

Lewis, S. C., Guzman, A. L., & Schmidt, T. R. (2019). Automation, Communication: Rethinking Roles and Relationships of Humans and Machines in News. *Digital Journalism (Abingdon, England)*, ●●●, 409–427. DOI:10.1080/21670811.2019.1577147

Li, Y. (2022, August 26). Research on the Application of Artificial Intelligence in the Film. *2022 International Conference on Science and Technology Ethics and Human Future (STEHF 2022)*, 1-6. EDP Sciences. DOI:10.1051/shsconf/202214403002

Li, G., Li, N., & Sethi, S. P. (2021). Does CSR reduce idiosyncratic risk? Roles of operational efficiency and AI innovation. *Production and Operations Management*, 30(7), 2027–2045. DOI:10.1111/poms.13483

Lighthill, J. (1972, July). *Artificial Intelligence: A General Survey*. Retrieved from CHILTON: https://www.chilton-computing.org.uk/inf/literature/reports/lighthill _report/p001.htm

Lim, W. M., Gunasekara, A., Pallant, J. L., Pallant, J. I., & Pechenkina, E. (2023). Generative AI and the future of education: Ragnarök or reformation? A paradoxical perspective from management educators. *International Journal of Management Education*, 21(2), 100790. DOI:10.1016/j.ijme.2023.100790

Lin, L., Gupta, N., Zhang, Y., Ren, H., Liu, C. H., Ding, F., . . . Hu, S. (2024). Detecting multimedia generated by large ai models: A survey. arXiv preprint arXiv:2402.00045. DOI:10.36227/techrxiv.170723324.44685515/v1

Linkon, A. A., Shaima, M., Sarker, M. S. U., Nabi, N., Rana, M. N. U., Ghosh, S. K., & Chowdhury, F. R. (2024). Advancements and Applications of Generative Artificial Intelligence And Large Language Models on Business Management: A Comprehensive Review. *Journal of Computer Science and Technology Studies*, 6(1), 225–232. DOI:10.32996/jcsts.2024.6.1.26

Lipton, Z. C. (2018). The Mythos of Model Interpretability: In machine learning, the concept of interpretability is both important and slippery. . *Queue*, 31-57.

Liu, Y., Du, H., Niyato, D., Kang, J., Xiong, Z., Kim, D. I., & Jamalipour, A. (2024). Deep Generative Model and Its Applications In Efficient Wireless Network Management: A Tutorial And Case Study. *IEEE Wireless Communications*, 31(4), 199–207. DOI:10.1109/MWC.009.2300165

Li, W., Su, Z., Li, R., Zhang, K., & Wang, Y. (2020). Blockchain-based data security for artificial intelligence applications in 6G networks. *IEEE Network*, 34(6), 31–37. DOI:10.1109/MNET.021.1900629

Lucci, S., Musa, S. M., & Kopec, D. (2022). *Artificial Intelligence in the 21st Century* (3rd ed.). Mercury Learning and Information. DOI:10.1515/9781683922520

Luger, G. F. (1998). *Artificial Intelligence: Structures and Strategies for Complex Problem Solving, 5/e*. Pearson Education India.

Luo, B., Lau, R. Y. K., Li, C., & Si, Y. W. (2022). A critical review of state-of-the-art chatbot designs and applications. In *Wiley Interdisciplinary Reviews: Data Mining and Knowledge Discovery* (Vol. 12, Issue 1). DOI:10.1002/widm.1434

Madaio, M. A., Stark, L., Wortman Vaughan, J., & Wallach, H. (2020). Proceedings of the 2020 CHI conference on human factors in computing systems. In Co-designing checklists to understand organizational challenges and opportunities around fairness in AI (pp. 1–14). ACM Digital Library.

Maddula, S. S. (2018). The Impact of AI and Reciprocal Symmetry on Organizational Culture and Leadership in the Digital Economy. *Engineering International*, 6(2), 201–210. DOI:10.18034/ei.v6i2.703

Maedche, A., Legner, C., Benlian, A., Berger, B., Gimpel, H., Hess, T., Hinz, O., Morana, S., & Söllner, M. (2019). AI-Based Digital Assistants: Opportunities, Threats, and Research Perspectives. *Business & Information Systems Engineering*, 61(4), 535–544. Advance online publication. DOI:10.1007/s12599-019-00600-8

Magas, M., & Kiritsis, D. (2022). Industry commons: An ecosystem approach to horizontal enablers for sustainable cross-domain industrial innovation. *International Journal of Production Research*, 60(2), 479–492. DOI:10.1080/00207543.2021.1989514

Magazine, S. (2018). Christie's is first to sell art made by artificial intelligence, but what does that mean? Available at: https://www.smithsonianmag.com/smart-news/christies-first-sell-art-made-artificial-intelligence-what-does-mean-180970642/ (Accessed: 25 May 2024).

Magistretti, S., Dell'Era, C., & Petruzzelli, A. M. (2019). How intelligent is Watson? Enabling digital transformation through artificial intelligence. *Business Horizons*, 62(6), 819–829. DOI:10.1016/j.bushor.2019.08.004

Makanadar, A., Shahane, S., & Patil, U. (2024). Tracing Algorithmic and AI-biased Data. *Economic and Political Weekly*, 59(7).

Makridakis, S. (2017). The forthcoming Artificial Intelligence (AI) revolution: Its impact on society and firms. *Futures*, 90, 46–60. DOI:10.1016/j.futures.2017.03.006

Mantymaki, M., Minkkinen, M., Birkstedt, T., & Viljanen, M. (2022). Defining organizational AI governance. *AI and Ethics*, 2(4), 603–609. DOI:10.1007/s43681-022-00143-x

Manyika, J., Silberg, J., & Presten, B. (2019). What do we do about the biases in AI? In *Harvard Business Review*.

Mariani, M., & Dwivedi, Y. K. (2024). Generative artificial intelligence in innovation management: A preview of future research developments. *Journal of Business Research*, 175, 114542. DOI:10.1016/j.jbusres.2024.114542

Marks, M., & Haupt, C. E. (2023). AI chatbots, health privacy, and challenges to HIPAA compliance. *Journal of the American Medical Association*, 2023(329), 1349–1350. DOI:10.1001/jama.2023.9458 PMID:37410450

Matytsin, D. E., Dzedik, V. A., Makeeva, G. A., & Boldyreva, S. B. (2023). "Smart" outsourcing in support of the humanization of entrepreneurship in the artificial intelligence economy. *Humanities & Social Sciences Communications*, 10(1), 1–8. PMID:36644399

McBride, R., Dastan, A., & Mehrabinia, P. (2022). How AI affects the future relationship between corporate governance and financial markets: A note on impact capitalism. *Managerial Finance*, 48(8), 1240–1249. DOI:10.1108/MF-12-2021-0586

McCarthy, J., Minsky, M. L., Rochester, N., & Shannon, C. E. (2006). A Proposal for the Dartmouth Summer Research Project on Artificial Intelligence, August 31, 1955. *AI Magazine*, •••, 12–14.

McCartney, G., & McCartney, A. (2020). Rise of the machines: Towards a conceptual service-robot research framework for the hospitality and tourism industry. *International Journal of Contemporary Hospitality Management*, 32(12), 3835–3851. DOI:10.1108/IJCHM-05-2020-0450

McCormack, J. (2023). Evolutionary Machine Learning in the Arts. In *Handbook of Evolutionary Machine Learning* (pp. 739–760). Springer Nature Singapore.

McCormack, J., & D'Inverno, M. (2012). *Computers and Creativity*. Springer-Verlag. DOI:10.1007/978-3-642-31727-9

McCulloch, W. S., & Pitts, W. (1943). A Logical Calculus of Ideas Immanent in Nervous Activity. *The Bulletin of Mathematical Biophysics*, 5(4), 115–133. DOI:10.1007/BF02478259

McKinney, S. M., Sieniek, M., Godbole, V., Godwin, J., Antropova, N., Ashrafian, H., Back, T., Chesus, M., Corrado, G. S., Darzi, A., Etemadi, M., Garcia-Vicente, F., Gilbert, F. J., Halling-Brown, M., Hassabis, D., Jansen, S., Karthikesalingam, A., Kelly, C. J., King, D., & Shetty, S. (2020). International evaluation of an AI system for breast cancer screening. *Nature*, 577(7788), 89–94. DOI:10.1038/s41586-019-1799-6 PMID:31894144

McKinsey & Company. (May 11, 2023). Extent of which commercial leaders feel their organizations are using machine learning (ML) or generative artificial intelligence (AI) in 2023 [Graph]. In *Statista*. Retrieved June 16, 2024, from https://www .statista.com/statistics/1411472/commerical-leader-use-of-gen-ai/

McKinsey & Company. (May 11, 2023). Extent of which commericial leaders feel their organizations should be using machine learning (ML) or generative artificial intelligence (AI) in 2023 [Graph]. In *Statista*. Retrieved June 16, 2024, from https:// www.statista.com/statistics/1411478/desired-commerical-leaders-use-of-generative -ai/

McKinsey Global Institute. (2017, December 1). *Jobs Lost, Jobs Gained: Workforce Transitions in a Time of Automation.* Retrieved from McKinsey & Company: https://www.mckinsey.com/~/media/mckinsey/industries/public%20and%20social %20sector/our%20insights/what%20the%20future%20of%20work%20will%20mean %20for%20jobs%20skills%20and%20wages/mgi-jobs-lost-jobs-gained-executive -summary-december-6-2017.pdf

McKinsey Global Institute. (2024, July 16). *Jobs lost, jobs gained: What the future of work will mean for jobs, skills, and wages.* Retrieved from McKinsey & Company: https://www.mckinsey.com/featured-insights/future-of-work/jobs-lost-jobs-gained -what-the-future-of-work-will-mean-for-jobs-skills-and-wages

McLennan, S. e., Fiske, A., Celi, L. A., Müller, R., Harder, J., Ritt, K., Haddadin, S., & Buyx, A. (2020). An embedded ethics approach for AI development. *Nature Machine Intelligence*, 2(9), 488–490. DOI:10.1038/s42256-020-0214-1

Mennella, C., Maniscalco, U., De Pietro, G., & Esposito, M. (2024). Ethical and regulatory challenges of AI technologies in healthcare:A narrative review. *Heliyon*, 10(4), e26297. DOI:10.1016/j.heliyon.2024.e26297 PMID:38384518

Mercader, V., Galván-Vela, E., Ravina-Ripoll, R., & Popescu, C. R. G. (2021). A focus on ethical value under the vision of leadership, teamwork, effective communication and productivity. *Journal of Risk and Financial Management*, 14(11), 522. DOI:10.3390/jrfm14110522

Mervosh, S. (2019, May 24). *Distorted Videos of Nancy Pelosi Spread on Facebook and Twitter, Helped by Trump.* Retrieved April 30, 2024, from The New York Times: https://www.nytimes.com/2019/05/24/us/politics/pelosi-doctored-video.html

Metz, R. (2021, August 6). *How a deepfake Tom Cruise on TikTok turned into a very real AI company.* Retrieved April 30, 2024, from CNN Business: https://edition.cnn .com/2021/08/06/tech/tom-cruise-deepfake-tiktok-company/index.html

Metzler, D., Tay, Y., Bahri, D., & Najork, M. 2021. Rethinking Search: Making Domain Experts Out of Dilettantes. In ACM SIGIR Forum, Vol. 55.

Miah, J., Cao, D. M., Sayed, M. A., & Haque, M. S. (2023). Generative AI Model for Artistic Style Transfer Using Convolutional Neural Networks. *arXiv preprint arXiv:2310.18237.*

Michel-Villarreal, R., Vilalta-Perdomo, E., Salinas-Navarro, D. E., Thierry-Aguilera, R., & Gerardou, F. S. (2023). Challenges and Opportunities of Generative AI for Higher Education as Explained by ChatGPT. *Education Sciences*, 13(9), 856. DOI:10.3390/educsci13090856

Microsoft. (2023). Responsible and trusted AI. Redmont https://learn.microsoft .com/en-us/azure/cloud-adoption-framework/innovate/best-practices /trusted-ai

Milton, J., & Al-Busaidi, A. (2023). New role of leadership in AI era: Educational sector. *In SHS Web of Conferences* (Vol. 156, p. 09005). EDP Sciences.

Minkkinen, M., Niukkanen, A., & Mäntymäki, M. (2022). What about investors? ESG analyses as tools for ethics-based AI auditing. *AI & Society*. Advance online publication. DOI:10.1007/s00146-022-01415-0

Minkkinen, M., Zimmer, M. P., & Mäntymäki, M. (2023). Co-shaping an ecosystem for responsible AI: Five types of expectation work in response to a technological frame. *Information Systems Frontiers*, 25(1), 103–121. DOI:10.1007/s10796-022-10269-2

Miranda, E. R. (2021). *Handbook of Artificial Intelligence for Music-Foundations, Advanced Approaches, and Developments for Creativity.* Springer.

Mitchell, M. (2021). Why AI is harder than we think. *arXiv preprint arXiv:2104.12871.* DOI:10.1145/3449639.3465421

Mittelstadt, B. (2019). Principles alone cannot guarantee ethical AI. *Nature Machine Intelligence*, 1(11), 501–507. DOI:10.1038/s42256-019-0114-4

Mittelstadt, B. D., Allo, P., Taddeo, M., Wachter, S., & Floridi, L. (2016). The ethics of algorithms: Mapping the debate. *Big Data & Society*, 3(2), 1–21. DOI:10.1177/2053951716679679

Mogaji, E., & Jain, V. (2024). How generative AI is (will) change consumer behaviour: Postulating the potential impact and implications for research, practice, and policy. *Journal of Consumer Behaviour*, cb.2345. DOI:10.1002/cb.2345

Mohav, Y. (2023). Ethical Leadership in AI-driven Organizations: Balancing Innovation and Responsibility. *International Journal of Advanced Engineering Technologies and Innovations*, 1(04), 204–220.

Mondal, S., Das, S., & Vrana, V. G. (2023). How to bell the cat? A theoretical review of generative artificial intelligence towards digital disruption in all walks of life. *Technologies*, 11(2), 44. DOI:10.3390/technologies11020044

Moor, J. (2006). The Nature, Importance, and Difficulty of Machine Ethics. *IEEE Intelligent Systems*, 21(4), 18–21. DOI:10.1109/MIS.2006.80

Moor, J. H. (1985). What is Computer Ethics? *Metaphilosophy*, 16(4), 266–275. DOI:10.1111/j.1467-9973.1985.tb00173.x

Morandín-Ahuerma, F. (2023, September 20). Montreal Declaration for Responsible AI: 10 Principles and 59 Recommendations. Retrieved from *OSF Preprints*: https://osf.io/preprints/osf/sj2z5DOI:10.31219/osf.io/sj2z5

Moulaei, K., Yadegari, A., Baharestani, M., Farzanbakhsh, S., Sabet, B., & Afrash, M. R. (2024). Generative artificial intelligence in healthcare: A Scoping Review on Benefits, Challenges And Applications. *International Journal of Medical Informatics*, 188, 105474. DOI:10.1016/j.ijmedinf.2024.105474 PMID:38733640

Mullins, M., Holland, C. P., & Cunneen, M. (2021). Creating ethics guidelines for artificial intelligence and big data analytics customers: The case of the consumer European insurance market. *Patterns (New York, N.Y.)*, 2(10), 100362. DOI:10.1016/j.patter.2021.100362 PMID:34693379

Multidisciplinary perspectives on opportunities, challenges and implications of generative conversational AI for research, practice, and policy, International Journal of Information Management, Volume 71, 2023, 102642, ISSN 0268-4012

Mungale, S. G., Mungale, N. G., Shaikh, M. S., Mungale, S. G., Wazalwar, S. S., Wanjari, M. M., & Jichkar, R. A. (2024). Safeguard Wrist: Empowering Women's Safety. In Ponnusamy, S., Bora, V., Daigavane, P., & Wazalwar, S. (Eds.), *Wearable Devices, Surveillance Systems, and AI for Women's Wellbeing* (pp. 192–205). IGI Global., DOI:10.4018/979-8-3693-3406-5.ch012

Murayama, M. (2021). Society 5.0 transformation: Digital strategy in Japan. In Management Education and Automation (pp. 7-29). Routledge.

Mustak, M., Salminen, J., Plé, L., & Wirtz, J. (2021). Artificial intelligence in marketing: Topic modeling, scientometric analysis, and research agenda. *Journal of Business Research*, 124, 389–404. DOI:10.1016/j.jbusres.2020.10.044

Myers, D., Mohawesh, R., Chellaboina, V. I., Sathvik, A. L., Venkatesh, P., Ho, Y. H., Henshaw, H., Alhawawreh, M., Berdik, D., & Jararweh, Y. (2024). Foundation and large language models: Fundamentals, challenges, opportunities, and social impacts. *Cluster Computing*, 27(1), 1–26. DOI:10.1007/s10586-023-04203-7

Nagarathinam, D. (2020). Leadership styles, qualities, and characteristics of the world great leaders with constitutional and judicial flavors. *Issue 4 Int'l JL Mgmt. &. Human.*, 3, 1428.

Narwani, K., Lin, H., Pirbhulal, S., & Hassan, M. (2022). *Towards AI-enabled approach for Urdu text recognition: A legacy for Urdu image apprehension.* IEEE., DOI:10.1109/ACCESS.2022.3203426

Nation, S. (2019). National Artificial Intelligence Strategy: Advancing our smart nation journey. Smart Nation and Digital Government Office https://www.smartnation.gov.sg/initiatives/artificial-intelligence/

Nazareno, L., & Schiff, D. S. (2021). The impact of automation and artificial intelligence on worker well-being. *Technology in Society*, 67, 101679. Advance online publication. DOI:10.1016/j.techsoc.2021.101679

Neubert, M. J., Kacmar, K. M., Carlson, D. S., Chonko, L. B., & Roberts, J. A. (2008). Regulatory focus as a mediator of the influence of initiating structure and servant leadership on employee behavior. *The Journal of Applied Psychology*, 93(6), 1220–1233. DOI:10.1037/a0012695 PMID:19025244

Ng, K. K., Chen, C. H., Lee, C. K., Jiao, J. R., & Yang, Z. X. (2021). A systematic literature review on intelligent automation: Aligning concepts from theory, practice, and future perspectives. *Advanced Engineering Informatics*, 47, 101246. DOI:10.1016/j.aei.2021.101246

Nicolae, M., & Nicolae, E. E. (2018). Leadership in Higher Education–coping with AI and the turbulence of our times. *In Proceedings of the International Conference on Business Excellence* (Vol. 12, No. 1, pp. 683-694). DOI:10.2478/picbe-2018-0061

Nikolenko, S. I., Koltcov, S., & Koltsova, O. (2017). Topic modelling for qualitative studies. *Journal of Information Science*, 43(1), 88–102. DOI:10.1177/0165551515617393

Nishant, R., Kennedy, M., & Corbett, J. (2020). Artificial intelligence for sustainability: Challenges, opportunities, and a research agenda. *International Journal of Information Management*, 53, 102104. Advance online publication. DOI:10.1016/j.ijinfomgt.2020.102104

Noble, S. U. (2018). *Algorithms of Oppression: How Search Engines Reinforce Racism*. NYU Press. DOI:10.18574/nyu/9781479833641.001.0001

O'Neil, C. (2016). *Weapons of Math Destruction: How Big Data Increases Inequality and Threatens Democracy*. Crown Publishing Group.

O'Regan, G. (2008). Artificial Intelligence and Expert Systems. In *A Brief History of Computing* (pp. 149–177). Springer London. DOI:10.1007/978-1-84800-084-1_5

Obermeyer, Z., Powers, B., Vogeli, C., & Mullainathan, S. (2019). Dissecting racial bias in an algorithm used to manage the health of populations. *Science*, 366(6464), 447–453. DOI:10.1126/science.aax2342 PMID:31649194

OECD. (2019). Recommendation of the council on artificial intelligence. Organization for Economic Cooperation and Development https://legalinstruments. oecd .org/en/instruments/oecd-legal-0449

Ojha, Nitish & Pandita, Archana & Ramkumar, J. (2024). Cyber Security Challenges and Dark Side of AI: Review and Current Status. .DOI:10.4018/979-8-3693-0724-3.ch007

Okonkwo, C. W., & Ade-Ibijola, A. (2021). Chatbots applications in education: A systematic review. In *Computers and Education* (Vol. 2). Artificial Intelligence., DOI:10.1016/j.caeai.2021.100033

Oliveira, J. D. S., Espindola, D. B., Barwaldt, R., Ribeiro, L. M. I., & Pias, M. (2019). IBM Watson Application as FAQ Assistant about Moodle. *Proceedings - Frontiers in Education Conference, FIE, 2019-October*. DOI:10.1109/FIE43999.2019.9028667

Oniani, D., Hilsman, J., Peng, Y., Poropatich, R. K., Pamplin, C. O. L., Legault, L. T. C., & Wang, Y. (2023). From Military to Healthcare: Adopting and Expanding Ethical Principles for Generative Artificial Intelligence. arXiv preprint arXiv:2308.02448.

Oniani, D., Hilsman, J., Peng, Y., Poropatich, R. K., Pamplin, J. C., Legault, G. L., & Wang, Y. (2023). Adopting and expanding ethical principles for generative artificial intelligence from military to healthcare. *NPJ Digital Medicine*, 6(1), 225. DOI:10.1038/s41746-023-00965-x PMID:38042910

Ooi, K. B., Tan, G. W. H., Al-Emran, M., Al-Sharafi, M. A., Capatina, A., Chakraborty, A., Dwivedi, Y. K., Huang, T.-L., Kar, A. K., Lee, V.-H., Loh, X.-M., Micu, A., Mikalef, P., Mogaji, E., Pandey, N., Raman, R., Rana, N. P., Sarker, P., Sharma, A., & Wong, L. W. (2023). The potential of generative artificial intelligence across disciplines: Perspectives and future directions. *Journal of Computer Information Systems*, 1–32. DOI:10.1080/08874417.2023.2261010

Oremus, W. (2014, March 17). *The First News Report on the L.A. Earthquake Was Written by a Robot*. Retrieved April 30, 2024, from Slate.com: https://slate.com/technology/2014/03/quakebot-los-angeles-times-robot-journalist-writes-article-on-la-earthquake.html

Ossa, L. e. (2024). Integrating ethics in AI development: A qualitative study. *BMC Medical Ethics*. PMID:38262986

Owczarczuk, M. (2023). Ethical and regulatory challenges amid artificial intelligence development. *Ekonomia i Prawo. Economics and Law*, 295-310.

Pai, V., & Chandra, S. (2022). Exploring factors influencing organizational adoption of artificial intelligence (AI) in corporate social responsibility (CSR) initiatives. *Pacific Asia Journal of the Association for Information Systems*, 14(5), 4. DOI:10.17705/1pais.14504

Pal, S., Rabehaja, T., Hitchens, M., & Hill, A. (2019). On the design of a flexible delegation model for the Internet of Things using blockchain. *IEEE Transactions on Industrial Informatics*, 16(5), 3521–3530. DOI:10.1109/TII.2019.2925898

Pant, A. e. (2023). Ethics in the Age of AI: An Analysis of AI Practitioners' Awareness and Challenges. *Journal of the Association for Computing Machinery*.

Pan, Z., Yu, W., Yi, X., Khan, A., Yuan, F., & Zheng, Y. (2019). Recent progress on generative adversarial networks (GANs): A survey. *IEEE Access : Practical Innovations, Open Solutions*, 7, 36322–36333. DOI:10.1109/ACCESS.2019.2905015

Papagiannidis, E., Enholm, I. M., Dremel, C., Mikalef, P., & Krogstie, J. (2023). Toward AI governance: Identifying best practices and potential barriers and outcomes. *Information Systems Frontiers*, 25(1), 123–141. DOI:10.1007/s10796-022-10251-y PMID:35464171

Parikh, P. (2019). *AI Film Aesthetics: A Construction of a New Media Identity for AI Films*. Chapman University. doi., https://digitalcommons.chapman.edu/film_studies_theses/8/

Paris, B., & Donovan, J. (2019, September 18). *Deepfakes and Cheapfakes- the manipulation of audio and video evidence*. Retrieved from Data & Society's Media Manipulation research: https://datasociety.net/library/deepfakes-and-cheap-fakes/

Partnership on AI. (2016). https://partnershiponai.org

Paul, R. K., & Sarkar, B. (2023) Generative AI and Ethical Considerations For Trustworthy Ai Implementation. *Journal ID, 2157*, 0178.

Peiser, J. (2019, February 5). *The Rise of the Robot Reporter*. Retrieved April 30, 2024, from The New York Times: https://www.nytimes.com/2019/02/05/business/media/artificial-intelligence-journalism-robots.html

Perkins, J. (2022). Immersive metaverse experiences in decentralized 3d virtual clinical spaces: Artificial intelligence-driven diagnostic algorithms, wearable internet of medical things sensor devices, and healthcare modeling and simulation tools. *American Journal of Medical Research (New York, N.Y.)*, 9(2), 89–104. DOI:10.22381/ajmr9220226

Peters, D., Vold, K., Robinson, D., & Calvo, R. A. (2020). Responsible AI—Two frameworks for ethical design practice. *IEEE Transactions on Technology and Society*, 1(1), 34–47. DOI:10.1109/TTS.2020.2974991

Pham, K. T., Nabizadeh, A., & Selek, S. (2022). Artificial Intelligence and Chatbots in Psychiatry. In *Psychiatric Quarterly* (Vol. 93, Issue 1). DOI:10.1007/s11126-022-09973-8

Poggi, M., Tosi, F., Batsos, K., Mordohai, P., & Mattoccia, S. (2021). On the synergies between machine learning and binocular stereo for depth estimation from images: A survey. *IEEE Transactions on Pattern Analysis and Machine Intelligence*, 44(9), 5314–5334. DOI:10.1109/TPAMI.2021.3070917 PMID:33819150

Popescu, G. H. (2022). K. Valaskova, J. Horak, Augmented reality shopping experiences, retail business analytics, and machine vision algorithms in the virtual economy of the metaverse, J. Self-Governance Manage. *Econ.*, 10(2), 67–81.

Po, R., Yifan, W., Golyanik, V., Aberman, K., Barron, J. T., Bermano, A., Chan, E., Dekel, T., Holynski, A., Kanazawa, A., Liu, C. K., Liu, L., Mildenhall, B., Nießner, M., Ommer, B., Theobalt, C., Wonka, P., & Wetzstein, G. (2024, May). State of the art on diffusion models for visual computing. *Computer Graphics Forum*, 43(2), e15063. DOI:10.1111/cgf.15063

Posner, T., & Fei-Fei, L. (2020). AI will change the world, so it's time to change AI. *Nature*, 588(7837), S118–S118. DOI:10.1038/d41586-020-03412-z

Press, G. (2016, December 30). *A Very Short History Of Artificial Intelligence (AI)*. Retrieved May 4, 2024, from Forbes: https://www.forbes.com/sites/gilpress/2016/12/30/a-very-short-history-of-artificial-intelligence-ai/?sh=68b5d5776fba

Rąb-Kettler, K., & Lehnervp, B. (2019). Recruitment in the times of machine learning. *Management Systems in Production Engineering*, 27(2), 105–109. DOI:10.1515/mspe-2019-0018

Rae, S. (2018). *Moral choices: An introduction to ethics*. Zondervan Academic.

Raisch, S., & Krakowski, S. (2021). Artificial intelligence and management: The automation–augmentation paradox. *Academy of Management Review*, 46(1), 192–210. DOI:10.5465/amr.2018.0072

Rajaraman, V. (2023). From ELIZA to ChatGPT: History of Human-Computer Conversation. *Resonance*, 28(6), 889–905. Advance online publication. DOI:10.1007/s12045-023-1620-6

Raji, I. D., Smart, A., White, R. N., Mitchell, M., Gebru, T., Hutchinson, B., Smith-Loud, J., Theron, D., & Barnes, P. (2020). In proceedings of the 2020 conference on fairness, accountability, and transparency. In *Closing the AI accountability gap: Defining an end-to-end framework for internal algorithmic auditing* (pp. 33–44). ACM Digital Library.

Raman, R., Meenakshi, R., Ramya, R., Jayaprakash, S., & Srinivasan, C. (2024). IoT-Based Magnetic Field Strength Monitoring for Industrial Applications. *2nd International Conference on Smart Technologies for Smart Nation, SmartTechCon 2023*, pp. 132–136.

Ramesh, A. e. (2022, April 13). *Hierarchical Text-Conditional Image Generation with CLIP Latents.* Retrieved from ARXIV: https://arxiv.org/abs/2204.06125

Ramya, R., & Ramamoorthy, S. (2022). Analysis of machine learning algorithms for efficient cloud and edge computing in the IoT, Challenges and Risks Involved in Deploying 6G and NextGen Networks, pp. 72–90.

Ramya, R. (2024). *Analysis and Applications Finding of Wireless Sensors and IoT Devices With Artificial Intelligence/Machine Learning. AIoT and Smart Sensing Technologies for Smart Devices.* IGI Global., DOI:10.4018/979-8-3693-0786-1.ch005

Ramya, R., & Ramamoorthy, S. (2022). Development of a framework for adaptive productivity management for edge computing based IoT applications. *AIP Conference Proceedings*, 2519, 030068. DOI:10.1063/5.0111710

Ramya, R., & Ramamoorthy, S. (2022). Survey on Edge Intelligence in IoT-Based Computing Platform. *Survey on Edge Intelligence in IoT-Based Computing Platform, Lecture Notes in Networks and Systems, Springer*, 356, 549–556. DOI:10.1007/978-981-16-7952-0_52

Ramya, R., & Ramamoorthy, S. (2023). Lightweight Unified Collaborated Relinquish Edge Intelligent Gateway Architecture with Joint Optimization. *IEEE Access : Practical Innovations, Open Solutions*, 11, 90396–90409. DOI:10.1109/ACCESS.2023.3307808

Ramya, R., & Ramamoorthy, S. (2024). Hybrid Fog-Edge-IoT Architecture for Real-time Data Monitoring. *International Journal of Intelligent Engineering and Systems*, 17(1), 2024. DOI:10.22266/ijies2024.0229.22

Ramya, R., & Ramamoorthy, S. (2024). QoS in multimedia application for IoT devices through edge intelligence. *Multimedia Tools and Applications*, 83(3), 9227–9250. DOI:10.1007/s11042-023-15941-6

Rane, N. (2023). ChatGPT and Similar Generative Artificial Intelligence (AI) for Smart Industry: role, challenges and opportunities for industry 4.0, industry 5.0 and society 5.0. Challenges and Opportunities for Industry, 4.

Rane, N. (2023). Role and challenges of ChatGPT and similar generative artificial intelligence in arts and humanities. Available at *SSRN* 4603208. DOI:10.2139/ssrn.4603208

Ray, P. P. (2023). *ChatGPT: a comprehensive review on background, applications, key challenges, bias, ethics, limitations and future scope, Internet Things Cyber-Phys.* Syst.

Reddy, S., Fox, J., & Purohit, M. P. (2021). Artificial intelligence-enabled healthcare delivery. *Journal of the Royal Society of Medicine*, ●●●, 372–376. PMID:30507284

Renieris, E. M., Kiron, D., & Mills, S. (2022). Should organizations link responsible AI and corporate social responsibility?It's Complicated. MIT Sloan https://sloanreview.mit.edu/article/should-organizations-link-responsible-ai-and-corporate-social-responsibility-its-complicated/

Renieris, E. M., Kiron, D., & Mills, S. (2022). Should organizations link responsible AI and corporate social responsibility?It's Complicated. MIT Sloan https://sloanreview.mit.edu/article/should-organizations-link-responsible-ai-and-corporate-social-responsibility-its-complicated/

Ribeiro, J., Lima, R., Eckhardt, T., & Paiva, S. (2021). Robotic process automation and artificial intelligence in industry 4.0–a literature review. *Procedia Computer Science*, 181, 51–58. DOI:10.1016/j.procs.2021.01.104

Rikap, C. (2023). The expansionary strategies of intellectual monopolies: Google and the digitalization of healthcare. *Economy and Society*, 52(1), 110–136. DOI:10.1080/03085147.2022.2131271

Robbins, S. P., Judge, T. A., & Vohra, N. (2013). *Organizational behaviour* (15th ed.). Pearson India.

Roberts, H., Cowls, J., Morley, J., Taddeo, M., Wang, V., & Floridi, L. (2021). The Chinese approach to artificial intelligence: an analysis of policy, ethics, and regulation. Ethics, governance, and policies in artificial intelligence, 47-79.

Roberts, H., Cowls, J., Morley, J., Taddeo, M., Wang, V., & Floridi, L. (2021). The Chinese Approach to Artificial Intelligence: An Analysis of Policy, Ethics, and Regulation. In Floridi, L. (Ed.), *Ethics, Governance, and Policies in Artificial Intelligence* (pp. 47–79). Springer. DOI:10.1007/978-3-030-81907-1_5

Robertson, E. G. (2022). *Video Games and Politics: An exploratory analysis of how narrative frames in video games can influence political perception.* Auckland, New Zealand: The University of Auckland. Retrieved April 30, 2024, from https://researchspace.auckland.ac.nz/bitstream/handle/2292/60892/Robertson-2022-thesis.pdf?sequence=4&isAllowed=y

Rodriguez-Barroso, N., Stipcich, G., Jiménez-Lopez, D., Ruiz-Millán, J. A., Martínez-Cámara, E., González-Seco, G., Luzón, M. V., Veganzones, M. A., & Herrera, F. (2020). Federated learning and differential privacy: Software tools analysis, the Sherpa. Ai FL framework and methodological guidelines for preserving data privacy. *Information Fusion*, 64, 270–292. DOI:10.1016/j.inffus.2020.07.009

Romao, M., Costa, J., & Costa, C. J. (2019). Robotic process automation: A case study in the banking industry. In 2019 14th Iberian conference on information systems and technologies (CISTI) (pp. 1–6). IEEE. DOI:10.23919/CISTI.2019.8760733

Rosário, A. T. (2024). Generative ai and generative pre-trained transformer applications: Challenges and opportunities. Making Art With Generative AI Tools, 45-71.

Rosenblatt, F. (1958). The Perceptron: A Probabilistic Model for Information Storage and Organization in the Brain. *Psychological Review*, 65(6), 386–408. DOI:10.1037/h0042519 PMID:13602029

Ross, C., & Swetlitz, I. (2017, September 5). *IBM pitched its Watson supercomputer as a revolution in cancer care. It's nowhere close.* Retrieved from STAT News: https://www.statnews.com/2017/09/05/watson-ibm-cancer/

Rost, J. C. (1995). Leadership: A discussion about ethics. *Business Ethics Quarterly*, 5(1), 129–142. DOI:10.2307/3857276

Rowe, W. G., & Guerrero, L. (2010). *Cases in leadership (2nd* ed.). New Delhi. Sage Publication India Pvt. Ltd.

Rumelhart, D. E., Hinton, G. E., & Williams, R. J. (1986). Learning representations by back-propagating errors. *Nature*, 323(6088), 533–536. DOI:10.1038/323533a0

Russell, S. J., & Norvig, P. (2016). *Artificial intelligence: A modern approach*. Pearson Education Limited.

Russell, S., & Norvig, P. (2020). *Artificial Intelligence: A Modern Approach*. Pearson.

Ryan, M., & Stahl, B. C. (2021). Artificial intelligence ethics guidelines for developers and users: clarifying their content and normative implications. *Journal of Information, Communication and Ethics in Society*, 61-86.

S. Murugesan and A. K. Cherukuri, "The Rise of Generative Artificial Intelligence and Its Impact on Education: The Promises and Perils," in Computer, vol. 56, no. 5, pp. 116-121, May 2023, Doi: .DOI:10.1109/MC.2023.3253292

Sachan, S., Yang, J. B., Xu, D. L., Benavides, D. E., & Li, Y. (2020). An explainable AI decision-support-system to automate loan underwriting. *Expert Systems with Applications*, 144, 113100. DOI:10.1016/j.eswa.2019.113100

Samant, R. M., Bachute, M. R., Gite, S., & Kotecha, K. (2022). Framework for deep learning-based language models using multi-task learning in natural language understand-ing: A systematic literature review and future directions. *IEEE Access : Practical Innovations, Open Solutions*, 10, 17078–17097. DOI:10.1109/ACCESS.2022.3149798

Sambasivan, N. (2021). Re-imagining Algorithmic Fairness in India and Beyond. *FAccT '21: Proceedings of the 2021 ACM Conference on Fairness, Accountability, and Transparency* (pp. 315-328). New York City: ACM. DOI:10.1145/3442188.3445896

Sankararaman, K. A., Wang, S., & Fang, H. 2022. BayesFormer: Transformer with Uncertainty Estimation. arXiv preprint arXiv:2206.00826 (2022).

Santos, B. S. (2002). *Towards a New Legal Common Sense: Law, Globalization, and Emancipation*. Butterworths.

Santosh, K. C., & Wall, C. (2022, January 1). *AI and Ethical Issues*. SpringerBriefs in Applied Sciences and Technology. DOI:10.1007/978-981-19-3935-8_1

Sarah, A. Chauncey, H. Patricia McKenna, A framework and exemplars for ethical and responsible use of AI Chatbot technology to support teaching and learning, Computers and Education: Artificial Intelligence, Volume 5, 2023, 100182, ISSN 2666-920X

Sargeant, H. (2023). Algorithmic decision-making in financial services: Economic and normative outcomes in consumer credit. *AI and Ethics*, 3(4), 1295–1311. DOI:10.1007/s43681-022-00236-7

Satra, H. S. (2021). A framework for evaluating and disclosing the ESG related impacts of AI with the SDGs. *Sustainability (Basel)*, 13(15), 8503. DOI:10.3390/su13158503

Saurabh, K., Arora, R., Rani, N., Mishra, D., & Ramkumar, M. (2022). AI led ethical digital transformation: Framework, research and managerial implications. Journal of Information. *Communication and Ethics in Society*, 20(2), 229–256. DOI:10.1108/JICES-02-2021-0020

Sautoy, M. d. (2019). *The Creativity Code-Art and Innovation in the age of AI*. Harvard University Press. DOI:10.2307/j.ctv2sp3dpd

Saxena, T. (2023, November 26). *The deepfake gender paradox in AI's moral maze*. Retrieved April 30, 2024, from Deccan Herald: https://www.deccanherald.com/specials/the-deepfake-gender-paradox-in-ai-s-moral-maze-2783639

Sayeed, M. S., Mohan, V., & Muthu, K. S. (2023). BERT: A Review of Applications in Sentiment Analysis. In *HighTech and Innovation Journal* (Vol. 4, Issue 2). DOI:10.28991/HIJ-2023-04-02-015

Schiuma, G., Santarsiero, F., Carlucci, D., & Jarrar, Y. (2024). Transformative leadership competencies for organizational digital transformation. *Business Horizons*, 67(4), 425–437. Advance online publication. DOI:10.1016/j.bushor.2024.04.004

Schmidhuber, J. (2015). Deep Learning in Neural Networks: An Overview. *Neural Networks*, 61, 85–117. DOI:10.1016/j.neunet.2014.09.003 PMID:25462637

Schneider, J., Abraham, R., Meske, C., & Vom Brocke, J. (2022). Artificial intelligence governance for businesses. Information Systems Management. /arXiv.2011.10672DOI:<ALIGNMENT.qj></ALIGNMENT>10.48550

Schwab, K. (2017). *The fourth industrial revolution*. Crown Business.

Schwartz, R., Dodge, J., Smith, N. A., & Etzioni, O. (2020). Green AI. *Communications of the ACM*, 63(12), 54–63. DOI:10.1145/3381831

Seaborn, K., Sawa, Y., & Watanabe, M. (2024). Coimagining the Future of Voice Assistants with Cultural Sensitivity. *Human Behavior and Emerging Technologies*, 2024, 2024. DOI:10.1155/2024/3238737

Selamat, M. A., & Windasari, N. A. (2021). Chatbot for SMEs: Integrating customer and business owner perspectives. *Technology in Society*, 66, 101685. Advance online publication. DOI:10.1016/j.techsoc.2021.101685

Sengar, S. S., Hasan, A. B., Kumar, S., & Carroll, F. (2024). Generative Artificial Intelligence: A Systematic Review and Applications. *arXiv preprint arXiv:2405.11029*.

Sethi, K., Sharma, M., & Gusain, A. (2023, June). The Acceptance of Machine module in the Hospitality Industry with Prospects and Challenges: A Review. In 2023 2nd International Conference on Computational Modelling, Simulation and Optimization (ICCMSO) (pp. 283-288). IEEE.

Shahriar, S., Allana, S., Hazratifard, S. M., & Dara, R. (2023). A survey of privacy risks and mitigation strategies in the Artificial Intelligence life cycle. *IEEE Access : Practical Innovations, Open Solutions*, 11, 61829–61854. DOI:10.1109/ACCESS.2023.3287195

Shaikh, M. S., Ali, S. I., Deshmukh, A. R., Chandankhede, P. H., Titarmare, A. S., & Nagrale, N. K. (2024). AI business boost approach for small business and shopkeepers: Advanced approach for business. In Ponnusamy, S., Assaf, M., Antari, J., Singh, S., & Kalyanaraman, S. (Eds.), *Digital Twin Technology and AI Implementations in Future-Focused Businesses* (pp. 27–48). IGI Global., DOI:10.4018/979-8-3693-1818-8.ch003

Shaikh, M. S., Chandrawat, U. B., Choudhary, S. M., Ali, S. I., Ponnusamy, S., Khan, R. A., & Sheikh, A. G. (2024). Harnessing Logistic Industries and Warehouses With Autonomous Carebot for Security and Protection: A Smart Protection Approach. In Ponnusamy, S., Assaf, M., Antari, J., Singh, S., & Kalyanaraman, S. (Eds.), *Harnessing AI and Digital Twin Technologies in Businesses* (pp. 239–257). IGI Global., DOI:10.4018/979-8-3693-3234-4.ch017

Shama, M. (2024). Generative Artificial Intelligence in Finance. Artificial Intelligence (Ai) and Business, 70.

Sharma, M., & Singh, A. (2024). Enhancing Competitive Advantages Through Virtual Reality Technology in the Hotels of India. In Utilizing Smart Technology and AI in Hybrid Tourism and Hospitality (pp. 243-256). IGI Global. DOI:10.4018/979-8-3693-1978-9.ch011

Sharma, M., Bathla, G., Kaushik, A., & Rana, S. (2023, June). A Study on Impact of Adaptation of AI-Artificial intelligence services on Business Performance of Hotels. In 2023 2nd International Conference on Computational Modelling, Simulation and Optimization (ICCMSO) (pp. 28-32). IEEE.

Sharma, R. (2024). A Comprehensive Guide to HIPAA Compliance in the Age of AI. https://www.protecto.ai/blog/hipaa-compliance-ai-comprehensive-guide

Sharma, S., Singh, G., Gaur, L., & Afaq, A. (2022). Exploring customer adoption of autonomous shopping systems. *Telematics and Informatics*, 73, 101861. DOI:10.1016/j.tele.2022.101861

Sharples, J. (2021). 'The machine being set in motion': The automaton chess-player in urban and literary culture, 1839–1851. *Sport in History*, 41(2), 181–212. DOI:10.1080/17460263.2021.1906310

Sheikh, M. S.. (2024). Harnessing Logistic Industries Using Autonomous Carebot for Smart Surveillence, Protection and Security. In Al-Turjman, F. (Ed.), *The Smart IoT Blueprint: Engineering a Connected Future. AIoTSS 2024. Advances in Science, Technology & Innovation*. Springer., DOI:10.1007/978-3-031-63103-0_20

Sherry, T. L. (2019). AI and Emotional Intelligence: Understanding the Impact on Human Relationships. *Journal of Human-Computer Interaction*, 303-320.

Shum, H. yeung, He, X. dong, & Li, D. (2018). From Eliza to XiaoIce: challenges and opportunities with social chatbots. In *Frontiers of Information Technology and Electronic Engineering* (Vol. 19, Issue 1). DOI:10.1631/FITEE.1700826

Siau, K., & Wang, W. (2020). Artificial Intelligence (AI) Ethics: Ethics of AI and Ethical AI. *Journal of Database Management*, 31(2), 74–87. DOI:10.4018/JDM.2020040105

Sidaoui, K., Mahr, D., & Odekerken-Schröder, G. (2024). Generative AI in Responsible Conversational Agent Integration: Guidelines for Service Managers. *Organizational Dynamics*, 53(2), 101045. DOI:10.1016/j.orgdyn.2024.101045

Silva, F. A. P., Velez, M. J. P., & Borrego, P. J. V. B. M. (2024). Perceived leadership effectiveness and turnover intention in remote work: The mediating role of communication. *Caderno Pedagógico*, 21(3), e3200–e3200. DOI:10.54033/cadpedv21n3-093

Silva, R. L., Canciglieri, O.Junior, & Rudek, M. (2022). A road map for planning-deploying machine vision artifacts in the context of industry 4.0. *Journal of Industrial and Production Engineering*, 39(3), 167–180. DOI:10.1080/21681015.2021.1965665

Silver, D. e., Huang, A., Maddison, C. J., Guez, A., Sifre, L., van den Driessche, G., Schrittwieser, J., Antonoglou, I., Panneershelvam, V., Lanctot, M., Dieleman, S., Grewe, D., Nham, J., Kalchbrenner, N., Sutskever, I., Lillicrap, T., Leach, M., Kavukcuoglu, K., Graepel, T., & Hassabis, D. (2016). Mastering the Game of Go with Deep Neural Networks and Tree Search. *Nature*, 529(7587), 484–489. DOI:10.1038/nature16961 PMID:26819042

Singapore Personal Data Protection Commission. (2020). *Singapore's Approach to AI Governance*. Retrieved from Singapore Personal Data Protection Commission: https://www.pdpc.gov.sg/help-and-resources/2020/01/model-ai-governance-framework

Singer, N. (2019, May 14). *San Francisco Bans Facial Recognition Technology*. Retrieved from NY Times: https://www.nytimes.com/2019/05/14/us/facial-recognition-ban-san-francisco.html

Singh, H., Kaur, K., & Singh, P. P. (2023). Artificial Intelligence as a facilitator for Film Production Process. *2023 International Conference on Artificial Intelligence and Smart Communication (AISC)*, (pp. 969-972). DOI:10.1109/AISC56616.2023.10085082

Sleaman, W. K., Hameed, A. A., & Jamil, A. (2023). Monocular vision with deep neural networks for autonomous mobile robots navigation. *Optik (Stuttgart)*, 272, 170162. DOI:10.1016/j.ijleo.2022.170162

Smith, J. (2024). *AI for Cyber Guardians: Revolutionizing Ethical Hacking*. Bod Third Party Titles.

Smuha, N. A. (2019). The EU approach to ethics guidelines for trustworthy artificial intelligence. *Computer Law Review International*, 20(4), 97–106. DOI:10.9785/cri-2019-200402

Sohn, J. J., Guo, N., & Chung, Y. (2024, June). Sustainable E-commerce Marketplace: Reshaping Consumer Purchasing Behavior Through Generative AI (Artificial Intelligence). In International Conference on Human-Computer Interaction (pp. 254-269). Cham: Springer Nature Switzerland. DOI:10.1007/978-3-031-61315-9_18

Spears, L. C. (2004). Practicing servant-leadership. *Leader to Leader*, 2004(34), 7–11. DOI:10.1002/ltl.94

Spirling, A. (2023, April). Why open-source generative AI models are an ethical way forward for science. *Nature*, 616(7957), 413. DOI:10.1038/d41586-023-01295-4 PMID:37072520

Stahl, B. C. (2021, January 1). *Addressing Ethical Issues in AI*. SpringerBriefs in Research and Innovation Governance. DOI:10.1007/978-3-030-69978-9_5

Stahl, B. C. (2021, January 1). *Ethical Issues of AI*. SpringerBriefs in Research and Innovation Governance. DOI:10.1007/978-3-030-69978-9_4

Stahl, B. C., & Stahl, B. C. (2021). Ethical issues of AI. Artificial Intelligence for a better future: An ecosystem perspective on the ethics of AI and emerging digital technologies, 35-53.

Stahl, B. C., Timmermans, J., & Mittelstadt, B. D. (2016). The Ethics of Computing: A Survey of the Computing-Oriented Literature. *ACM Computing Surveys*, 48(4), 1–38. DOI:10.1145/2871196

Stainer, L. (2003). Ethical dimensions of management decision-making: a stakeholder values approach to performance and strategy. https://doi.org/DOI:10.18745/th.14114

Stamboliev, E., & Christiaens, T. (2024). How empty is Trustworthy AI? A discourse analysis of the Ethics Guidelines of Trustworthy AI. *Critical Policy Studies*, 1–18. DOI:10.1080/19460171.2024.2315431

Sterne, J. (2017). *Artificial intelligence for marketing: Practical applications.* John Wiley and Sons. DOI:10.1002/9781119406341

Stine, R. A. (2019). Sentiment analysis. *Annual Review of Statistics and Its Application*, 6(1), 287–308. DOI:10.1146/annurev-statistics-030718-105242

Stone, P. Brooks, et al. (2022) Artificial Intelligence and Life in 2030: The One Hundred Year Study on Artificial Intelligence. Rep. 2015–2016 Study Panel. Available online: https://ai100.stanford.edu

Strickland, E. (2019, April 1). *IBM Watson, Heal Thyself: How IBM Overpromised and Underdelivered on AI Health Care.* Retrieved from IEEE Spectrum: https://read.nxtbook.com/ieee/spectrum/spectrum_na_april_2019/ibm_watson_heal_thyself.html

Strickland, E. (2019). IBM Watson, heal thyself: How IBM overpromised and underdelivered on AI health care. *IEEE Spectrum*, 56(4), 24–31. DOI:10.1109/MSPEC.2019.8678513

Strubell, E., Ganesh, A., & McCallum, A. (2019). Energy and Policy Considerations for Deep Learning in NLP. *Proceedings of the 57th Annual Meeting of the Association for Computational Linguistics* (pp. 3645-3650). PKP Publishing Services Network. DOI:10.18653/v1/P19-1355

Strubell, E., Ganesh, A., & McCallum, A. (2020). Energy and Policy Considerations for Deep Learning in NLP. *Proceedings of the AAAI Conference on Artificial Intelligence* (pp. 13693-13696). Burnaby, B.C., Canada: PKP Publishing Services.

Suchman, L. (2007). *Human-Machine Reconfigurations: Plans and Situated Actions.* Cambridge University Press.

Su, J., & Yang, W. (2023). Unlocking the power of ChatGPT: A framework for applying generative AI in education. *ECNU Review of Education*, 6(3), 355–366. DOI:10.1177/20965311231168423

Sushina, T., & Sobenin, A. (2020, May). Artificial intelligence in the criminal justice system: leading trends and possibilities. In 6th International Conference on Social, economic, and academic leadership (ICSEAL-6-2019) (pp. 432-437). Atlantis Press. DOI:10.2991/assehr.k.200526.062

Tahiru, F. (2008). Pannu (2020), A. Artificial Intelligence and Its Application in Different Areas. *Int. J. Eng. Innov. Technol.*, 4, 79–84.

Taj, Y. (2023). Ethical Dilemmas in Leadership: Navigating Difficult Situations with Integrity. *International Journal For Multidisciplinary Research*, 5(4), 4222. DOI:10.36948/ijfmr.2023.v05i04.4222

Teixeira, N., & Pacione, M. (2024). *Implications of Artificial Intelligence on Leadership in Complex Organizations: An Exploration of the Near Future.* (Master's Thesis)

Thorp, H. H. (2023). ChatGPT is fun, but not an author. *Science*, 379(6630), 313. DOI:10.1126/science.adg7879 PMID:36701446

Thrun, S. e., Montemerlo, M., Dahlkamp, H., Stavens, D., Aron, A., Diebel, J., Fong, P., Gale, J., Halpenny, M., Hoffmann, G., Lau, K., Oakley, C., Palatucci, M., Pratt, V., Stang, P., Strohband, S., Dupont, C., Jendrossek, L.-E., Koelen, C., & Mahoney, P. (2006). Stanley: The Robot That Won the DARPA Grand Challenge. *Journal of Field Robotics*, 23(9), 661–692. DOI:10.1002/rob.20147

Tiago, I. R. (2021). The influence of artificial intelligence leadership on employee ethical decision-making (Doctoral dissertation).

Tong, Z., & Zhang, H. (2016, May). A text mining research based on LDA topic modelling. In *International conference on computer science, engineering and information technology* (pp. 201-210). DOI:10.5121/csit.2016.60616

Topol, E. J. (2019). *Deep Medicine: How Artificial Intelligence Can Make Healthcare Human Again.* Basic Books.

Treasury Board of Canada Secretariat. (2023, April 25). *Directive on Automated Decision-Making.* Retrieved from Policies, directives, standards and guidelines: https://www.tbs-sct.gc.ca/pol/doc-eng.aspx?id=32592

Troise, C., & Camilleri, M. A. (2021). The use of digital media for marketing, CSR communication and stakeholder engagement. In *Strategic corporate communication in the digital age*. Emerald Publishing Limited. DOI:10.1108/978-1-80071-264-520211010

Tseng, P. E., & Wang, Y. H. (2021). Deontological or utilitarian? An eternal ethical dilemma in outbreak. *International Journal of Environmental Research and Public Health*, 18(16), 8565. DOI:10.3390/ijerph18168565 PMID:34444311

Tulshan, A. S., & Dhage, S. N. (2019). Survey on virtual assistant: Google assistant, Siri, Cortana, Alexa. *Communications in Computer and Information Science*, 968, 190–201. Advance online publication. DOI:10.1007/978-981-13-5758-9_17

Turing, A. M. (1950). Computing Machinery and Intelligence. *Mind- a Quarterly Review of Psychology and Philosophy*, 433-460.

Turkle, S. (2011). *Alone Together: Why We Expect More from Technology and Less from Each Other*. Basic Books.

Uddin, A. S. M. (2023). The Era of AI: Upholding ethical leadership. *Open Journal of Leadership*, 12(04), 400–417. DOI:10.4236/ojl.2023.124019

UNESCO. "Ethics of Artificial Intelligence" (2022): https://www.unesco.org/en/ artificialintelligence / recommendation-ethics.

US Department of Defense. (2020, February 24). *DOD Adopts Ethical Principles for Artificial Intelligence*. Retrieved from US Department of Defense: https://www .defense.gov/News/Releases/Release/Article/2091996/dod-adopts-ethical-principles -for-artificial-intelligence/

Vaja, J. R. (2017). Ethical Leadership in the Digital Age: Assessing the Role of Leaders in Nurturing Ethical Behavior in Technology-Driven Organizations. *International Journal of Management and Development Studies*, 6(10), 118–126. DOI:10.53983/ijmds.v6n10.013

Van Dierendonck, D. (2011). Servant leadership: A review and synthesis. *Journal of Management*, 37(4), 1228–1261. DOI:10.1177/0149206310380462

van Dis, E. A. M., Bollen, J., Zuidema, W., van Rooij, R., & Bockting, C. L. (2023, February). ChatGPT: Five priorities for research. *Nature*, 614(7947), 224–226. DOI:10.1038/d41586-023-00288-7 PMID:36737653

Vartiainen, H., & Tedre, M. (2024). How Text-to-Image Generative AI is Transforming Mediated Action. *IEEE Computer Graphics and Applications*, 44(2), 12–22. DOI:10.1109/MCG.2024.3355808 PMID:38285567

Vaswani, A. e. (2017). Attention is All You Need. *Proceedings of 31st Conference on Neural Information Processing Systems*. Glasgow: Curran Associates.

Vimalkumar, M., Sharma, S. K., Singh, J. B., & Dwivedi, Y. K. (2021). 'Okay google, what about my privacy?': User's privacy perceptions and acceptance of voice based digital assistants. *Computers in Human Behavior*, 120, 106763. Advance online publication. DOI:10.1016/j.chb.2021.106763

Vogel, K. M., Reid, G., Kampe, C., & Jones, P. (2021). The impact of AI on intelligence analysis: Tackling issues of collaboration, algorithmic transparency, accountability, and management. *Intelligence and National Security*, 36(6), 827–848. Advance online publication. DOI:10.1080/02684527.2021.1946952

Voigt, P., & Von dem Bussche, A. (2017). *The EU General Data Protection Regulation (GDPR)*. Springer International Publishing. DOI:10.1007/978-3-319-57959-7

Wach, K., Duong, C. D., Ejdys, J., Kazlauskaitė, R., Korzynski, P., Mazurek, G., Paliszkiewicz, J., & Ziemba, E. (2023). The dark side of generative artificial intelligence: A critical analysis of controversies and risks of ChatGPT. *Entrepreneurial Business and Economics Review*, 11(2), 7–30. DOI:10.15678/EBER.2023.110201

Wakunuma, K., & Eke, D. (2024). Africa, ChatGPT, and Generative AI Systems: Ethical Benefits, Concerns, and the Need for Governance. *Philosophies*, 9(3), 80. DOI:10.3390/philosophies9030080

Wallace, R. S. (2009). *The anatomy of ALICE*. Springer.

Wallach, W., & Allen, C. (2009). *Moral Machines-Teaching Robots right from wrong*. Oxford University Press. DOI:10.1093/acprof:oso/9780195374049.001.0001

Walumbwa, F. O., Hartnell, C. A., & Oke, A. (2010). Servant leadership, procedural justice climate, service climate, employee attitudes, and organizational citizenship behavior: A cross-level investigation. *The Journal of Applied Psychology*, 95(3), 517–529. DOI:10.1037/a0018867 PMID:20476830

Walumbwa, F. O., & Schaubroeck, J. (2009). Leader personality traits and employee voice behavior: Mediating roles of ethical leadership and work group psychological safety. *The Journal of Applied Psychology*, 94(5), 1275–1286. DOI:10.1037/a0015848 PMID:19702370

Wamba-Taguimdje, S. L., Fosso Wamba, S., Kala Kamdjoug, J. R., & Tchatchouang Wanko, C. E. (2020). Influence of artificial intelligence (AI) on firm performance: The business value of AI-based transformation projects. *Business Process Management Journal*, 26(7), 1893–1924. DOI:10.1108/BPMJ-10-2019-0411

Wang, T. "Navigating Generative AI (ChatGPT) in Higher Education: Opportunities and Challenges," (2023) in C. Anutariya, D. Liu, Kinshuk, A. Tlili, J. Yang, M. Chang (eds.), Smart Learning for A Sustainable Society, ICSLE 2023, Lecture Notes in Educational Technology, Singapore: Springer: https://doi.org/DOI:10.1007/978-981-99-5961-7_28

Warwick, K. (2013). *The future of artificial intelligence and cybernetics. There's a future: visions for a better world. BBVA Open Mind.* TF Editores.

Watson, R. (2022). The virtual economy of the metaverse: Computer vision and deep learning algorithms, customer engagement tools, and behavioral predictive analytics. *Linguistic Philos. Investig.*, (21), 41–56.

Watts, J., & Adriano, A. (2021). Uncovering the sources of machine-learning mistakes in advertising: Contextual bias in the evaluation of semantic relatedness. *Journal of Advertising*, 50(1), 26–38. DOI:10.1080/00913367.2020.1821411

Weber, M., Beutter, M., Weking, J., Böhm, M., & Krcmar, H. (2022). AI startup business models: Key characteristics and directions for entrepreneurship research. *Business & Information Systems Engineering*, 64(1), 91–109. DOI:10.1007/s12599-021-00732-w

Weisz, J. D., He, J., Muller, M., Hoefer, G., Miles, R., & Geyer, W. (2024, May). Design Principles for Generative AI Applications. In *Proceedings of the CHI Conference on Human Factors in Computing Systems* (pp. 1-22).

Weizenbaum, J. (1966). ELIZA-A computer program for the study of natural language communication between man and machine. *Communications of the ACM*, 9(1), 36–45. Advance online publication. DOI:10.1145/365153.365168

Weizenbaum, J. (1976). *Computer Power and Human Reason: From Judgment to Calculation.* W.H. Freeman.

Weng, Y. H., & Hirata, Y. (2018). Ethically Aligned Design for Assistive Robotics. *2018 International Conference on Intelligence and Safety for Robotics, ISR 2018.* DOI:10.1109/IISR.2018.8535889

West, D. M. (2018). *The Future of Work: Robots, AI, and Automation.* Brookings Institution Press.

What ChatGPT means for universities: Perceptions of scholars and students. (2023). Journal of Applied Learning and Teaching, 6(1). DOI:10.37074/jalt.2023.6.1.22

White, M. "A Brief History of Generative AI," Medium (2023): https://matthewdwhite.medium.com/a-brief-historyof-generative-ai-cb1837e67106

WhiteHouse. (2022). Blueprint for an AI bill of rights: Making automated systems work for the American people. The White House Washington DC https:// www .whitehouse.gov/ostp/ai-bill-of-rights/

Whyte, C. (2020). Deepfake news: AI-enabled disinformation as a multi-level public policy challenge. *Journal of Cyber Policy*, 5(2), 199–217. Advance online publication. DOI:10.1080/23738871.2020.1797135

Wong, P. H. (2020). Cultural differences as excuses? Human rights and cultural values in global ethics and governance of AI. *Philosophy & Technology*, 33(4), 705–715. DOI:10.1007/s13347-020-00413-8

Wood, M. S. (2005). Determinants of shared leadership in management teams. *International Journal of Leadership Studies*, 1(1), 64–85.

Wooldridge, M. (2021). *A Brief History of Artificial Intelligence: What It Is, Where We Are, and Where We Are Going*. Macmillan Publishers.

World Economic Forum. (2023, April 30). *The Future of Jobs Report 2023*. Retrieved from World Economic Forum: https://www.weforum.org/publications/the-future-of -jobs-report-2023/

Wright, J. (2024). The Development of AI Ethics in Japan: Ethics washing Society 5.0? *East Asian Science, Technology and Society*, 18(2), 117–134. DOI:10.1080/1 8752160.2023.2275987

Wu, T., He, S., Liu, J., Sun, S., Liu, K., Han, Q. L., & Tang, Y. (2023). A Brief Overview of ChatGPT: The History, Status Quo and Potential Future Development. *IEEE/CAA Journal of Automatica Sinica, 10*(5). DOI:10.1109/JAS.2023.123618

Wu, Y. C., & Wang, X. (2024). Balancing Innovation and Regulation in the Age of Generative Artificial Intelligence. Journal of Information Policy, 14. Regulatory Framework.

Wu, L., Dodoo, N. A., Wen, T. J., & Ke, L. (2022). Understanding twitter conversations about artificial intelligence in advertising based on natural language processing. *International Journal of Advertising*, 41(4), 685–702. DOI:10.1080/0 2650487.2021.1920218

Wu, W., Huang, T., & Gong, K. (2020). Ethical principles and governance technology development of AI in China. *Engineering (Beijing)*, 6(3), 302–309. DOI:10.1016/j. eng.2019.12.015

Wu, X., Guan, F., & Xu, A. (2020). Passive ranging based on planar homography in a monocular vision system. *Journal of Information Processing Systems*, 16(1), 155–170.

Wyden, R. (2022, March 2). *Algorithmic Accountability Act of 2022*. Retrieved from Congress.gov: https://www.congress.gov/bill/117th-congress/senate-bill/3572

Xu, D., Zhang, D., Yang, G., Yang, B., Xu, S., Zheng, L., & Liang, C. (2024). Survey for Landing Generative AI in Social and E-commerce Recsys—the Industry Perspectives. arXiv preprint arXiv:2406.06475.

Xu, J., Wu, B., Huang, J., Gong, Y., Zhang, Y., & Liu, B. (2024). Practical Applications of Advanced Cloud Services and Generative AI Systems in Medical Image Analysis. arXiv preprint arXiv:2403.17549.

Yafei, X., Wu, Y., Song, J., Gong, Y., & Lianga, P. (2024). Generative AI in Industrial Revolution: A Comprehensive Research on Transformations, Challenges, and Future Directions. Journal of Knowledge Learning and Science Technology ISSN: 2959-6386 (online), 3(2), 11-20.

Yang, S., Lee, J., Sezgin, E., Bridge, J., & Lin, S. (2021). Clinical advice by voice assistants on postpartum depression: Cross-sectional investigation using apple siri, amazon Alexa, google assistant, and microsoft cortana. *JMIR mHealth and uHealth*, 9(1), e24045. Advance online publication. DOI:10.2196/24045 PMID:33427680

Yeralan, S., & Lee, L. A. (2023). Generative AI: Challenges to higher education. *Sustainable Engineering and Innovation*, 5(2), 107–116.

Yeung, L. K. C., Tam, C. S. Y., Lau, S. S. S., & Ko, M. M. (2023). Living with AI personal assistant: An ethical appraisal. *AI & Society*. Advance online publication. DOI:10.1007/s00146-023-01776-0

Yin, M., Jiang, S., & Niu, X. (2024). Can AI really help? The double-edged sword effect of AI assistant on employees' innovation behavior. *Computers in Human Behavior*, 150, 107987. Advance online publication. DOI:10.1016/j.chb.2023.107987

Yukl, G. A. (2013). *Leadership in organizations*. Pearson Education India.

Zafar, M. R., & Khan, N. M. (2019). DLIME: A deterministic local interpretable model-agnostic explanations approach for computer-aided diagnosis systems. arXiv preprint arXiv:1906.10263.

Zemčík, M. (2019). A brief history of chatbots. DEStech Transactions on Computer Science and Engineering, 10.

Zhai, C., & Wibowo, S. (2022). A systematic review on cross-culture, humor and empathy dimensions in conversational chatbots: The case of second language acquisition. *Heliyon*, 8(12), e12056. Advance online publication. DOI:10.1016/j.heliyon.2022.e12056 PMID:36531630

Zhai, X. ChatGPT user experience: Implications for education. *SSRN* 2022. [CrossRef] DOI:10.2139/ssrn.4312418

Zhang, C., Zhang, C., Zheng, S., Qiao, Y., Li, C., Zhang, M., & Hong, C. S. (2023). A Complete Survey On Generative A*i* (Aigc): is Chatgpt From Gpt-4 To Gpt-5 All You Need? *arXiv preprint arXiv:2303.11717*.

Zhang, B., Zhu, J., & Su, H. (2023). Toward the third generation artificial intelligence. *Science China. Information Sciences*, 66(2), 1–19. DOI:10.1007/s11432-021-3449-x

Zhang, C., & Lu, Y. (2021). Study on artificial intelligence: The state of the art and future prospects. *Journal of Industrial Information Integration*, 23, 100224. DOI:10.1016/j.jii.2021.100224

Zhang, E. Y., Cheok, A. D., Pan, Z., Cai, J., & Yan, Y. (2023). From Turing To Transformers: A Comprehensive Review and Tutorial On The Evolution And Applications of Generative Transformer Models. *Sci*, 5(4), 46. DOI:10.3390/sci5040046

Zhang, P., & Kamel Boulos, M. N. (2023). Generative AI in medicine and healthcare: Promises, opportunities and challenges. *Future Internet*, 15(9), 286. DOI:10.3390/fi15090286

Zhao, B. (2023). Analysis on the Negative Impact of AI Development on Employment and Its Countermeasures. In SHS Web of Conferences (Vol. 154, p. 03022). EDP Sciences. DOI:10.1051/shsconf/202315403022

Zhu, T., Ye, D., Wang, W., Zhou, W., & Philip, S. Y. (2020). More than privacy: Applying differential privacy in key areas of artificial intelligence. *IEEE Transactions on Knowledge and Data Engineering*, 34(6), 2824–2843. DOI:10.1109/TKDE.2020.3014246

Zlateva, P., Steshina, L., Petukhov, I., & Velev, D. (2024). A Conceptual Framework for Solving Ethical Issues in Generative Artificial Intelligence. In *Electronics, Communications and Networks* (Vol. 381, pp. 110–119). IOS Press. DOI:10.3233/FAIA231182

Zuboff, S. (2019). *The Age of Surveillance Capitalism: The Fight for a Human Future at the New Frontier of Power*. PublicAffairs.

About the Contributors

Loveleen Gaur is currently working as an adjunct professor with Taylor University, Malaysia & University of South Pacific, Fiji and academic consultant with Australian School of Graduate Studies. Before moving to USA, she was working as Professor with Amity University, India. She has supervised several PhD scholars, Post Graduate students, mainly in Artificial Intelligence and Data Analytics for business and healthcare. Under her guidance, the AI/Data Analytics research cluster has published extensively in high impact factor journals and has established extensive research collaboration globally with several renowned professionals. She is a senior IEEE member and Series Editor with CRC and Wiley. She has high indexed publications in SCI/ABDC/WoS/Scopus and has several Patents/copyrights on her account, edited/authored many research books published by world-class publishers. She has excellent experience in supervising and co-supervising postgraduate and PhD students internationally. An ample number of Ph.D. and master's students graduated under her supervision. She is an external Ph.D./Master thesis examiner/ evaluator for several universities globally. She has also served as Keynote speaker for several international conferences, presented several Webinars worldwide, chaired international conference sessions. Prof. Gaur has significantly contributed to enhancing scientific understanding by participating in many scientific conferences, symposia, and seminars, by chairing technical sessions and delivering plenary and invited talks. She has specialized in the fields of Artificial Intelligence, Machine Learning, Pattern Recognition, Internet of Things, Data Analytics and Business Intelligence. She has chaired various positions in International Conferences of repute and is a reviewer with top rated journals of IEEE, SCI and ABDC Journals. She has been honored with prestigious National and International awards. She has introduced courses related to Artificial Intelligence specialization including, Predictive Analytics, Deep and Reinforcement learning etc. She has vast experience teaching advanced-era specialized courses, including Predictive Analytics, Data Visualization, Social Network Analytics, Deep Learning, Power BI, Digital Marketing and Digital Innovation etc., besides other undergraduate and postgraduate courses, graduation projects, and thesis supervision.

Munir Ahmad is a seasoned professional in the realm of Spatial Data Infrastructure (SDI), Geo-Information Productions, Information Systems, and Information Governance, boasting over 25 years of dedicated experience in the field. With a PhD in Computer Science, Dr. Ahmad's expertise spans Spatial Data Production, Management, Processing, Analysis, Visualization, and Quality Control. Throughout his career, Dr. Ahmad has been deeply involved in the development and deployment of SDI systems specially in the context of Pakistan, leveraging his proficiency in Spatial Database Design, Web, Mobile & Desktop GIS, and Geo Web Services Architecture. His contributions to Volunteered Geographic Information (VGI) and Open Source Geoportal & Metadata Portal have significantly enriched the geospatial community. As a trainer and researcher, Dr. Ahmad has authored over 50 publications, advancing the industry's knowledge base and fostering innovation in Geo-Tech, Data Governance, and Information Infrastructure, and Emerging Technologies. His commitment to Research and Development (R&D) is evident in his role as a dedicated educator and mentor in the field.

Syed Ibad Ali is a distinguished and accomplished electronics professor, renowned for their significant contributions to the field of electrical engineering and technology. With an illustrious career spanning several years, Dr. Syed Ibad Ali has emerged as a thought leader, educator, and innovator in the realm of electronics.

Aravind Ganesan is a Programmer Analyst Trainee at Cognizant, embarking on his professional career in the field of technology. He holds a degree in Artificial Intelligence and Machine Learning from Hindusthan College of Engineering & Technology, Coimbatore, India. His academic pursuits and research interests span across Artificial Intelligence, Machine Learning, Data Analytics, and the Internet of Things (IoT). Aravind has co-authored a chapter titled Internet of Things & Smart Cities in the book Modern Research in Engineering & Technology. With a foundation in mentoring and lecturing on AI and ML, he is now transitioning from academia to the tech industry.

Ajay Gangele is working as an Assistant Professor in School of Management. He has more than 18 years of experience in the field of academic administration. He also has 3 Years of Industrial Experience. Dr. Ajay Gangele obtained his Bachelor's degree in Science, Master's degree in Business Administration; he has completed Ph.D in the field of management. He has qualified UGC- NET. He obtained all degree with first division. He has many research papers in the field of management in his credit including UGC approved Journal of Management. He has guided more than 120 students at graduation/post graduation level in field research work. He has participated and presented research papers in numerous conferences, seminars at national and international level. He has also participated in a couple of AICTE sponsored QIP at BHU Banaras. He has special interest in the field of Management and Behavioral Games. He works with a burning intension of creating future leaders possessing excellent communication skills and commendable sense of self confidence. Dr. Ajay chaired various sessions in national and international conference.

Rubén González Vallejo has focused his research on legal translation and the environment in the Italian-Spanish combination, as well as on language teaching. With a background in philology and in translation and interpreting, he is a member of various research groups/centres and scientific and editorial committees. He has also coordinated monographs in scientific journals. As a teacher, he has carried out mobilities in national and foreign universities and has taught at Bachelor's and Master's degree levels. He is currently an Associate Professor.

Sunil Kumar, currently an Associate Professor at Faculty of Management Sciences, Shoolini University, possesses extensive academic and research experience. With a background in Business Administration, Sunil's research endeavors span multiple domains, including human resource management, e-training, and the application of artificial intelligence in business. His prolific contributions are evidenced by numerous research papers, book chapters, copyrights, and patents. Dr. Kumar has a knack for teaching with expertise in subjects like HR Analytics, Talent Management, Industrial Psychology, and more. He has also successfully guided PhD students and actively contributes to various academic administrative roles. Dr. Kumar's vast research landscape includes topics like AI-based chatbots in recruitment, e-training impact, and more. His papers have been published in esteemed journals like Journal of Workplace Learning and International Journal of Law and Management. Moreover, he has authored books and edited volumes touching upon themes related to the contemporary business environment. Dr. Sunil has been a dynamic participant in numerous workshops, ranging from data analysis to stress

Anand M is a Research Scholar at the Department of Computing Technologies, SRM Institute of Science and Technology, Kattankulathur, India. His research focuses on Machine Learning, Deep Learning, Image Processing, IOT, and Cloud Computing. He has completed his M.E Computer Science and Engineering from Anna University, Chennai, India. He has 6+ years of teaching and industry experience.

Kiran Macwan is a distinguished Computer Science researcher renowned for her extensive contributions to the fields of Internet of Things (IoT) and security. With a career spanning over 14 years in academia, Macwan has established herself as a leading authority in her domain. Macwan commenced her academic journey by earning her Bachelor's degree in Technology (B.Tech) in Computer Science, laying the groundwork for her illustrious career. She further solidified her expertise by completing a Master's degree (M.Tech) in Computer Science, where she delved deeper into advanced concepts and methodologies within the field. Driven by an insatiable thirst for knowledge and a passion for pioneering research, Macwan embarked on her doctoral journey in Computer Engineering, with a focus on addressing the intricacies of IoT and security. Her ongoing pursuit of a Ph.D. underscores her commitment to pushing the boundaries of knowledge and spearheading innovative solutions to contemporary challenges. Throughout her tenure in academia, Macwan has not only cultivated a profound understanding of theoretical frameworks but has also demonstrated a remarkable ability to translate these theories into practical applications. Her research endeavours have been instrumental in enhancing the efficiency, reliability, and security of IoT systems, thereby laying the groundwork for the future evolution of interconnected technologies. As a faculty member at Parul University, Macwan plays a pivotal role in shaping the next generation of computer scientists and engineers. Her dynamic teaching methodologies, coupled with his wealth of industry experience, empower students to navigate the complex landscape of modern technology with confidence and ingenuity. Macwan's prolific research output, comprising numerous publications in esteemed journals and conference proceedings, serves as a testament to his scholarly prowess and unwavering dedication to advancing the frontiers of knowledge. Her groundbreaking contributions continue to garner widespread acclaim within the academic community, positioning her as a trailblazer in the realm of Computer Science. Driven by a relentless pursuit of excellence, Kiran Macwan remains steadfast in her commitment to unraveling the mysteries of IoT and security, thereby paving the way for a future defined by innovation, resilience, and technological advancement.

Bodhibrata Nag (Ph.D./Fellow, Operations Research and System Analysis, IIM Calcutta & Bachelor of Technology in Electrical Engineering, IIT Madras) is a Professor of Operations Management at the Indian Institute of Management (IIM) Calcutta. He has held the position of Acting Director, Dean (Academic), Provost & Chairperson of Student Affairs, Chairperson of Doctoral Programs and Research, National Coordinator for Post Graduate Program for Executives for Visionary Leadership in Manufacturing at IIM Calcutta, as well as Director of IIM Calcutta Innovation Park. He holds certifications in various areas of Operations Management, Logistics, Supply Chain, and Data Sciences from The World Bank, Massachusetts Institute of Technology (MIT), Georgia Institute of Technology, ESSEC Business School, University of Virginia, University of Amsterdam, John Hopkins University, Delft University of Technology and Technical University of Munich. He has received training in the case methods of teaching and case development at Harvard Business School and power projects design at Electricite de France. He has 21 years of transportation and energy industry experience and 18 years of teaching and research

Thamizhikkavi P. is currently an Assistant Professor at SRM Institute of Science and Technology, Chennai, India. he received a Master's Degree in Information Security and Cyber Forensics from SRM Institute of Science and Technology in 2015. His research interest includes Cryptography and Ethical hacking.

Shweta Bhattacharjee Porna She has a BSc in computer science from Ahsanullah University of Science and Technology. Currently, she works as a Web developer interested in AI Research and web design.

Ramya R. Completed her B.E and M.E degree in Computer science and engineering. She had 14 years of teaching experience in various Engineering colleges. She pursuing PhD in computer science and engineering and published many articles and book chapters in her domain. She specialises in edge computing, cloud computing, Machine learning, artificial intelligence, Quality of services and resource management.

Md. Asifur Rahman is a distinguished postgraduate student in the Department of Anthropology at the University of Dhaka. He completed his Bachelor of Social Sciences (BSS) in Anthropology with exceptional grades, earning the Dean's Award for academic excellence. Asif's research focuses on the intersection of gender and technology, particularly in relation to children, youth, and adolescents (CAY). His monograph, "Silhouettes of the Digital Minds: Exploring Emotional and Intellectual Interaction between Youth and Artificial Intelligence," has been well-received in the academic community for its rigorous analysis and innovative perspectives on the role of artificial intelligence in youth development. Asif's scholarly work reflects a deep engagement with contemporary issues and contributes meaningfully to the field of anthropology.

Smita Shahane, a passionate researcher in the field of electronics and telecommunication, holds a Master's degree in the discipline. With a focus on networking, electronics, and communication, her expertise shines through in her dedication to advancing these domains. Smita's unwavering passion for research fuels her drive to explore new frontiers and innovate within her field, shaping the future of technology.

Mohammad Shahnawaz Shaikh awarded Ph.D. in Electronics from Rani Durgavati University Jabalpur (M.P.) in the year 2021. He completed M. Tech. in Digital Communication in year 2014 and B.E. in Electroincs and Comunication Engineering in the year 2008 from Rajiv Gandhi Technical University Bhopal (M.P.). He is working as Associate Professor in Department of CSE, Parul University, Vadodra, Gujrat. He has 15 years of experience in teaching and research with remarkable achievements in academics in the from of publications in internarional journal, international and national connferences, book chapters, patents and copyrights . My research area is AI, ML, IoT, Signal Processing and Image Processing and Advanced Communication Networks

Irim Shehzadi is a dedicated and passionate individual with a strong educational background, holding a Master's degree in Environmental Sciences from PMAS Arid Agricultural University in Rawalpindi.

Amrik Singh is working with Lovely Professional University, Punjab, India. He is PhD in Hospitality Management from Kurukshetra University, India. His area of expertise in tourism and hospitality, human resource management, sustainability, waste management, green practices, restaurant and resort management. He has published more than 42 scopus indexed research articles and book chapters. He is associated as editorial board members and reviewers with reputed and scopus indexed journals.

Index

A

AI assistant 296, 299, 315

AI Governance 58, 70, 71, 73, 78, 79, 82, 104, 105, 106, 111, 115, 130, 131, 132, 135, 136, 140, 151, 154, 155, 157, 158, 163, 269, 271, 277, 375, 377, 378, 381

AI Literacy 111, 132, 134, 136, 155, 157, 158, 160, 227, 232, 233

AI Regulations 69, 72, 131, 135, 154, 157

Artificial Intelligence 1, 4, 6, 19, 20, 26, 29, 30, 32, 35, 36, 39, 40, 41, 42, 46, 47, 49, 50, 51, 53, 54, 61, 63, 67, 69, 70, 73, 75, 76, 77, 78, 79, 81, 82, 83, 85, 86, 87, 90, 91, 92, 93, 94, 95, 96, 100, 101, 102, 103, 104, 105, 106, 107, 108, 109, 110, 111, 112, 118, 130, 145, 162, 163, 164, 165, 166, 170, 171, 179, 180, 184, 186, 187, 188, 190, 191, 195, 196, 197, 199, 200, 201, 202, 208, 211, 212, 215, 216, 217, 218, 219, 220, 221, 223, 224, 227, 228, 230, 232, 233, 236, 237, 239, 240, 242, 243, 244, 245, 246, 247, 249, 250, 254, 255, 256, 262, 264, 265, 266, 271, 272, 273, 282, 283, 284, 285, 286, 287, 288, 289, 290, 292, 294, 295, 308, 309, 310, 311, 312, 313, 314, 322, 329, 332, 333, 334, 335, 336, 337, 338, 339, 340, 341, 343, 346, 347, 349, 351, 352, 353, 355, 358, 360, 364, 365, 367, 368, 369, 370, 373, 374, 375, 376, 377, 379, 380, 381, 382

Augmented reality 31, 43, 185, 210, 293, 367, 368, 369, 373, 374, 377

B

Bias 1, 5, 21, 37, 43, 48, 51, 52, 53, 54, 62, 64, 65, 68, 73, 75, 76, 91, 98, 100, 109, 116, 117, 118, 123, 126, 131, 134, 138, 140, 153, 154, 156, 162, 163, 169, 175, 176, 177, 180, 191, 199, 211, 241, 243, 249, 259, 260, 261, 270, 277, 278, 279, 291, 293, 300, 311, 317, 318, 319, 320, 321, 322, 323, 324, 325, 326, 327, 328, 329, 331, 332, 337, 338, 340, 344, 345, 347, 348, 349, 350, 351, 356, 357, 364, 365, 370, 376, 377

Bias Mitigation 140, 177, 331

C

Chatbots 16, 18, 101, 119, 120, 163, 174, 209, 225, 233, 240, 254, 291, 292, 293, 294, 295, 296, 298, 299, 302, 303, 308, 309, 310, 312, 313, 314, 315, 368, 369, 370, 373

ChatGPT 4, 22, 24, 25, 27, 29, 30, 31, 40, 43, 100, 109, 166, 170, 171, 186, 187, 188, 192, 224, 225, 226, 227, 228, 229, 230, 231, 233, 235, 240, 241, 243, 244, 245, 246, 247, 254, 264, 288, 298, 313, 314, 315, 333, 334, 335, 336, 380

D

Data Security 30, 60, 94, 105, 124, 125, 180, 212, 278, 281, 291, 293, 299, 323, 360, 361

Deepfake 60, 68, 77, 80, 172, 187, 193, 204, 205, 206, 207, 217, 219, 220, 308, 315

Deep Learning 2, 15, 30, 31, 32, 37, 38, 39, 41, 42, 43, 46, 51, 52, 55, 78, 79, 81, 82, 94, 101, 104, 118, 134, 190, 191, 193, 223, 252, 253, 254, 262, 267, 284, 293, 298, 305, 341, 352

E

E-commerce 172, 244, 289, 317, 318, 319, 320, 321, 322, 323, 324, 325, 326, 327, 328, 329, 330, 331, 332, 333, 334, 335, 336

Ethical AI practices 19, 59, 136, 154, 155, 156, 158, 159, 160, 276

192, 201, 207, 208, 218, 224, 251,
254, 293, 305, 342, 354, 357

P

Pillars of AI Ethics 85, 95, 96, 98, 99
Predictive Policing 60, 117, 162, 337, 338,
    343, 344, 345, 346, 351, 358, 360
Privacy Laws 94, 361, 377
Public Engagement 111, 113, 127, 132,
    136, 146, 150, 157, 160, 261

R

responsibility and accountability 45, 49,
    64, 176

S

Sentiment Analysis 210, 274, 275, 280,
    288, 309, 314, 371
Social responsibility 101, 102, 104, 106,
    108, 257, 258, 381
Societal Impact 47, 59, 112, 126, 178

T

Transparency 5, 21, 30, 45, 51, 52, 53,
    55, 57, 58, 59, 60, 61, 62, 64, 65, 66,
    69, 70, 71, 73, 74, 75, 76, 81, 85, 86,
    87, 88, 89, 90, 91, 92, 93, 94, 95, 96,
    97, 99, 100, 103, 107, 111, 115, 116,
    117, 118, 121, 122, 123, 124, 129,
130, 131, 133, 139, 140, 142, 143,
145, 146, 147, 150, 151, 152, 153,
154, 155, 156, 158, 160, 169, 176,
177, 178, 183, 186, 189, 191, 193,
195, 196, 198, 199, 200, 203, 212,
216, 250, 255, 257, 258, 260, 261,
270, 271, 272, 276, 277, 278, 280,
281, 291, 293, 299, 300, 301, 302,
303, 306, 308, 311, 315, 317, 323,
327, 328, 330, 331, 332, 337, 343,
346, 348, 349, 350, 352, 354, 359,
360, 361, 376, 377
Trust 5, 20, 21, 45, 52, 53, 57, 64, 65, 66,
    69, 71, 72, 74, 87, 88, 90, 91, 92, 95,
    96, 112, 122, 124, 126, 133, 136, 139,
    142, 143, 150, 151, 153, 156, 159,
    160, 177, 180, 193, 198, 207, 208,
    209, 210, 211, 231, 250, 255, 258,
    260, 268, 269, 271, 272, 291, 292,
    299, 300, 301, 308, 312, 318, 322,
    323, 326, 329, 330, 331, 332, 352,
    353, 355, 374, 377

U

User Autonomy 291, 292, 301, 312

V

Virtual Reality 31, 41, 119, 210, 367, 368,
    369, 373, 374, 377, 381